Yearbook of the Maimonides Centre for Advanced Studies 2019

Yearbook of the Maimonides Centre for Advanced Studies

Chief Editor
Giuseppe Veltri

Yearbook of the Maimonides Centre for Advanced Studies 2019

Volume Editor
Yoav Meyrav

DE GRUYTER

The Yearbook is published on behalf of
the Maimonides Centre for Advanced Studies

ספקנות
Maimonides Centre for Advanced Studies
JEWISH SCEPTICISM

ISBN 978-3-11-060332-3
e-ISBN (PDF) 978-3-11-061883-9
e-ISBN (EPUB) 978-3-11-060737-6

Library of Congress Control Number: XXX

Bibliographic information published by the Deutsche Nationalbibliothek
The Deutsche Nationalbibliothek lists this publication in the Deutsche Nationalbibliografie; detailed bibliographic data are available on the Internet at http://dnb.dnb.de.

Cover image: Staats- und Universitätsbibliothek Hamburg, Ms Cod. Levy 115, fol. 158r: Maimonides, More Nevukhim, Beginn von Teil III.
Printing and binding: CPI books GmbH, Leck

www.degruyter.com

In Memoriam Roi Benbassat

Contents

Editorial

The *Yearbook of the Maimonides Centre for Advanced Studies 2019* is the fourth volume of articles stemming from research conducted by MCAS's fellows and affiliates. It also includes a section with an overview of the activities and events conducted at the centre during its fourth academic year and a report on its library.

The twelve articles contained in the present volume exhibit the extent to which MCAS scholarship advances the understanding of Judaism and scepticism, both individually and together, among different thinkers and within different areas of study. Whether a short note or a full-fledged analysis, each paper has something fresh to offer and something new to say, opening new directions of research and inviting the scholarly community to join the discussion.

Aryeh Botwinick's opening essay outlines a new and thought-provoking approach to Western thought and religion's perception of itself. Instead of a narrative of conflict, Botwinick's historiosophy raises the notion of God's unknowability as the mediating thread that renders monotheism and scepticism fundamentally intertwined.

Michela Torbidoni's study adds an important chapter to the as-yet understudied history of the persona and function of Socrates within the Jewish tradition. Focusing on the cases of Moses Mendelssohn and Simone Luzzatto, who operated at different times and in different contexts, she tantalisingly argues that the figure of Socrates can be considered as a vehicle for Jewish emancipation. The comparative discussion adds evidence to the thesis that modern Jewish Italian culture was part of the intellectual basis of the programme to reform the status of European Jews in the late eighteenth century.

Behnam Zolghadr tackles one of the cornerstones of Aristotelian logic—namely, the principle of non-contradiction—by presenting critiques of it as reported by the Muslim theologian Faḫr al-Dīn al-Rāzī (d. 1210). Zolghadr argues that these critiques, which can be traced back to the Muʿtazilite *mutakallimūn*, function as part of its consolidation of the theory of states (*aḥwāl*), which can consequently be regarded as a dialetheic theory. As such, Zolghadr devises the first steps in modelling it logically using tools of paraconsistent logic developed in the late twentieth century, more than a thousand years after the theory's inception.

Máté Veres enriches the understanding of ancient scepticism by exploring Clement of Alexandria's discussion of *epochē* (suspension of judgment) in his *Stromateis*. He examines and refines arguments that support the claim that Clement is presenting a variety of scepticism that predates the well-known version expounded by Sextus Empiricus and considers different explanations as to why a Christian theologian of Clement's standing would express interest in sceptical arguments, let alone recognise their potential legitimacy.

 https://doi.org/10.1515/9783110618839-001

Ze'ev Strauss offers the first comprehensive scholarly work in English on Meister Eckhart's employment of Ibn Gabirol's *Fons vitae*. He shows the vast extent to which Eckhart draws on this work, explicitly and also implicitly, in order to account for his own Christian philosophy, primarily gravitating towards Ibn Gabirol's metaphysics of the One. Strauss shows that many of the provocative and unusual elements of Eckhart's Christian thought correspond to Eckhart's favourable view of *Fons vitae*.

Giada Coppola elucidates Obadiah Sforno's (?1475–1550) philosophical interpretation of the individual human soul in his *Light of the Nations*, with emphasis on his defence of its immortality. Coppola shows how exegetical interpretation and the text's philosophical background are intrinsically connected, identifies Sforno's sources common to the Hebrew and Latin traditions, and displays the interesting dynamics between Sforno's philosophical composition and his biblical commentaries.

José María Sánchez de León Serrano offers a new reading of the Flemish thinker Arnold Geulincx (1624–69) based on an audacious interpretation of his epistemological views. By tracing the central tenets of Geulincx's thought back to his method of self-inspection, Sánchez de León Serrano illuminates the unity and systematic character of this multifaceted philosopher—who at first glance seems to offer an incoherent blend of scepticism and metaphysics—and opens up new vistas in the study of this still widely overlooked thinker.

Avraham Rot's paper offers an original and challenging interpretation of Spinoza's philosophical project, arguing that if one shifts the focus of analysis from his theory of knowledge to his theory of affects, Spinoza emerges as a philosopher whose scepticism is significantly more radical than its Cartesian counterpart. Rot claims that for Spinoza, the doubt that matters is not doubt as a methodological device employed to attain certainty, but rather the doubt that human beings constantly experience insofar as they are finite, embodied, and uncertain of their future.

Timothy Franz offers a highly detailed, in-depth analysis of Solomon Maimon's seldom-studied *Weltseele* (1790), which he describes as a proto-dialectical metaphysics of the absolute. Beyond exploring the *Weltseele* on its own terms, Franz identifies it as a key stage in Maimon's philosophical development and argues that it functions as an attempt to overcome the self-conceived failure of Maimon's well-known *Essay on Transcendental Philosophy*. In turn, the ideas in the *Weltseele* were abandoned, although not before they exposed the need to construct a logic of the absolute, which was ultimately expressed in Maimon's *Logik* (1794).

Michah Gottlieb's contribution does much to clarify the debate surrounding Judaism and dogma as reflected in German-speaking Jewish authors in the eighteenth and nineteenth centuries, showing that it is not merely a theoretical question, but also has much to do with concrete considerations concerning Jewish society. He begins with a thorough analysis of Moses Mendelssohn's view of Judaism, explaining in what sense it should be understood as anti-dogmatic and also how it reflects Mendelssohn's aspiration for a self-conception of Judaism that is distinct from Christianity. Gottlieb then turns to discuss leading nineteenth-century German-speaking Jewish thinkers and shows how they transform the problem from an intrareligious one to an interreligious one. Each response is informed by a concept of Judaism that is intimately connected to the problem of Jewish communal unity.

Anne Fiebig's article is the first to offer a systematic account of the concept of tolerance in the thought of Yeshayahu Leibowitz (1903–94), one of the most important—and certainly most controversial—Israeli philosophers, as it emerges from his religious scepticism. Analysing Leibowitz's epistemology, Fiebig presents Leibowitz as a model against religious dogmatism, which Leibowitz regards as corruptive to a genuinely faithful position.

Finally, through an analysis of an early seventeenth-century Jesuit report about religious minorities in China, Giuseppe Veltri introduces the novel notion of "strategic scepticism," which is a form of abusing sceptical argumentation aimed at preserving and reinforcing political power. At the present time, strategic scepticism as an analytical category is becoming pressingly relevant.

We would like to take this opportunity to thank Bill Rebiger, who edited the first three volumes of the Yearbook and who has done an excellent job of making it a cornerstone of the yearly scholarly activities at MCAS. Special thanks are also due to Katharine Handel for her superb copy- and language editing of all the papers in the present volume. We would also like to thank Maria Wazinski, the centre's academic coordinator, for compiling this year's list of activities and events; Silke Schaeper, the centre's librarian, for preparing the report on the library; Sarah Wobick-Segev for her perceptive comments; and Mikheil Kakabadze for his valuable assistance in the editorial process.

Hamburg, October 2019

Giuseppe Veltri (Chief Editor)
Yoav Meyrav (Volume Editor)

Postscript

As this volume was going to press, we were deeply saddened to learn of the untimely passing of our dear friend and colleague Dr. Roi Benbassat.

Dr. Benbassat, a scholar of Jewish existentialism and a gifted musician, belonged to the first generation of MCAS fellows. He earned his bachelor's degree in philosophy and music composition from Tel Aviv University, his master's degree in philosophy from the University Sorbonne-Pantheon Paris 1 (with a *mémoire* entitled "The Religiosity of Spinoza"), and his PhD in Philosophy from Tel Aviv University in 2011, with a dissertation entitled "The Relation between the Philosophical and the Religious: Reconsidering Kierkegaard's Confrontation with Philosophy."

After a Minerva Postdoctoral Fellowship (Max Planck Institute) at the Institute of Comparative Ethics at Freie Universität Berlin, Benbassat arrived at MCAS as a postdoctoral fellow in 2015–16 with a research project entitled "Yeshayahu Leibowitz—Strict Orthodox Practice and Unbound Scepticism." One of the fruits of his research at MCAS was his paper entitled "Jewish Faith and Scepticism—The Example of Yeshayahu Leibowitz," which was published in the 2017 edition of the *Yearbook of the Maimonides Centre for Advanced Studies*. Until recently, he was a research associate at the Goethe Universität in Frankfurt am Main with a research project entitled "Tolerance and Intolerance in the Dialogues between Conflicting Jewish Positions in the 20th Century."

Benbassat had a wide array of research interests that was reflected in his publications, including the work of the Israeli thinker Yeshayahu Leibowitz, the relations between religious and moral-philosophical standpoints, religious existentialism, Jewish–Christian interreligious conflicts and dialogues, religion and scepticism, and religious positioning in the conflicts and dialogues within Judaism in Israel. He was also a talented singer and songwriter, releasing three albums: *Into Your Craft* (2017), *Footloose* (2018), and *Hard Nut* (2019).

Roi Benbassat was not only a talented scholar, but also a kind, gentle, and humble human being. He was also very funny. His radiant personality filled the Centre's corridors from the day of his arrival. Those who became closer to him learned that underneath the ironic smile and quiet presence was a huge heart and a generous soul.

Roi was supposed to return to MCAS in 2021 with a research project entitled "Yeshayahu Leibowitz's Concept of Faith—Between Intellectualism and Sentimentalism." We were all eagerly awaiting his arrival. This volume is dedicated to his memory.

Part I: **Articles**

Aryeh Botwinick

The Simultaneous Genesis of Monotheism and Scepticism in the Jewish Religion

In this paper, I will argue that the conceptual dichotomy between religion and scepticism cannot adequately capture what biblical monotheism inaugurated in the world. In a number of key respects that I will attempt to analytically plot, I will aim to show that Jewish scepticism and its cultural correlative of secularism are co-extensive with Jewish monotheism. I will try to plot the general outlines of an argument that will support the claim that both Western monotheism and Western scepticism have their roots in negative theology. What I will attempt to show is that the logical impediments that block a fully coherent statement of the tenets of both negative theology and scepticism point to a common origin and a common destiny for both doctrines.

The traditional narrative of Western religion and its relationship to Western scepticism needs to be radically revised. The traditional story is both structured around and invoked to lend support to the idea of irreconcilable conflict between religion and sceptical modes of thought. The conflicts between religion and scepticism are taken to apply both to the substantive doctrines and to the methodologies of argument of religious and sceptical discourse. With negative theology (the idea that we can only say what God is not, but not what He is) as our organising perspective, we can argue that the areas of continuity and overlap between religion and scepticism are much more prominent and decisive than the areas of tension and discontinuity. The practice of religion is in no way threatened or undermined by this realisation. If anything, it is strengthened and reinforced by it. Where the vista of distance between the human community and God is as steadily maintained as it is by negative theology, the prospect of our succumbing to idolatry and worshipping our own handiwork is considerably diminished.

My argument is formulated as a kind of thought experiment. Working with the Maimonidean assumptions in *Sefer Ha-Madaʿ* and *The Guide of the Perplexed* (including the annexation of God as a necessary First Cause in the genealogy of monotheism and the historical assumption that the patriarch Abraham was the discoverer of the monotheistic principle), what relationship emerges between monotheism and scepticism?

The implicit conception of the relationship between religion and scepticism derivable from negative theology is a function of the central place that Platonic and Neoplatonic arguments occupy in its arsenal. Throughout this paper, I am using the term "scepticism" to refer to the philosophical doctrine that states that words are underdetermined by things; that no item in the "furniture" of the world or the human psyche can be framed in only one way, but that openness and flexibility always prevail with regard to designating the identity of things.

https://doi.org/10.1515/9783110618839-003

From a Platonic perspective later fleshed out by Aristotle, the One (which was later called the Western monotheistic God) emerges as a response to an antecedent question: How can we bring the explanatory quest to a halt? What drives this pursuit is simply the sense that most of us aspire to reach "rock bottom" in the search for an explanation: that is, that we have not overlooked any general factor in dealing with the situation or phenomenon at hand. In order to achieve internal reassurance that our explanation in any given circumstance has been along the right path, we seek confirmation that we have exhausted all possibilities of going further. If there is no move left to us but to invoke the infinite (and if this move could be shown to be logically legitimate), we feel that our explanation has been exhaustive. Where can the search for more abstract and comprehensive explanations than those we are currently invoking end?

In his dialogue *Parmenides*, Plato argues that it can only end with the invocation of a name: the One, the forerunner of the Western God.[1] In *Parmenides*, the One is just a name, and if anything could correspond to this name, that would undermine our ability to use it, because then there would be two entities rather than one: the name and the object that it names. Radical oneness presupposes the absence of both sense and reference, connotation as well as denotation. The idea of God is mystical at its inception, and it returns to this baseline again and again throughout the course of Western religious and intellectual history.

In the crucial senses that I have just described, modernism (and even postmodernism) are with us from the beginnings of the Western story (that is, the Western philosophical and theological narratives). As it turns out, a nominalistic reading of God, just like a nominalistic reading of knowledge (statements and claims), may be found in ancient thought, and Plato is the author of both. In *Cratylus* and *Theaetetus*, he offers us a nominalistic reading of epistemology, and in *Parmenides*, he presents us with an acutely nominalistic version of the One.

A cogent approach for delineating the tenets of negative theology links it to the formulation of the monotheistic principle itself. This principle—the assertion of God's absolute oneness and otherness—represents a desire to bring the philosophical/commonsensical/scientific quest to a halt. As long as the subject matter to which the *explanans* (the statement doing the explaining) makes reference is in any way continuous with the material to which the *explanandum* (the statement that needs to be explained) refers, the pursuit of reasons and causes can proceed indefinitely in the search for a primary explanation. The only way that the philosophical/commonsensical/scientific quest can be brought to a satisfactory arrest and that philosophical repose can be achieved is by postulating an *explanans* that invokes an entity totally unassimilable to human conceptual schemata. The origin of the monotheistic God who is depicted as

1 Plato, *Parmenides*, 130B–132B in Plato, *The Republic and Other Works*, trans. Benjamin Jowett (Garden City, NY: Doubleday/Anchor, 1973), 388. See also the discussion in Aryeh Botwinick, *Skepticism, Belief, and the Modern: Maimonides to Nietzsche* (Ithaca: Cornell University Press, 1997), chapters 1 and 2.

being wholly other than man can plausibly be reconstructed as arising out of the need to achieve release from philosophical anxiety by postulating as First Cause an entity that embodies absolute difference.

There is an important price that the monotheistic believer who conceives of his belief in these terms has to pay. On the one hand, by postulating God as being so distant, the believer has latched on to an entity that can appropriately be conceptualised as God. On the other hand, however, by being so distant and untranslatable into human things, He ceases to be explicable and to have the capacity to serve as a source of explanation altogether. The quest for an ultimate explanatory factor is resolved not by coming up with a solution for it, but in effect by de-legitimising the quest by dramatising its unattainability.

A deepened appreciation of this problematic offers a ray of hope for salvaging God, but only negatively as a possibility, not positively as an actuality. The negative theological critique of the popular concept of God, which ascribes attributes to Him and therefore propels the explanatory quest ever forward, cannot be formulated consistently. Although from a negative theological perspective, none of the traditional theological terms literally applies to God—He does not see, hear, or even exist in the way traditional religious understanding projects—the adherent of negative theology still wants all of his negative criticisms to apply to Him: the subject of all of his negative, paring-down work is the traditional (mono)theistic God. In order for the negative theological critical apparatus to have a viable target, it needs to maintain in some form, to some extent, whatever it is that this critical apparatus shoots down.

The negative theological conceptualisation of God is therefore contradictory. It states that it is only by denying God in any humanly comprehensible way that we have God. Ironically, it can be argued that it is precisely negative theology's embracing of contradiction that gives its adherents their only access to God. Karl Popper, in his essay "What Is Dialectic?", argued that according to the rules of inference enshrined in traditional Aristotelian logic, toleration of contradiction means that no possibilities can be ruled out, so that a generalised agnosticism (which theoretically leaves the door open to all possibilities) becomes one of the major corollaries of the acceptance of contradiction. As Popper succinctly states his thesis: "If two contradictory statements are admitted, any statement whatever must be admitted; for from a couple of contradictory statements any statement whatever can be validly inferred."[2] Since no possibility can be ruled out in the aftermath of contradiction, ontological space has been created for God.

A further paradox emerges that it is only negative theological monism (that is to say, a monism that locates scepticism and God on the same ontological plane by identifying analogous problems of consistency with regard to the formulation of both positions) that is able to sustain a place for God. Only an approach which emphasises that

---

2 Karl Popper, *Conjectures and Refutations: The Growth of Scientific Knowledge*, 2nd ed. (London: Routledge and Kegan Paul, 1965), 317.

attempts to coherently formulate scepticism confront the same self-defeating prospects as attempts to coherently formulate the principles of monotheism—that efforts to sustain an intellectual universe without God founder in much the same way as efforts to sustain an intellectual universe with God—can manage, in this crucial negative sense, to keep the prospect of God alive. In order to emerge as properly consistent, sceptical doctrine must encompass a reflexive manoeuvre whereby sceptical critical canons are turned against the very tenets of scepticism, forcing them to recoil and thereby inhibiting their adequate formulation. Analogously, the utter conceptual removal of the monotheistic God, which renders Him entirely unlike anything human, requires some kind of grammatical subject concerning whom the continual divestiture of predicates can take place. Scepticism is a doctrine of radical critique that both presupposes and denies a stable subject (in a grammatical sense), which are tenets of scepticism. Monotheism is also a doctrine of radical critique that simultaneously presupposes and denies a stable grammatical subject; namely, God.

There is a third level of paradox here which relates to the fact that through the extremeness of its critique, negative theology preempts a literal translation even of itself. To know with certainty that we cannot literally grasp or pin down God's attributes because of His transcendence of known things, so that we can translate those attributes only metaphorically, is already to have breached the unbridgeable distance separating us from God. Negative theology undermines itself by the thoroughgoing nature of its critique, as well as the official targets of its attack. In the end, it cannot foreclose the possibility that on a superficial level, it is most adamantly concerned with refuting: namely, that the literal God exists, literally understood. Negative theology propels us to go full circle from the jettisoning of literalism to the literal affirmation of God.

The approach that I am outlining here will open new sociological, historical, and theoretical vistas for re-conceptualising the relationship between monotheism and scepticism in Western thought. From sociological and historical perspectives as well as logical ones, monotheistic religion (with negative theology as its dominant exegetical decoder) needs to be viewed as a sceptical response to the primitive animistic religion that was practised in the cultures in which monotheism emerged. Monotheism seeks to discredit the immanentist deities affirmed by animistic religion. It counterposes to them a single, transcendent, infinite deity who both created and governs the world. In His aloofness and loftiness, this deity can appropriately be viewed as flagging the limitations of human reason with regard to the search for ultimate explanations, rather than as a cognitively accessible and substantively recognisable and identifiable God.

The God of monotheistic religion has to be understood in terms of what He is blocking and ruling out and not in relation to what He is ostensibly affirming. He is blocking the divinisation of natural and historical forces rather than presenting Himself as their literal replacement. In terms of the strategic factors at stake here, He cannot do what they do in the manner in which they do it without becoming them and forfeiting His status as the one transcendent God.

In order to serve as a tool for critiquing animistic religion, God has to be conceived as being utterly transcendent, as occupying a level of ultimacy of explanation beyond which it is not possible to penetrate. Animistic religion sells the quest for explanation short by arresting the search at a superficial naturalistic level which it is clearly possible to transcend. Positing God as unique, singular, and transcendent means that by definitional fiat, it is not possible to go beyond Him in the search for explanation.

There is another more narrowly logical factor that renders the concept of God rationally impenetrable that is incisively captured in a verse in the Psalms. Psalms 119:138 reads: "You commanded the justice of your testimonies, and great faithfulness." The first part of the verse appears highly paradoxical, and the second seems deeply ambiguous. Instead of justice being implicitly or intrinsically known or knowable, the first part of the sentence emphasises that the justice of God's laws is a function of His commands. We are to imagine the Psalmist addressing God and saying: "You commanded and stipulated that your laws are just; [there is nothing inherently just in them/they are not inherently just] by virtue of their content." One way of reading the second half of the sentence is that the opening words of the verse, "You commanded," are applicable to the second half as they are to the first. The key word in the second half of the sentence in Hebrew is *emunah*, which can be translated as "faith," as well as "faithfulness." Read in this light, the Psalmist is reminding God, as it were, that "we are believers because you commanded us to believe." We do not follow your commandments because we are believers. We believe because we follow your commandment to have faith.

From the perspective of this construal of the verse in the Psalms, we might say that faith evokes the spectre of an infinite regress. Why do we have faith? Because we have faith in the idea of having faith. Why do we have faith in the idea of having faith? Because we have faith in the idea of having faith in the idea of having faith, and so on ad infinitum. The infinite regress can only be broken by acknowledging or stipulating that God commands us to have faith.

The notion of "command" conjures up the same infinite regress as the idea of "faith." A command can only be authorised by a prior command that requires and/or legitimates compliance with the first command. But this second command in turn has to be grounded in a previous command, and so on indefinitely, so that the prospect of an infinite regress emerges with regard to "command" just as it does in relation to "faith." In the end, both "faith" and "commandment" loom as circular categories parasitically feeding off each other's circularity. The Psalmist thus envisages faith as a circle. There is no rational way to enter into it; one has to be in it in order to be in it. Negative theology reinforces the circular character of faith, and the attainment of ultimacy which the introduction of the idea of the infinite God achieves forecloses the possibility of explanation.

The implicit logic of the concept of the biblical God leads directly into negative theology. To ensure that one achieves finality in one's explanatory quest, one has to switch dimensions from the finite to the infinite. This switching of dimensions means that the concept of the biblical God can only serve as a critique of other theistic conceptions that arrest the search for explanation at the level of the finite. The infinite, transcendent biblical God, being deliberately designed not to intersect with human things, cannot serve as an explanatory factor Himself. All the descriptive epithets projected onto Him can only be construed as metaphors, capturing what God literally is not. The biblical God paradoxically only encodes the limits to reason that prevent the explanatory quest from being consummated.

The explanatory line that links human beings to God thus has to be reconfigured as a circle. The blockages that we confront in trying to access and validate Him insinuate to us that perhaps it is only by classifying the limits of our rational understanding of God as God that we have God. Only by re-conceptualising the limits at the outer perimeters of our lines of explanations as a circle (whether a circle of doubt such as scepticism or a circle of belief such as faith in the monotheistic God) are we able to sustain the lines within the circle that we choose. In its inability to sustain itself as a straight line linking premise to conclusion, the circle of doubt (just like the circle of belief) bespeaks in a negative sense the possibility of God. It is the circular character of both faith and doubt that conjures up a domain beyond either of them.

The above-quoted Psalm 119 contains intimations about how the affirmation of God's infinity leads into a circular conceptualisation of monotheistic faith. Verse 160 reads: "Your very first utterance is truth, and every ordinance of your righteousness is valid for all time." One way of translating this sentence is to say: "Your first statement which announced your infinity is true, and this is what establishes the righteousness of your laws." Judaism, then, is envisioned by the Psalmist as a self-contained, self-justifying deductive system of faith and practice.

The fate of monotheistic religion understood from a negative theological perspective is duplicated in virtually all of the fields of human inquiry that have developed alongside religion, or in self-conscious opposition to it. All acts of theorising can be sceptically questioned, at least in terms of interrogating how they privilege their premises. In order to remain consistent, scepticism in turn also needs to be sceptically assessed. What emerges as a result is either a generalised agnosticism—where an uncertain "maybe" is assigned to all human knowledge claims, including scepticism—or a full-scale rehabilitation of the original targets of scepticism, which, however, lacks the resources to preclude the resumption of the cycle of sceptical attacks, instantaneous rehabilitation of the targets of those attacks, and revivals of the attacks themselves. Either way, secure, fully reliable knowledge has been blocked in all of the intellectual areas that are successors to religion, as fully as negative theology's self-undermining blocks the attainment of secure religious knowledge. Religion prefigures what befalls the successor modes of knowledge after the historical eclipse of religious worldviews and sensibility has occurred. These successor modes of sensibility and practice do not extricate us from the quandaries and dilemmas that surround religion itself, but rather, on a purely formal, structural level, re-encode them.

For example, the empiricist approach to everyday knowledge is bogged down in a problematic that resembles the problematic of monotheism. The empiricist seeks to substantiate knowledge statements and claims in terms of their pattern of derivation from sense experience. What the great empiricists almost immediately notice is that sense experience does not come pre-packaged or pre-labelled and needs to be assembled into a usable and reliable basis for knowledge on the basis of an antecedent naming process that makes the sense data valuable and relevant for the justificatory project on which the empiricist is embarked. The neutral and unbiased character of the sense data is thus called into question, with the result that our everyday statements remain unanchored. The empiricist who, on one level, believes that he has the antidote to scepticism—unadorned, irrefutable sense data—paradoxically ends up reinforcing it by the unavoidable need to supplement sense data through endlessly evolving human naming processes. The logical problematic that confronts scepticism itself means that the empiricist's work remains systematically unresolved—and perhaps unresolvable.

Monotheism interpreted from a negative theological perspective constitutes the first run of Western scepticism to have had a pervasive cultural influence. The most influential forms of scepticism in Western thought and experience appear to have been first monotheism and then, more explicitly and self-consciously, the scepticism of the seventeenth and eighteenth centuries.

Conceiving the relationship between monotheism and scepticism as I have theorised it in this essay opens new approaches for interpreting the Western intellectual tradition and the relationship between modernity and postmodernity. The traditional Western intellectual story needs to be revised. Western intellectual history does not consist in a dramatic supplanting of a religious sensibility and outlook by sceptical self-understandings and modes of being in the world. Rather, Western thought seems to move from more inchoate and less self-conscious forms of scepticism to more fully explicit and articulated versions of it. If modernity is to be theoretically and sociologically identified with the emergence of scepticism into full prominence, I believe that postmodernity can be most persuasively envisioned as the emergence of the limitations and dilemmas surrounding scepticism into an augmented self-consciousness. The unexitable square one of philosophy, to which the early forms of Western self-consciousness in Plato alerted us, can now be seen as our common habitation.

The upshot of my reading is that "modernity" has been with us from the start of Western intellectual history. An awareness of scepticism and its limits and dilemmas has been present in Western thought since at least the time of Plato, and it is integral to the reception of Greek ideas in both the medieval and post-medieval periods. In an important sense, a proper appreciation of the origins and development of negative theology clues us in to the fact that all of Western thought consists of one vast continual stationary moment characterised by preoccupation with scepticism and the dilemmas that it engenders. One of the most cogent readings of postmodernism is that it highlights the incomplete and uncompletable character of what precedes it in Western thought and therefore constitutes modernity's most elevated moment of reflectiveness and self-consciousness. Beneath postmodernism's glorying in the fragmentary and the transitory lies an appreciation of the unity of Western thought that perhaps goes deeper than earlier modernist perceptions of it.

Bibliography

Botwinick, Aryeh. *Skepticism, Belief, and the Modern: Maimonides to Nietzsche*. Ithaca: Cornell University Press, 1997.

Plato. *The Republic and Other Works*. Translated by Benjamin Jowett. Garden City, NY: Doubleday/Anchor, 1973.

Popper, Karl. *Conjectures and Refutations: The Growth of Scientific Knowledge*. 2nd edition. London: Routledge and Kegan Paul, 1965.

Michela Torbidoni

Socratic Impulse, Secular Tendency, and Jewish Emancipation: A Comparison between Simone Luzzatto and Moses Mendelssohn

1 Introduction

Since antiquity, the legendary figure of Socrates has embodied the human desire for knowledge, and since the earliest theological controversies, he has symbolised the first authentic free thinker as well as a victim of social and religious intolerance. The role of Socrates as a representative of secular knowledge within the framework of the struggle between philosophy and revelation was greatly thematised during the eighteenth century, which celebrated him as a paradigm of reason, a model of secular virtue, "certainly not an atheist one, but one which was independent of the dogmas of the priests, [...] the archetype of the free thinker in matters of faith."[1] For this reason, the character of Socrates was imbued with a new secular *pietas*, which must be considered a significant expression of the Enlightenment's concerns about the question of religion.

According to Leo Strauss, the Socratic philosophical method reflects the original meaning of the term as a way of life devoted to the realm of knowledge and a search for the truth. Strauss makes a distinction between this concept of philosophy and that of philosophy as a mass of doctrines and a system established by history.[2] In his view, Socrates's philosophy embodies the term's original meaning because it consists on the one hand of an awareness of human ignorance towards fundamental issues and on the other of the *eros* which prompts inquiry and the search for the truth. And indeed, an awareness of the uncertainty of any solution and of our ignorance of the main aspects of philosophical inquiry is the main feature of the philosopher's search for truth and his way of life; namely, a way of life that is incompatible with the biblical one. As part of the great discussion of the opposition between Jerusalem and Athens within Strauss's thought, there is an ambivalence between the Bible and Greek philosophy concerning how to deal with morality and which is the best way of life: philosophers attempt to answer to these questions by relying on their reason and not merely obeying the laws of a particular tradition, thereby opposing biblical hu-

1 Raymond Trousson, *Socrates devant Voltaire, Diderot et Rousseau: la conscience en face du mythe* (Paris: Lettres modernes, 1967), 18.

2 See Leo Strauss, "The Mutual Influence of Theology and Philosophy," *Independent Journal of Philosophy* 3 (1979): 114.

https://doi.org/10.1515/9783110618839-004

mility, which is grounded in fear and piety.[3] Strauss expounded the two different approaches to life through the story of Abraham and the example of Socrates in the following way: the former obeys an unintelligible command without hesitation; the latter, Socrates, despite his belief that Apollo was commanding him, nevertheless did not react "in unhesitating obedience, but in examining an unintelligible saying of Apollo."[4] Socrates questions the truth of the Delphic oracle's statement that he is the wisest man and in doing so, his rational inquiry replaces the blind obedience to divine orders requested by the biblical tradition. In light of this example, the Socratic form of inquiry must be interpreted as the authentic expression of human wisdom in opposition to the divine one.

Despite the radical opposition between Jerusalem and Athens defended by Strauss's view, his speculation about Socrates has the merit not only of having highlighted the secular tendency of Socratic philosophy—namely, Socrates's attempt to free the mind from the fetters of dogmatism and his impact on the social order—but also, and even more surprising, of having linked Mendelssohn's enthusiasm for the figure of Socrates and Simone Luzzatto's *Socrates, Or On Human Knowledge* (1651). In a small passage in his early writing on Mendelssohn's *Phaedon*, Strauss quoted a sentence written by Mendelssohn in a letter to his friend Thomas Abbt at a time when his *Phaedon* (1767) was almost complete:

> I am putting my reasons into Socrates' mouth. [...] I need a pagan so as not to have to get involved with revelation.[5]

The above sentence taken from Mendelssohn's letters is completed by a handwritten marginal note in which Strauss adds a reference to Luzzatto and his *Socrates.*[6] The similarities that Strauss had perhaps seen between Luzzatto's book and Mendelssohn's *Phaedon* seem to be mostly found in Mendelssohn's philosophical decision to speak through Socrates's voice. According to Strauss, this fact reflected not only Mendelssohn's struggle against the Scholastic tradition, but also his struggle against the very concept of tradition; namely, "prejudices," especially those of the revealed religions. In Strauss's opinion, Socrates and Plato attracted Mendelssohn's interest because they were philosophers who did not believe in revelation; that is, they were "rational pagans."[7]

---

3 See Chiara Adorisio, *Leo Strauss lettore di Hermann Cohen: dalla filosofia moderna al ritorno agli antichi* (Florence: Giuntina, 2007), 170–74.

4 Leo Strauss, *Jewish Philosophy and the Crisis of Modernity: Essays and Lectures in Modern Jewish Thought*, ed. Kenneth Hart Green (Albany, NY: SUNY Press, 1997), 111.

5 Letter from Moses Mendelssohn to Thomas Abbt of July 22, 1766, in Leo Strauss, *On Moses Mendelssohn*, ed. and trans. Martin D. Yaffe (Chicago: University of Chicago Press, 2012), 34.

6 Yaffe mentions this handwritten note and directs the reader to Leo Strauss's *Gesammelte Schriften, Band II: Philosophie und Gesetz—Frühe Schriften*, ed. Heinrich Meier (Stuttgart: J.B. Metzler Verlag, 1997). See Strauss, *On Moses Mendelssohn*, 34 n. 40.

7 Strauss, *On Moses Mendelssohn*, 33–34.

Strauss's interpretation suggests that the choice of the pagan Socrates may have been methodological in order to fulfil Mendelssohn's wish to speak free from religious concerns and thus also to raise criticism of traditional prejudices. This consideration, when viewed in connection with the marginal quotation, thus also invites us to reflect on the role of the pagan Socrates within Luzzatto's book, and most of all to what extent this figure may be bound to the history of Jewish emancipation. The cases of Moses Mendelssohn and Simone Luzzatto, as representatives of the German Jewish Enlightenment and Baroque Italy respectively, provide meaningful evidence of the close connection between the use of Socrates and his critical approach and their struggle for the rights of Jews in society.

The present comparison is a preliminary work aiming to provide some hints of a wider topic that will not be fully realised here. However, this brief overview of the Socratic impulses of Mendelssohn and Luzzatto may be considered an interesting starting point in order to examine not only their philosophical projects, but also the social and religious claims behind them. Starting from the better-known case of Mendelssohn's rational Socrates and proceeding to Luzzatto's sceptical version of him, the great complexity of the Jews' relationship to their surrounding culture will be explored. This will help to add some further considerations on the state of the Jews in early modern Italy and the German *Aufklärung*.

As we will see, under Socrates's flag, both Mendelssohn and Luzzatto pursue a common aim; namely, the foundation of a universalistic human community. However, what is interesting and more problematic in this respect is the different methodological approaches to such a figure. While Mendelssohn uses Socrates as a means of disseminating his original optimism and trust into the universalistic value of reason, careful to conform to the cultural tendencies of the Enlightenment, Luzzatto makes Socrates the mediator of his provocative sceptical pessimism towards human certainties. These two contrary methodological approaches invite us to reflect on the freedom of speech already achieved by this early modern Venetian rabbi, on the multi-level challenge issued by his philosophical writing in which Socrates was the main character, and above all on the real status of the Jews in Venice, where a rabbi could state that his book had been written by a "Venetian Jew."

2 Mendelssohn's Socrates of Reason

Within eighteenth-century Jewish thought, however, Socrates must be seen as an ambivalent figure of reason, and Moses Mendelssohn's commitment to achieving a true and coherent synthesis between rationalist philosophy and Jewish religion properly shows this ambivalence.

We should not overlook the fact that within the framework of intellectual history, the figure of Socrates also embodied crucial religious and political issues during the early German *Haskalah* (1680–1720): in an era of intense criticism of Lutheran orthodoxy and its relationship to the Aristotelian philosophical tradition in German universities,[8] Socrates had usually been taken into account independently of the Platonic tradition and praised as the philosopher of moral questions who had turned his attention from the contemplation of natural things to human customs and behaviours.[9] However, according to the German Lutheran theologian Johann Franz Budde (1667–1729), "the figure of Socrates assumes an important role: his struggle against the Sophists more or less foreshadowed the struggle of modern philosophy against the empty metaphysical debates and arrogance of the Scholastics (*supercilium scholasticum*)."[10]

A similar attitude may also be found in Mendelssohn's decision to put his reasoning into the mouth of Socrates, a pagan, in his *Phaedon* (1767): this was meant to fulfil his wish to speak free from religious concerns and thus reflected his attempt to underline the universalistic foundation of philosophy, theology, and above all reason as the common ground of every religion. *Phaedon* was an essay about what reason can know regarding the immortality of the soul when it is not guided by revelation; namely, by counting only on its own capacities. Mendelssohn's decision to write such a work was prompted by his admiration for Plato and Socrates, which was rooted in his opposition to the dogmatism and pedantic authority of Scholastic philosophy. In putting forth strictly philosophical arguments for the immortality of the soul, the dialogues contend that such reasonings must take place independently of the demands of religion.

Mendelssohn's portrayal of Socrates greatly idealised his wisdom, his patience and impartial philanthropy, his devotion to the laws of his homeland until his death, his liberal attitude towards religious traditions, and his decisive opposition to prejudices and superstitions.[11] Although Socrates lived in a time that did not share the oppressive theological circumstances of Mendelssohn's century, nevertheless Mendelssohn described him by focusing less on his personality and more on his cultural context, which seemed to reflect Mendelssohn's own dilemma:

> He had on the one hand to conquer the prejudices of his own education, to enlighten the ignorance of others, to refute sophistry and calumny of his adversaries, to bear poverty, to contend against established power, and, what was still more hard than either, to dissipate the dark terrors of superstition. On the other hand, the feeble understanding of his fellow citizens required

---

8 From 1660 in Germany, and more specifically following the foundation of the University of Halle, a religious circle that was highly critical of Lutheran orthodoxy began disseminating its ideas: its promoters, such as Christian Thomasius, Johann Franz Budde, and some members of the Pietist community, challenged Lutheran traditions and encouraged a new study of the relationship between Christianity and classical philosophy. On this topic, see Guido Bartolucci, "Jewish Scepticism in Christian Eyes: Jacob F. Reimmann and the Transformation of Jewish Philosophy," in *Yearbook of the Maimonides Centre for Advanced Studies 2018*, ed. Bill Rebiger (Berlin: De Gruyter, 2018), 145–64.

9 Gregorio Piaia and Giovanni Santinello, eds., *Models of the History of Philosophy. Volume II: From the Cartesian Age to Brucker* (Dordrecht and New York: Springer, 2011), 357.

10 Piaia and Santinello, 357.

11 See Alexander Altmann, *Moses Mendelssohn: A Biographical Study* (Philadelphia: Jewish Publication Society of America, 1973), 157–58.

> to be managed with great tenderness, that he might draw no reproaches upon himself, or lessen the influence which the poorest religion has on the morals of weak minds.[12]

The goal of Mendelssohn's Socrates is to eradicate "the dark terrors of superstition" in favour of a broader horizon made by an enlightened concept of a universal God which unifies reason and religion and thus may be shared by different cultures. The philosophical project of *Phaedo*'s Socrates is thus intimately linked to Mendelssohn's personal intellectual experience and desire for a space in which speculative thought could be independent from any religious group:

> Wisdom knows a universal fatherland, a universal religion, and if it tolerates divisions, it does not tolerate the unkindness, misanthropy of such divisions, which you have made the foundation of your political institutions.—This is how, I think, a man like Socrates would think in our days and seen from this angle, the mantle of modern philosophy, which I have given him, would not seem entirely unfitting.[13]

In his use of a pagan speaker, Mendelssohn was attempting to overcome the particularity of a specific religious creed and to join Enlightenment's universalist project; however, the entire book must also be read as an answer to a Christian readership. According to Miriam Leonard, Mendelssohn engaged in this dialogue in order to appease the long-standing Christian critique that accused the Jews of having neglected the immortality of the soul:

> From the Christian perspective, the question of immortality was just one more indication of the essential lack of Judaism. Paradoxically, immortality is a Greek not a Jewish idea. Jewish philosophers like Philo had to turn to Plato to fill in the gaps of Jewish theology. In this version, Platonism is not so much the antithesis as the supplement to Judaism. Plato hardly represented the dangerous lure of paganism. Jews needed Plato to become better Jews.[14]

Socrates is anything but Mendelssohn's mouthpiece; he leads Mendelssohn's opinion to the conclusion that this world has been conceived so as to make those of its inhabitants who are endowed with reason continue in their progress towards perfection and happiness. From this assumption, he deduced that it was impossible to conceive of a divine wisdom which would allow the interruption of this progress and cause them to lose the results of their efforts.[15] Mendelssohn strengthens this argu-

---

12 Mendelssohn, *Phaedon*, in *Schriften zur Philosophie und Ästhetik*, ed. Fritz Bamberger and Leo Strauss, vol 3.1 of *Gesammelte Schriften. Jubiläumsausgabe*, ed. Alexander Altmann et al. (Berlin: Akademie Verlag, 1972), 15–16. From now on, I will refer to this edition as *JubA*. The English translation is taken from Willi Goetschel, *Spinoza's Modernity: Mendelssohn, Lessing, and Heine* (Madison: University of Wisconsin Press, 2003), 119–20.
13 *JubA*, 3.1:151–2; Goetschel, *Spinoza's Modernity*, 121.
14 Miriam Leonard, "Greeks, Jews, and the Enlightenment: Moses Mendelssohn's Socrates," *Cultural Critique* 74 (2010): 194.
15 See Mendelssohn, *JubA*, 3.1:115.

ment, which is in itself sufficient to demonstrate the immortality of the human soul, and underlined that human beings have a need to believe in the existence of a life after death.[16]

Based on Plato's *Phaedo*, Mendelssohn's dialogue on the immortality of the soul put forth a persuasive plea for the recognition that human individuality is grounded in every person's inalienable right. His use of Socrates to expound his philosophy—namely, a Greek philosopher who represented a common universal point of reference in Western philosophy—was the key to the immense success of his *Phaedon*.

Mendelssohn's essay is more than simply evidence of the conflict between the Enlightenment and religion: on the one hand, the figure of Socrates reflected Mendelssohn's philosophical search for the acknowledgment of the universal value of reason, and through him, he was able to present his philosophical project to a greater public. On the other, the Socratic impulse which permeated Mendelssohn's life and work was sometimes misinterpreted by the Christian side and seen as an attempt to distance himself from Judaism and come closer to Christianity. For this reason, behind the epithet "the German Socrates,"[17] which was bestowed on him by his contemporaries, there was not only the great acknowledgment that Mendelssohn received from the Christian intellectual world, but also persistent prejudice concerning his Judaism and the expectation of his conversion.

An example of this concern is the caustic letter exchange between Mendelssohn and Johann Kaspar Lavater. As soon as Mendelssohn had established himself in the republic of letters, he was obliged to face Lavater's public challenge, which had a devastating effect on his life. In his translation of Charles Bonnet's *La palingénésie philosophique*,[18] the Swiss philosopher and theologian addressed a dedicatory epistle to Mendelssohn. Since Lavater had interpreted Bonnet's arguments as a demonstration of the truth of Christianity meant for non-Christians, he invited Mendelssohn to either refute these arguments or else to do "what Socrates would have done if he had read [Bonnet's work] and found it irrefutable";[19] namely, to convert to the religion of truth and in so doing, to do what, in his opinion, reason would demand.

Mendelssohn, deeply offended by this public provocation, was also compelled to react, because his failure to answer to Lavater's attack could have been seen as an incapacity to dismiss Bonnet's arguments and could also have raised suspicion to-

---

16 See Allan Arkush, *Moses Mendelssohn and the Enlightenment* (Albany, NY: SUNY Press, 1994), 57.

17 On this topic, see Christoph Schulte, *Die Jüdische Aufklärung: Philosophie, Religion, Geschichte* (Munich: C.H. Beck, 2002), 199–206.

18 Charles Bonnet's book, *La palingénésie philosophique, ou, Idées sur l'état passé et sur l'état futur des êtres vivans*, was published in Geneva in 1769 by Claude Philibert and Barthelemi Chirol.

19 Mendelssohn, *Zueignungsschreiben Johann Caspar Lavaters an Moses Mendelssohn* in *Schriften zum Judentum I*, ed. Simon Rawidowicz, vol 7.1 of *Gesammelte Schriften. Jubiläumsausgabe*, ed. Alexander Altmann et al. (Berlin: Akademie Verlag, 1930. The English translation is taken from Moses Mendelssohn, *Writings on Judaism, Christianity, and the Bible*, ed. Michah Gottlieb, trans. Curtis Bowman, Elias Sacks, and Allan Arkush (Waltham, MA: Brandeis University Press, 2011), 3.

wards his sincere commitment to Judaism. At the same time, his denial of Bonnet's arguments could have been interpreted as an implicit attack on Christianity. He was thus obliged to take a public stand with regard to Judaism, which was what he had above all wished to avoid. In this way, Lavater created a situation which undermined the openness and fairness of the free speech of the republic of letters. The solution that Mendelssohn devised masterfully allowed him to avoid such an insidious trap. In an open letter to Lavater, he focused on Judaism's tolerant attitude towards non-Jews by arguing that it was a religion based on the universal laws of rational morality, the Noahide Laws, which do not demand the conversion of others, contrary to Christianity.[20]

Mendelssohn's involvement in the emancipation of the Jews consisted, as is well-known, in his mediating role between two apparent opposites, the Enlightenment and Judaism. His main goal was to create the foundations for a Judeo-German literature by propagating an image of Judaism that could suit the contemporary conception of reason. The uneasy compromise between reason and revelation that preoccupies Mendelssohn in his *Phaedon* is also at the heart of his most famous work, *Jerusalem, Or On Religious Power and Judaism* (1783), in which he attempted to show that Judaism was not only reconcilable to the Enlightenment, but also that it provided the most convincing model of a life ruled by reason.

According to David Sorkin, Mendelssohn's wished-for synthesis may be situated within what he calls the "religious Enlightenment," a term that points out the complexity of his relationship to the larger culture.[21] On the one hand, there was his support of an ideal compatibility between reason and divine revelation based on the cultivation of intellectual perfection and the practice of universal ethics; an ideal prompted by "an egalitarian impulse, holding that all human beings (not just élite philosophers) could know metaphysical truth"[22] and by the refusal to search for an ultimate truth. On the other, there was his reliance on what Bernard Septimus has defined as the "Andalusian tradition,"[23] meaning a flexible approach to Judaism within medieval Jewish philosophy which did not reject rationalism itself, but established boundaries to it by subordinating philosophy to piety and limiting the horizon of human knowledge. For this reason, according to Sorkin, Mendelssohn seems to use novel means to achieve conservative ends.

---

20 See Mendelssohn's open letter to Lavater, Mendelssohn, *JubA*, 7.1:7–17.

21 David Sorkin, *Moses Mendelssohn and the Religious Enlightenment* (London: Halban, 2012), 25.

22 Sorkin, xx–xxi.

23 Bernard Septimus, "Nahmanides and the Andalusian Tradition," in *Rabbi Moses Nahmanides (Ramban): Explorations in His Religious and Literary Virtuosity*, ed. Isadore Twersky (Cambridge, MA: Harvard University Press, 1983), 34.

3 Luzzatto's Sceptical Socrates

Little more than a century earlier, a Venetian rabbi, fully committed to the defence of the Jews, had adopted the figure of Socrates as a spokesperson for his philosophical, religious, and political ideas: Simone Luzzatto. One of the chief rabbis of Venice, Luzzatto is known for having published a plea to prevent the expulsion of his people from the city, *Discourse on the State of the Jews*, in 1638. Among the many arguments put forth in this work, Luzzatto also underlined the universalistic foundation of Judaism. He stated that the human community must be grounded in the observance of the precepts of natural morality and in some cognition of the superior cause;[24] in this way, he showed that he shared the widespread belief of his time that humanity consisted in acknowledging the existence of a superior cause and in sharing the precepts of natural morality.[25] In this work, Judaism is presented as being deeply bound to this universal community thanks to its traditional struggle against misbeliefs due to its inner adherence to rational thinking:

> No other precept is reiterated as frequently in the Scripture as the prohibition against magic, the judicial magic of the times, necromancy, and all other vain observances and abominable idolatries. For they are all the offshoots of superstition, which is nothing but an abuse, or worse, a waste product of true religion and the legitimate worship of God. Indeed, Moses was not so worried about chasing impiety from the world as he was about rooting out superstition. For a regular series of events, a chain of connected causes, or even the anatomy of a base, ugly little animal is proof of the existence of God and a confutation of atheism.[26]

In his response to Tacitus's denigration of Judaism, Luzzatto wants not only to demonstrate that Scripture itself is against idolatry, but also to underline the rational nature of the Jewish religion:

> I cannot fully understand what he wishes to infer when he states that the ancient Jews were oppressed by superstition, while he himself affirms first that they "conceive of one god only, and that with the mind alone."[27]

However, the idea of Judaism having a universal rational foundation is only briefly mentioned in the *Discourse*. After all, how could this book, which was an official defence of the Jews in light of dangers to their lives in Venice, be the right place to un-

24 See Simone Luzzatto, *Discourse on the State of the Jews of Venice*, ed. and trans. Giuseppe Veltri and Anna Lissa (Berlin and Boston: De Gruyter, 2019), 134.

25 See Alessandro Guetta, *Italian Jewry in the Early Modern Era: Essays in Intellectual History* (Brighton, MA: Academic Studies Press, 2014), 97.

26 Luzzatto, *Discourse*, 171–73.

27 Luzzatto, 172–73. The quotation from Tacitus is taken from his *Histories*, 5.5.18–19, in Tacitus, *Histories: Books 4–5. Annals: Books 1–3*, trans. Clifford H. Moore and John Jackson (Cambridge, MA: Harvard University Press, 1931), 183.

fold such a philosophical argument? Luzzatto would only properly articulate this idea more than ten years later in his second book, *Socrates, Or On Human Knowledge*, in which the famous Socrates and his critical attitude are allied to the pattern provided by sceptical philosophy. This combination of elements thus created the ideal premise for an inquiry able to delve into what Luzzatto defined as the "human community" described above.

Mendelssohn's Socrates is like a saint, a positive character who entirely relies on the certainty of reason. His *Phaedo* amply documented Mendelssohn's reliance on the abilities of reason, because it was an essay meant to show what reason can know regarding the immortality of the soul when it is not guided by revelation; namely, by counting only on its own capacities. Luzzatto's Socrates is rather an anti-hero; he is insolent, restless in his inquisitive approach to life, and substantially sceptical. In the extended title of *Socrates, Or On Human Knowledge*, the premises of Mendelssohn's *Phaedon* seem to be reversed, as it states that it is "a book that shows how deficient human understanding can be when it is not led by revelation." He belongs to the sceptical tradition going back to Michel de Montaigne's philosophy, with which Luzzatto was very conversant. Montaigne's use of Socrates's attitude and irony[28] was meant to overcome the borders established by rituals, superstitions, and prejudices. His goal was to discover the universality of rational thinking and at the same time to build the autonomy of the conscience, because according to Montaigne, every human being bears in himself the entire human condition.[29] Like Montaigne, Luzzatto, through the figure of Socrates, defines the guidelines of his philosophical project: the possibility of overcoming the mere singularity of his Jewish identity and achieving universality through his speculative thought.

The pattern followed by Luzzatto's Socrates was strongly inspired by Sextus Empiricus's *Outlines of Scepticism*: the ten *tropoi* lead his inquiry through the subjectivity of human perception, a source of continual illusion, or the inconstancy of the state of the whole up to the final acknowledgment of the inscrutability of any object by itself.[30] Like Sextus, Luzzatto's Socrates is persuaded that the suspension of judgement (*epochē*) is the correct attitude towards the world, because every statement is subject to irreconcilable conflicts (*diaphōnia*). Equipollence, which demands the suspension of judgement which they believe restores peace of mind or tranquil-

28 See Alexander Nehamas, *The Art of Living: Socratic Reflections from Plato to Foucault* (Berkeley, Los Angeles, and London: University of California Press, 1998), 101–27.
29 See Montaigne, *The Complete Essays*, ed. and trans. M.A. Screech (London: Penguin, 2003), 3:2.
30 On the impact of Sextus Empiricus's *Outlines of Scepticism* on Luzzatto's *Socrates*, see Michela Torbidoni, "Il metodo del dubbio nel *Socrate* di Simone Luzzatto," in *Filosofo e rabbino nella Venezia del Seicento: Studi su Simone Luzzatto*, ed. Giuseppe Veltri (Rome: Aracne, 2015), 183–245, and Giuseppe Veltri, *Alienated Wisdom: Enquiry into Jewish Philosophy and Scepticism* (Berlin and Boston: De Gruyter, 2018), 217–24.

lity (*ataraxia*), reigns in every inquiry that sceptics undertake.[31] Socrates is here not only masterfully managing Sextus's *Outlines*, but he is also well acquainted with the Academic scepticism of Carneades, whereof Cicero's *Academics* is the chief source. The "probable" (from the Greek *pithanon*) plays a major role in Luzzatto's book, in which it is defined as the "North Star in his journey through life," showing thus that it is consistent with the sceptic's instruction to follow the probable in the absence of truth.

Socrates here embodies the restless human curiosity which questions past and present beliefs without stopping to wonder what the future will bring. He questions anthropocentric cosmology by recalling disrupting thinkers such as Lucretius or Bruno and referring to the impious theory of the existence of an infinite number of worlds or the idea of the eternity of the world. Through Socrates, Luzzatto aims to break the illusion of mankind's superiority, and on the contrary, he upholds its marginalised position in the universe: this stance is also strengthened by his comparison between human beings and animals and the fact that he reduces the gap between them. Luzzatto's goal is to show the miserable state of every human being who arrogantly boasts of his wisdom or is absolutely convinced that he is following the right belief or custom.

Following modern scepticism, Luzzatto comes to acknowledge the relativity of the customs and beliefs of every civilisation to every epoch: his arguments lead to an unavoidable "isostheny" between all views, past and present. Luzzatto's Socrates champions the idea that no certainty exists, but rather that everything is true relative to its era and to the culture to which one belongs. An example of this is found in the following passage:

> However, in addition to this, is there any opinion that, despite having strongly persisted for a long time instead of ending its life and dying out in accordance with all other human things, turns out not to be temporary in relation to the infinite future and the undefined posterity to come? Rather, every day, we experience that many opinions held in high esteem and consideration by antiquity nowadays turn into mere fabulous fantasies and ridiculous tales.[32] The worshipped ox, deified by the Egyptians along with many other brutes, already begins to be considered as frenzied and mad by us Greeks.[33]

Luzzatto's Socrates fulfils a double task: there is an evident *pars destruens* meant to demolish traditional certainties in every field from astronomy to ethics. This

---

31 On the legacy of Pyrrhonism, see Richard H. Popkin, *The History of Scepticism from Savonarola to Bayle*, rev. ed. (Oxford: Oxford University Press, 2003) and *The High Road to Pyrrhonism*, ed. Richard A. Watson and James E. Force, repr. ed. (Indianapolis: Hackett, 1993).

32 Living by experience and without opinions plays a relevant role in Sextus's arguments against dogmatic assertions; see *Outlines of Scepticism* 2.246.

33 The reference to the Egyptians' worship of animals also occurs in *Discourse*, 168–71, where Luzzatto quotes Juvenal, *Satires* 15.1–4. The same topic, together with the same quotation, is found in Montaigne's *Essays* in 2:11, 462–64.

part, which is the most consistent in the whole of the *Socrates*, pursues the daring undertaking of undermining the traditional body of knowledge based on Aristotelian philosophy, because it is evidently no longer able to contain the immense changes and discoveries that arose during the seventeenth century. The reader of this first part of Luzzatto's book will find the Socratic celebration of ignorance—meant as an awareness of not knowing anything that may claim the name of truth and the critical use of reason—to be a tool for refusing everything dictated by tradition or blindly accepted as religious belief. Thanks to his Socrates, the rabbi expounded a refined criticism of the alleged truths or dogmata of his time established by an authorial tradition which demanded to be obeyed and not questioned. This part of the book, as has been seen, was made possible thanks to Luzzatto putting ancient and modern sceptical strategies into the mouth of Socrates, the most restless mind in Western culture.

However, the long *pars destruens* prepares the basis for the *pars construens*: the uncertainty disseminated throughout the book generates a sense of loss which creates bonds between human beings rather than separating them. Doctrine, social wealth, beauty, custom, religious ceremonies—namely, the major sources of human division—are, in Luzzatto's treatise, all subjected to doubt, and the human being ends up being stripped of all his convictions that are opposed to God and Nature. Indeed, the universality that Luzzatto's thinking aims to achieve is not atheistic, but is rather a theistically inspired religiosity which does not need any particular creed. Specifically, the final part of his *Socrates* shows that in his view, proper religious behaviour conforms to reason, is universal, and is against superstition and degenerate beliefs:

> For it is sufficient for my defence that you observe the public and private reasonings that I have always delivered concerning the reverence due to the first and worthiest cause, which moves and rules everything. Indeed, I have always promulgated that the cognition that one has of it and the veneration that is due to it come not only from subtle and wide-ranging deductions, but were also given to us along with milk by Nature itself. Hence, it follows that the human mind is so inclined and favourably disposed to religion and divine worship that if it were deprived of such a pursuit, he would not be very different from brute animals.
>
> I have never despised or omitted [to perform] ceremonies or institutions ordered by our city for the observance of religion, but I have always publicly offered sacrifices in accordance with the rites of my homeland, in appropriate places, at the right moment, and in a legitimate manner. And if I sometimes took a position against the ignorant by reprehending them for their ridiculous superstition or degenerate religion, I was not then attempting, as the Giants did, to expel Jupiter from the sky, but rather trying to remove those despicable concepts which disfigured the beauty and grace of the true religion in their minds.
>
> Therefore, I have often solemnly said to those who were truly prudent that they must protect themselves from the infection of superstition, an epidemic and serious disease of the people. They [must] be aware that often the religion of the vile common people is abominable blas-

> phemy for wise men and that the true temple of God is in the wise man's mind, where He is adored through offerings of love and sacrifices of veneration.[34]

Although Luzzatto's opinion is presented very prudently in this passage, it is nevertheless evident that he deploys reason as the true temple of God. Only there, in his opinion, does true religion seem to find shelter from "the infection of superstition" and "abominable blasphemy."

Luzzatto's conception of Nature is intended to increase this common ground shared by human beings: his sceptical Socrates promotes a naturalistic morality based on Nature as a "lover of equality" able to provide mankind with a profitable tool that should be the only reliable guide through life; namely, the probable, described as coming directly "from the hands of Nature." Luzzatto portrays a sort of genealogy in which Nature must be considered the true source of moral values:

> You must not even doubt that in discrediting its own judgement and accusing it of falsity, the court of conscience attended by the human mind would be debased and lacking in authority, because it will be replaced by the majesty of Nature, which will lead the way more decorously towards the good and remove the evil.[35]

Moral virtues are indeed, according to Luzzatto's Socrates, "daughters of the probable," and thus "legitimate grandchildren of Nature itself."[36] Nature is thus an impersonal entity free of any traditional features, which therefore offers Luzzatto a solution to the dizzying results of his free inquiry into dogmatic knowledge. Luzzatto's focus on Nature seems to continue to fulfil his attempt to bring the discussion to a universal level. According to him, humanity must be conceived "not as something abstracted and excluded from universal Nature, but as something included in it that should comply with it and be ably ruled by it like the other mundane things."[37]

Luzzatto's arguments concretely erode dogmatic beliefs as a source of separation and division in society and indirectly put forth the premises of unity and equality among human beings; namely, the acknowledgment of what he had already defined as the "human community" in his *Discourse*. In his view, this larger community shares the uncertainty of knowledge and beliefs as the only possible certainty and may rely only on a natural morality based on the probable and on the faith in God which dwells in the mind and not in religious ceremonies, securing itself from any form of degeneration. In this way, Luzzatto constructed an indirect apologetic discourse which fosters tolerance and extends the borders of the concept of

---

34 Luzzatto, *Socrates, Or On Human Knowledge*, ed. and trans. Giuseppe Veltri and Michela Torbidoni (Berlin and Boston: De Gruyter, 2019), 478–79.
35 Luzzatto, 470–71.
36 Luzzatto, 476–77.
37 Luzzatto, 306–7.

community by creating a new space for more acceptance of social and religious pluralism.

Luzzatto could develop this *pars construens* only thanks to the Greek Socrates. After reading Luzzatto's *Socrates*, one may note that most of the author's speech is delivered by Socrates himself. It may furthermore be seen that the text displays a rich variety of classical doctrines, including gnoseology, ethics, psychology, medicine, mathematics, geometry, and astronomy, embellished by passages and verses taken from classical poems and plays. With the exception of a few direct quotations from medieval and Renaissance Italian literature—such as Dante's *Divine Comedy*, Ariosto's *Orlando Furioso*, and Tasso's *The Liberation of Jerusalem*, which were typical for a learned Italian Jew[38]—and two from the Bible (Ecclesiastes and Proverbs), Luzzatto's book openly prefers the Greeks and Latins as speakers; namely, Greek philosophers and mythological and legendary figures or politicians from this ancient world, as well as Greek and Latin literature for the quotations between the paragraphs. Questioning Luzzatto's choice of the classical world for the setting of his philosophical scepticism therefore requires us to engage with the long-lasting relationship between Jewish thought and religion and classical philosophy, which began during the Middle Ages thanks to Jewish encounters with texts from the Greek philosophical traditions[39] and was revived during the Renaissance, strongly impacting the passage to modernity. During the Middle Ages, when the teachings of the two textual traditions seemed to contradict one another, a variety of methods for resolving contradictions and synthesising these traditions emerged, which spurred the development of medieval Jewish religious philosophy and gave rise to public controversies. Ancient philosophy, which emerged as a polemical middle ground on which to engage criticism during the Middle Ages, experienced a rebirth during the Renaissance, when many classical treatises were edited, translated, and commented on.[40] This revival was a challenging moment because it allowed a transition from a philosophical setting shaped by Scholastic Aristotelian doctrine to modernity. Ancient wisdom (Platonism, Neoplatonism, Stoic ethics, Atomism, or Scepticism) continued to be examined for its "neutrality" in religious matters and theological controversies which were deeply connected to Aristotelian authority and became extremely important during the sixteenth and seventeenth centuries. Classical philosophy was therefore seen as a safe zone for those intellectuals who were committed to free thinking or to the new experimental science, and for this reason, it was an excellent bridge of dialogue and comparison between Jews and Christians.

---

38 See Asher Salah, "A Matter of Quotation: Dante and the Literary Identity of Jews in Italy," in *The Italian Judaica Jubilee Conference*, ed. Shlomo Simonsohn and Joseph Shatzmiller (Leiden and Boston: Brill, 2013), 167–96; Guetta, *Italian Jewry in the Early Modern Era.*

39 See Adam Shear, "Science, Medicine, and Jewish Philosophy," in *The Cambridge History of Judaism. Volume 7: The Early Modern World, 1500–1815*, ed. Jonathan Karp and Adam Sutcliffe (Cambridge: Cambridge University Press, 2017), 522–49.

40 See Michael Heyd, "Introduction," *Science in Context* 15, no. 1 (March 2002): 1–8.

As in Mendelssohn's case, Luzzatto's erudition in secular and Christian learning also allowed him to engage in a solid dialogue with the learned people outside the ghetto and even to gain the respect and attention of those Christians with whom he was acquainted. There is indeed some evidence which clearly testifies to the regard in which he was held among non-Jewish Venetians and foreigners: the way he argued in defence of the Jews in his *Discourse* inspired and constituted the pattern of another plea on behalf of the Jews made between 1659 and 1660 by a Venetian senator of the Loredan family, twenty-one years after the publication of Luzzatto's book.[41] Recent studies have furthermore shown that scientists like the astronomer Ismaël Boulliau, who was in the service of the French librarian and erudite humanist Jacques Dupuy, enjoyed the company of the wisest rabbis of the Venice ghetto, such as Leone Modena and Simone Luzzatto, for book collection and theological debates.[42] Furthermore, there is evidence in Giulio Morosini's work *Via della fede mostrata a' gli ebrei* (1683) that Luzzatto was asked to be the arbitrator in a public debate between two brothers concerning the prophecy of Daniel.[43]

The celebration of Luzzatto's erudition and wisdom among the Gentiles is evidence of the intense dialogue and intellectual exchange he undertook with the Christian world. However, as in Mendelssohn's case, this openness was sometimes misinterpreted by Christians: the Franciscan Luigi Benetelli, in his invective work *I dardi rabbinici* (*Rabbinical Arrows*) published in 1705, mentioned Simone Luzzatto and Leone Modena as being known for addressing their works only to a Christian readership. Benetelli interpreted Luzzatto's behaviour and work as a wish to convert or an inclination to Christianity. It is not surprising indeed that Benetelli should have classified Luzzatto and Modena among the so-called evangelical rabbis (*rabbini evangelici*), meaning those who follow the Gospel but do not do so openly in order to not offend their coreligionists.[44] The Franciscan strengthened his statement by mentioning some rumours that Cardinal Gregorio Barbarigo reported to him; namely, that Luzzatto wanted to convert to Christianity on his deathbed, but that his coreligionists

---

41 See Giuseppe Veltri and Gianfranco Miletto, "Difesa inedita del senatore veneziano Loredan in favore degli ebrei nel 1659–60 basata sul *Discorso* di Simone Luzzatto," in *Filosofo e rabbino*, 249–74.

42 See Giuseppe Veltri and Evelien Chayes, *Oltre le mura del ghetto: Accademie, scetticismo e tolleranza nella Venezia barocca* (Palermo: New Digital Press, 2016), 121–46.

43 See Giulio Morosini, *Via delle fede mostrata a' gli ebrei* (Rome: Stamperia della Sacra Congregazione de Propaganda Fide, 1683), 13–14. Morosini reports that Luzzatto stated: "As you know, the text we are currently discussing leaves all the most excellent rabbis in a state of perplexity, and they are so upset that they almost do not know whether they are in Heaven or on earth." Furthermore, Morosini wrote that after a few similar words, Luzzatto puts a finger to his lips and said: "Be quiet, if I may ask, because if we continue speculating on the prophecy of Daniel, we should all become Christians. I cannot deny that it shows very clearly that the Messiah arrived some time ago. However, whether this was Jesus of Nazareth or not, I do not want to state my opinion too hastily" (my translation).

44 See Luigi Benetelli, *I dardi rabbinici* (Venice: Domenico Lovisa, 1705), 19.

did not allow any Christian, regardless of their social standing, to be present during the last moments before he died.[45]

Contrary to Mendelssohn's case, the scant amount of information concerning Luzzatto's intellectual life in Venice does not provide any evidence concerning his opinions or reactions in this regard. However, this short overview may be considered helpful to initiate an exploration of the meaning of this Jewish Socratic impulse—namely, the critical and anti-dogmatic approach to established knowledge—and how it was perceived from a Christian perspective.

4 Conclusion

The comparison between Mendelssohn and Luzzatto is useful here to understand the many implications involved in the use of Socrates by Jewish authors. For both of them, speaking through Socrates becomes a way to make their philosophical views reach a wider readership and reflects their wish to be seen only as far as their intellectual value is concerned. Mendelssohn and Luzzatto share the portrait of Socrates as a free thinker and free speaker beyond any boundaries; they both celebrate his critical attitude towards degenerated forms of beliefs, against superstition, in favour of a spirituality deeply rooted in reason and natural morality. For both of them, the figure of Socrates implicitly becomes a means of fostering their struggle for Jewish emancipation: according to Mendelssohn, Socrates is the mediator of his strong belief in the power of reason to build a community of human beings as estimators of speculative thinking which overcomes religious, social, or ethnic boundaries.

As far as Luzzatto's Socrates is concerned, he embodies the sceptic who inquires and demolishes every certainty, causing the apparently solid castle of beliefs established as undoubtable certainties over the centuries to collapse. From the ashes of this, his sceptical Socrates leads the way for a common ground based on natural morality and a faith in God which, in order to maintain its integrity and avoid blasphemy, can only be celebrated in the mind and not with external ceremonies.

Through the figure of Socrates and Socratic dialogue, the two philosophers achieve not only the goal of saving religion from degenerated forms of beliefs, but also that of fostering critical and free inquiry into dogmatic traditional knowledge and furthering one's self-knowledge in dialogue with others. Socrates therefore becomes a means of overcoming the borders of religious division and preparing the ground for a larger community; he was thus a medium for removing Judaism's particularistic features and instead stressing its universal content. Most of all, the Greek philosopher embodies the cultural role of philosophy and "secular" knowledge within the framework of Judaism; namely, the emergence of a group of intellectuals who gave

45 Benetelli, 19.

priority to such study, at least in their private lives, over the traditional Halakhic focus of the rabbinic élite.

It is also crucial to underline the revival of Socrates's critical spirit among Jewish thinkers because it may be considered as a possible beginning of Jewish philosophy, which may be identified in the very moment when the traditional bonds holding the community together begin to collapse, occurring when tradition starts to be perceived as the antithesis to progress and progressive thinking. Jewish philosophy may therefore be interpreted as a response to this perception and also as the emergence of new questions usually articulated by modern philosophers concerning individuality and community, the commensurability of reason and revelation, and the historisation of tradition.[46]

In his great work on Jewish philosophy, Daniel Frank specified that "there are problems that arise at the end of the tradition, at a time when a certain distancing from the tradition has occurred"[47] and also that this was not an issue of interpreting or understanding tradition, but of questioning it. The use of Socrates may document the way Jewish philosophy interweaved with emancipation: it refers to the intellectual experience of those Jewish thinkers who were challenging the monolithic ground of tradition and inquiring into their own identity not only as a community, but as individual citizens in a Christian society. These Jews were those who developed personal relationships with Christians through the medium of philosophy: by doing so, they began to emancipate the individual from the sole identity that the Jewish community offered to him and started to gain a double acknowledgment.

Finally, it must be pointed out that while Mendelssohn made Socrates the interpreter of his ideas in search of a way to find harmony between his religion and a more general education, Italian Jewish thinkers like Simone Luzzatto were already able to link their Jewish religion to humanistic education and thus to express themselves via the figure of Socrates and celebrate his genius.[48] The use of Socrates made by the Venetian rabbi Luzzatto substantially supports the thesis that early modern Italian Jewish culture must be explored more carefully as an important intellectual foundation of the programme to improve the status of the Jews of Europe in the late eighteenth century. As has been underlined by Adam Sutcliffe, "in many different early modern European contexts, Jews found ways to establish themselves on a firmer footing in the communities in which they lived without making any appeal to ideal of legal normalization or political equality, which would have been almost entirely anachronistic until the 1780s."[49] Through the maintenance of community

46 Daniel H. Frank and Oliver Leaman, eds., *History of Jewish Philosophy* (London and New York: Routledge, 1997), 4.

47 Frank and Leaman, 5.

48 See Jehuda Bergman, "Sokrates in der Jüdischen Literatur," *Monatsschrift für Geschichte und Wissenschaft des Judentums* 80 (1936): 13.

49 Adam Sutcliffe, "Toleration, Integration, Regeneration, and Reform," in *The Cambridge History of Judaism*, ed. Jonathan Karp and Adam Sutcliffe (Cambridge: Cambridge University Press, 2017), 1063.

concord, shared spaces, and inter-confessional interactions, the early modern period experienced a concrete Jewish–Christian coexistence in addition to the political and theological stances which only affected these exchanges in a collateral way.[50] Furthermore, one should reconsider the paradoxical impact of ghettoisation on Jewish culture, because rather than "eroding its distinctiveness, it provided, behind the ghetto walls, a secure home for its autonomous development, influenced by the dominant culture while also to some extent sealed from it. [...] Far from making Jews and Christians invisible to each other, this in fact provided the basis for more robust and self-conscious encounters across this boundary of difference."[51]

Bibliography

Adorisio, Chiara. *Leo Strauss lettore di Hermann Cohen: dalla filosofia moderna al ritorno agli antichi.* Florence: Giuntina, 2007.

Altmann, Alexander. *Moses Mendelssohn: A Biographical Study.* Philadelphia: Jewish Publication Society of America, 1973.

Arkush, Allan. *Moses Mendelssohn and the Enlightenment.* Albany, NY: SUNY Press, 1994.

Bartolucci, Guido. "Jewish Scepticism in Christian Eyes: Jacob F. Reimmann and the Transformation of Jewish Philosophy." In *Yearbook of the Maimonides Centre for Advanced Studies 2018*, edited by Bill Rebiger, 145–64. Berlin: De Gruyter, 2018.

Benetelli, Luigi. *I dardi rabbinici.* Venice: Domenico Lovisa, 1705.

Bergman, Jehuda. "Sokrates in der Jüdische Literatur." *Monatsschrift für Geschichte und Wissenschaft des Judentums* 80 (1936): 3–13.

Bonnet, Charles. *La palingénésie philosophique, ou, Idées sur l'état passé et sur l'état futur des êtres vivans.* Geneva: Claude Philibert et Barthelemi Chirol, 1769.

Frank, Daniel H., and Oliver Leaman, eds. *History of Jewish Philosophy.* London and New York: Routledge, 1997.

Goetschel, Willi. *Spinoza's Modernity: Mendelssohn, Lessing, and Heine.* Madison: University of Wisconsin Press, 2004.

Guetta, Alessandro. *Italian Jewry in the Early Modern Era: Essays in Intellectual History.* Brighton, MA: Academic Studies Press, 2014.

Heyd, Michael. "Introduction." *Science in Context* 15, no. 1 (March 2002): 1–8.

Israel, Jonathan. *European Jewry in the Age of Mercantilism, 1550–1750.* 3rd ed. London: Littman Library of Jewish Civilization, 1998.

Kaplan, Benjamin J. *Divided by Faith: Religious Conflict and the Practice of Toleration in Early Modern Europe.* Cambridge, MA: Harvard University Press, 2007.

Leonard, Miriam. "Greeks, Jews, and the Enlightenment: Moses Mendelssohn's Socrates." *Cultural Critique* 74 (2010): 183–99.

Luzzatto, Simone. *Discourse on the State of the Jews of Venice.* Edited and translated by Giuseppe Veltri and Anna Lissa. Berlin and Boston: De Gruyter, 2019.

---

50 See Benjamin J. Kaplan, *Divided by Faith: Religious Conflict and the Practice of Toleration in Early Modern Europe* (Cambridge, MA: Harvard University Press, 2007).

51 Sutcliffe, "Toleration, Integration, Regeneration, and Reform," 1064. See also Jonathan Israel, *European Jewry in the Age of Mercantilism, 1550–1750*, 3rd ed. (London: Littman Library of Jewish Civilization, 1998), 67–98.

Luzzatto, Simone. *Socrates, Or On Human Knowledge*. Edited and translated by Giuseppe Veltri and Michela Torbidoni. Berlin and Boston: De Gruyter, 2019.
Mendelssohn, Moses. *Phaedon*. In *Schriften zur Philosophie und Ästhetik*. Edited by Fritz Bamberger and Leo Strauss. Volume 3.1 of *Gesammelte Schriften. Jubiläumsausgabe*, edited by Alexander Altmann et al. Stuttgart: Friedrich Frommann, 1972.
Mendelssohn, Moses. *Writings on Judaism, Christianity, and the Bible*. Edited by Michah Gottlieb. Translated by Curtis Bowman, Elias Sacks, and Allan Arkush. Waltham, MA: Brandeis University Press, 2011.
Mendelssohn, Moses. *Zueignungsschreiben Johann Caspar Lavaters an Moses Mendelssohn* in *Schriften zum Judentum I*. Edited by Simon Rawidowicz. Volume 7.1 of *Gesammelte Schriften. Jubiläumsausgabe*, edited by Alexander Altmann et al. Stuttgart: Friedrich Frommann, 1974.
Montaigne, Michel de. *The Complete Essays*. Edited and translated by M.A. Screech, London: Penguin, 2003.
Morosini, Giulio. *Via delle fede mostrata a' gli Ebrei*. Rome: Stamperia della Sacra Congregazione de Propaganda Fide, 1683.
Nehamas, Alexander. *The Art of Living: Socratic Reflections from Plato to Foucault*. Berkeley, Los Angeles, and London: University of California Press, 1998.
Piaia, Gregorio, and Giovanni Santinello, eds. *Models of the History of Philosophy. Volume II: From the Cartesian Age to Brucker*. Dordrecht and New York: Springer, 2011.
Popkin, Richard H. *The High Road to Pyrrhonism*. Edited by Richard A. Watson and James E. Force. Reprint ed. Indianapolis: Hackett, 1993.
Popkin, Richard H. *The History of Scepticism: From Savonarola to Bayle*. Revised ed. Oxford: Oxford University Press, 2003.
Salah, Asher. "A Matter of Quotation: Dante and the Literary Identity of Jews in Italy." In *The Italian Judaica Jubilee Conference*, edited by Shlomo Simonsohn and Joseph Shatzmiller, 167–96. Leiden and Boston: Brill, 2013.
Schulte, Christoph. *Die Jüdische Aufklärung: Philosophie, Religion, Geschichte*. Munich: C.H. Beck, 2002.
Septimus, Bernard. "Nahmanides and the Andalusian Tradition." In *Rabbi Moses Nahmanides (Ramban): Explorations in His Religious and Literary Virtuosity*, edited by Isadore Twersky, 11–34. Cambridge, MA: Harvard University Press, 1983.
Shear, Adam. "Science, Medicine, and Jewish Philosophy." In *The Cambridge History of Judaism. Volume 7: The Early Modern World, 1500–1815*, edited by Jonathan Karp and Adam Sutcliffe, 522–49. Cambridge: Cambridge University Press, 2017.
Sorkin, David. *Moses Mendelssohn and the Religious Enlightenment*. London: Halban, 2012.
Strauss, Leo. *Gesammelte Schriften, Band II: Philosophie und Gesetz—Frühe Schriften*. Edited by Heinrich Meier. Stuttgart: J.B. Metzler Verlag, 1997.
Strauss, Leo. *Jewish Philosophy and the Crisis of Modernity: Essays and Lectures in Modern Jewish Thought*. Edited by Kenneth Hart Green. Albany, NY: SUNY Press, 1997.
Strauss, Leo. *On Moses Mendelssohn*. Edited and translated by Martin D. Yaffe. Chicago: University of Chicago Press, 2012.
Strauss, Leo. "The Mutual Influence of Theology and Philosophy." *Independent Journal of Philosophy* 3 (1979): 111–18.
Sutcliffe, Adam. "Toleration, Integration, Regeneration, and Reform." In *The Cambridge History of Judaism. Volume 7: The Early Modern World*, edited by Jonathan Karp and Adam Sutcliffe, 1058–88. Cambridge: Cambridge University Press, 2017.
Tacitus. *Histories: Books 4–5. Annals: Books 1–3*. Translated by Clifford H. Moore and John Jackson. Cambridge, MA: Harvard University Press, 1931.

Torbidoni, Michela. "Il metodo del dubbio nel *Socrate* di Simone Luzzatto." In *Filosofo e rabbino nella Venezia del Seicento. Studi su Simone Luzzatto*, edited by Giuseppe Veltri, 183–245. Rome: Aracne, 2015.

Trousson, Raymond. *Socrates devant Voltaire, Diderot et Rousseau: la conscience en face du mythe*. Paris: Lettres modernes, 1967.

Veltri, Giuseppe. *Alienated Wisdom: Enquiry into Jewish Philosophy and Scepticism*. Berlin and Boston: De Gruyter, 2018.

Veltri, Giuseppe, and Evelien Chayes. *Oltre le mura del ghetto: Accademie, scetticismo e tolleranza nella Venezia barocca*. Palermo: New Digital Press, 2016.

Veltri, Giuseppe, and Gianfranco Miletto. "Difesa inedita del senatore veneziano Loredan in favore degli ebrei nel 1659–60 basata sul *Discorso* di Simone Luzzatto." In *Filosofo e rabbino nella Venezia del Seicento. Studi su Simone Luzzatto*, edited by Giuseppe Veltri, 249–74. Rome: Aracne, 2015.

Behnam Zolghadr

The Theory of *Aḥwāl* and Arguments against the Law of Non-Contradiction

Introduction

Following the rise of dialetheism—the view that some contradictions are true—in the 1980s, there was a growing interest in exploring and searching for the possible opponents of the law of non-contradiction throughout the history of philosophy, in both the East and the West. However, between these two—that is, philosophy in Europe and philosophy in India and East Asia—there was a philosophical tradition in the Islamicate world which, as far as I know, has been overlooked from this dialetheic point of view. In this paper, I will discuss some arguments against the law of non-contradiction (henceforth, LNC) set forth by some philosophers in the Islamicate world. The text in which these arguments can be found is Faḫr al-Dīn al-Rāzī's magnum opus *Muḥaṣṣal afkār al-mutaqaddimīn wa-l-muta'aḫḫirīn* (*The Compendium of the Thought of the Ancients and the Moderns*).

It should be noted that Rāzī (d. 1210) himself is not an opponent of the LNC or the law of excluded middle (henceforth, LEM). However, in a review of the ideas of his predecessors, he discusses the opposition to the LNC. The arguments against the LNC appear in the first part of the book in his discussion of assents (*taṣdīqāt*). According to Rāzī, some assents are based on sense-perception (henceforth, sensible assents), such as "the fire is hot," and some are self-evident, such as "negation and affirmation do not combine and cannot be denied" (*al-nafy wa-l-iṯbāt lā yağtami'ān wa-lā yartafi'ān*). As Rāzī explains, some philosophers deny sensible assents altogether and some accept some of them. Similarly, some philosophers deny all self-evident assents and some confirm that there are self-evident assents. Hence, four possibilities emerge, and thus there will be four groups of thinkers. The first is those who accept both sensible and self-evident assents. As Rāzī says, these constitute the majority. The second group contains those who accept self-evident assents, but deny sensible assents. Rāzī names Plato, Aristotle, Ptolemy, and Galen as members of this group and then discusses their arguments for rejecting sensible assents. The third group is made up of those who deny self-evident assents, but accept sensible assents. Rāzī does not explicitly tell us who they are, but he puts forward their arguments against self-evident assents, such as the LNC and the LEM. In fact, these arguments are the main concern of this paper. As we will see, they are based on Abū Hāšim al-Ğubbā'ī's theory of *aḥwāl*.[1] Finally, the fourth group is

1 Two comments are in order here. First, the Mu'tazilites are famous for advocating rationality and asking for the arguments for every claim. This third group of thinkers about whom Rāzī is speaking

https://doi.org/10.1515/9783110618839-005

formed of those who deny both self-evident assents and sensible assents. According to Rāzī, the Sophists belong to this group.

Since the arguments against the LNC are logically based on Abū Hāšim's theory of *aḥwāl* (i.e., states), before discussing the arguments, I should briefly explain the theory of states and its main claim: that there is a middle (*wāsiṭah*) between existence and non-existence, or, in other words, that some things are neither existent nor non-existent. Thus, according to Abū Hāšim, there are truth-value gaps.[2] In fact, it was not uncommon among the *mutakallimūn* (i.e., the Islamic theologians) to hold views which accepted some truth-value gaps or gluts.[3] Thus, in the first section, I will explain the theory of states which is the basis of the aforementioned arguments against the LNC. In the second part, we will discuss the arguments that Rāzī put forward. After this, in the third part, we will see the relationship between other so-called self-evident assents and the LNC. These are the three following assents: "The whole is bigger than its parts," "two objects identical with a third object are identical to each other," and "an object cannot be in two places at the same time." On behalf of the opponents of the LNC, Rāzī also argues that these so-called self-evident assents are based on the LNC; that is, they are true because of the truth of the LNC. Finally, in the fourth part, we will encounter a paraconsistent logic and

---

rejects all self-evident assents. Does this mean that they cannot be among the followers of Abū Hāšim; i.e., the Bahšamiyyah? The Bahšamiyyah were some of the Mu'tazilites, and thus it does not seem that this can be the group of the thinkers to whom Rāzī is referring. However, this would be jumping to conclusions. It is possible to violate the law of non-contradiction and the law of excluded middle and still ask for reasons and arguments, or, in other words, to remain logical and coherent. For one explanation, see section 4 of this paper, in which some paraconsistent logics are introduced. For a detailed discussion, see Graham Priest, *Doubt Truth to Be a Liar* (Oxford: Oxford University Press, 2006), part 3. Second, in this paper, I provide no historical evidence that this group of thinkers were among the Bahšamiyyah. I only focus on the relationship between their arguments against the LNC and the theory of states. In fact, Rāzī is infamous for arguing against many positions for the sake of argumentation. Thus, it may be the case that those arguments against the LNC belong to Rāzī himself and that they are nothing more than dialectical constructions. However, I will not take a side as to whether there was such a group of thinkers among the Bahšamiyyah or not; my concern here is only philosophical and not historical.

2 Truth-value gaps are propositions which are neither true nor false (or in other words, propositions which do not take any truth value). Truth-value gluts are propositions which are both true and false. Consider the proposition "states are existent." According to the theory of states, this proposition is neither true nor false. Thus, it is a truth-value gap.

3 For some examples, see Ahmed Alwishah and David Sanson, "The Early Arabic Liar: The Liar Paradox in the Islamic World from the Mid-Ninth to the Mid-Thirteenth Centuries CE," *Vivarium* 47 (2009): 97–127, and Ahad Qaramaleki, "Khafri and the Liar Paradox" [Persian], *Journal of Religious Thought of Shiraz University* 11 (2004): 33–44. For the debates between the *mutakallimūn* and Aristotelian logicians, see Khaled El-Rouayheb, "Theology and Logic," in *The Oxford Handbook of Islamic Theology*, ed. Sabine Schmidtke (Oxford: Oxford University Press, 2016), 409–35, and Josef van Ess, "The Logical Structure of Islamic Theology," in *Logic in Classical Islamic Culture*, ed. Gustave E. von Grunebaum (Wiesbaden: Otto Harrassowitz, 1970), 21–50.

the way it can express the truth-value gaps and gluts in the theory of states without falling into triviality.

1 The Theory of States

It is well-known that Abū Hāšim invented the theory of states to answer a problem concerning divine attributes.[4] However, the theory of states is not confined to this problem, as states also play an extensive role in Abū Hāšim's metaphysics. First, let us discuss the problem of divine attributes and then we will see the other aspects of the theory of states.

In the Qurʾan, different attributes are ascribed to God. The *mutakallimūn* were concerned about the nature of God in relation to these properties. The main question concerns the relationship between these attributes and the essence of God. There were two main rival views on this issue in early Kalām: (1) Abū al-Huḏayl al-ʿAllāf (d. 841) held the view that God is identical with His attributes. For instance, Him being knowledgeable (*ʿālim*) implies that being knowledgeable is not a distinct reality from His. Being knowledgeable is God Himself; thus, according to Abū al-Huḏayl, God's attributes are not distinct realities from Him and "God's being knowledgeable" refers to nothing but God Himself. (2) Ibn Kullāb (d. 859) advocated the view that God's attributes are distinct entities from Him and so they are not identical with God. In his opinion, "being knowledgeable" refers to a reality distinct from and alongside God.

Each of these views had its problems. On the one hand, if God were identical with His attributes, He could not be transcendent. Moreover, His attributes, by transitivity of identity relation, would be identical with each other and thus the differences between the attributes would disappear: for example, "being knowledgeable" would be the same as "being powerful." On the other hand, if God were distinct from His attributes, then He could not be one, for affirming eternal realities alongside God means admitting the existence of several Gods. Thus, neither case is acceptable. To answer this problem of divine attributes, Abū Hāšim came up with his theory of states.[5] According to this theory, attributes are *aḥwāl* (pl. of *ḥāl*)—that is, states—and

4 See Muḥammad Šahrastānī, *Nihāyah* (Beirut: Dār al-Kutub al-ʿIlmiyyah, 1956), 79; Allāmeh-Hellī, *Kašfu l-Murād* (Qom: Muʾassese Našr-e Eslami, 1995), 15; Jan Thiele, "Abū Hāšim al-Ǧubbāʾī's (d. 321/933) Theory of 'States' (*aḥwāl*) and Its Adaption by Ashʿarite Theologians," in *The Oxford Handbook of Islamic Theology*, 368; and Richard Frank, *Beings and Their Attributes: The Teaching of the Basrian School of the Muʿtazila in the Classical Period* (Albany, NY: SUNY Press, 1978), 16.

5 It is worth mentioning that before Abū Hāšim, different theologians had tried to answer this problem, though all of them were faced with crucial objections. The most notable attempt was that of Abū Hāšim's father, Abū ʿAlī al-Ǧubbāʾī (d. 915), who was the head of the Basrian Muʿtazilite school at the time. For more, see Frank, *Beings and Their Attributes*, chapter 1.

aḥwāl are neither existent nor non-existent. But what are *aḥwāl* and how do they resolve the aforementioned problem?

Abū Hāšim borrowed the concept of *ḥāl* from grammarians. In Arabic grammar, one of the instances of *ḥāl* which Abū Hāšim took is the situation or state of the subject or object of the verb at the time of its occurrence.[6] For example, in the sentence *ǧāʾanī Zaydun māšīyan* ("Zayd came to me walking"), *māšīyan* ("walking") is a *ḥāl* ("state"); the state in which the verb occurs. Abū Hāšim paraphrases subject–predicate sentences in such a way that the predicate becomes a *ḥāl*. Consider the sentence *Zaydun ʿālimun* ("Zayd is knowledgeable"). For Abū Hāšim, this sentence means that Zayd *is* and that he is in the state of being knowledgeable. He paraphrases the sentence by adding the verb *kāna*[7] ("to be" or "to become"). Thus, *yakūnu Zaydun ʿālimun* is a paraphrase of *Zaydun ʿālimun*. Although they both have the same meaning, each implies a different proposition on the metaphysical level. The original sentence is a usual subject–predicate sentence, but the paraphrase is not. The latter implies the being of the subject and then the state in which the subject is. Abū Hāšim paraphrases all predicates as such. However, as we will see, his use of the concept of *ḥāl* is not restricted to predicates.

The Muʿtazilites were atomists and for them, the metaphysical categories consisted of God, atoms, and accidents.[8] Hence, there was no place for universals or attributes as such in their categories of beings. Abū Hāšim added *aḥwāl* to these categories. One of the distinct roles of *aḥwāl* was to explain the relationship between different accidents which go by the same title. Thus, the theory of *aḥwāl* completes the Muʿtazilites' metaphysical picture by explaining universals and attributes as a specific metaphysical category. Thus, *aḥwāl* are real. However, they are not objects (*ḏawāt*) and they are neither existent nor non-existent. There are five categories of *aḥwāl*:[9]

(1) The attribute of essence: this is the self-identity of things. Through this attribute, objects differ from each other. A thing being itself is the attribute of essence. For example, for an atom, being an atom is its attribute of essence which makes it what it is.
(2) Essential attributes: these are the attributes which are entailed as soon as objects become existent via the attribute of essence. For instance, the essential attribute of an atom is to occupy space (*taḥayyuz*). When an atom is non-existent, it is still an atom; however, it does not occupy space. But as soon as it becomes existent,

6 Frank, 20.
7 The verb *kāna* is sometimes a complete verb and sometimes incomplete. Here, the complete form of *kāna*, which means "to be" or "to become," is used.
8 Alnoor Dhanani, *The Physical Theory of* Kalām*: Atoms, Space, and Void in Basrian Muʿtazilī Cosmology* (Leiden: Brill, 1994), 17.
9 See Frank, *Beings and Their Attributes*, 27; Thiele, "Abū Hāšim al-Ǧubbāʾī's Theory of States," 370–76; and Sabine Schmidtke, "The Muʿtazilite Movement III: The Scholastic Phase," in *The Oxford Handbook of Islamic Theology*, 163–64.

it occupies space. Thus, Abū Hāšim and his followers, as well as some of the other Muʿtazilites, held the view that some objects do not exist. Their idea of existence and objecthood will become more evident in our discussion of the next category of *aḥwāl*.

(3) Attributes which are affected by an agent (*al-ṣifāt bi-l-fāʿil*): specifically, this attribute is *existence*. An agent, which can be either God or human, generates an object and then the object is existent. It is the agent's action which causes the existence of the object, and a non-existent object becomes existent through the action of an agent. Thus, for an object, being existent is a state in which the essential attributes of the object are actualised. As already mentioned, some objects are non-existent. For instance, a non-existent atom is still an atom: it is knowable and we can discuss its attributes as an atom. However, as long as it is not in the state of existence, its essential attributes, such as occupying space, are not actualised. What causes it to be in the state of existence is an act of creation performed by an agent, and what causes a non-existent (*maʿdūm*) to be an object is its knowability.[10] Anything that is knowable, whether existent or non-existent, is an object. However, existence itself is not an object: it is a state (*ḥāl*). Since only objects are knowable and states are not objects, states are not knowable; however, they are real.

(4) Attributes which are grounded in accidents: as already mentioned, the Muʿtazilites' metaphysical categories consisted of God, atoms, and accidents. Therefore, they could not explain the similarities and differences between objects by relying on these categories. One important aspect of this shortcoming is a coherent account of universals. What is the similarity between a raven and the night sky? The answer is easy; they are both black. But what is black? According to the Muʿtazilites, an accident (*ʿaraḍ*) of blackness inheres in a raven and another accident of blackness inheres in the night sky. These two are distinct accidents; for one thing, they are in different places. Thus, the Muʿtazilites' metaphysics cannot explain the similarity between two black objects. Abū Hāšim's fourth category of states is those states which are grounded in accidents. An object in which an accident of blackness inheres is in the state of being black. Thus, states also play the role of universals. Now we can say that the similarity between a raven and the night sky is the state of being black.

(5) Those states which do not gain actuality by the object itself or by another object, the most important example of which is "being perceiving." In contrast to his predecessors, Abū Hāšim takes perception not as an accident, but as a state.[11]

10 Dhanani, *The Physical Theory of* Kalām, 30.

11 Considering these five categories of *aḥwāl*, one might ask whether Abū Hāšim was under the influence of the Stoics, specifically, the theory of *hexeis*. The historical roots of the theory of states are not my concern here, but I can say by referring to Dhanani, *The Physical Theory of* Kalām, 5, that there is no historical evidence of any direct link between the *mutakallimūn* and Greek philosophers.

To sum up, states are neither existent nor non-existent. They are also not objects, which means that they are not knowable. However, Abū Hāšim explained different characteristics of states. Moreover, although states are not objects, they are real. These features of states help Abū Hāšim to reply to the problem of the divine attributes. According to the theory of states, attributes of God are states. States are real, and thus they can explain the attributes ascribed to God. However, states are not objects, and hence, in ascribing His attributes to Him, it does not follow that some objects *are* alongside Him. Thus, God's unity will not be violated. In this way, not only His unity, but also His transcendence and the reality of His attributes are preserved.

Let us see where the theory of states violates the LNC. That states are neither existent nor non-existent apparently entails truth-value gaps and thus violates the law of excluded middle.[12] Avicenna was well aware of the theory of states and calls the followers of the theory of states those who "are not among the assemblage of the discerning."[13] Some pages later, in his *Metaphysics* (*Ilāhiyyāt*) of *al-Šifā'*, while defending the LNC, he argues that any violation of the LEM entails a violation of the LNC. The LEM fails when both affirmation and negation are removed. Avicenna tells us how this leads to the violation of the LNC:

> **[T1]** If both [affirmation and negation] together are false with respect to one thing, then that thing would, for example, be "not man" and also "not not man." Hence, the thing which is "not man" and its negation (which is "not not man") would have been combined (Avicenna, *Metaphysics of the Healing*, 42–43).[14]

If this argument works, then the theory of states violates the LNC, and thus it is a dialetheic theory. There is a middle between existence and non-existence, or, in other words, states are neither existent nor non-existent. It follows that states are not existent and not not existent. Therefore, according to Avicenna's argument, the middle between existence and non-existence entails a contradiction.

On the other hand, although states are not knowable according to Abū Hāšim, he has told us about many of their features and we now know several characteristics of states as well as their roles in his metaphysical theory. This is also a contradiction and is mentioned by some opponents of the theory of states. Whether these are good reasons to take the theory of states as a dialetheic theory can be discussed further, but for our purposes, these would suffice. Let us now examine the arguments against the LNC which Rāzī ascribes to those who accept sensible propositions but deny self-evident propositions.

---

12 It is worth mentioning that this claim is based on the idea that existence and non-existence are the most general metaphysical categories. In other words, everything is either existent or non-existent and consequently, predicating existence/non-existence on states is not a categorical mistake.

13 Avicenna, *The Metaphysics of the Healing*, trans. Michael E. Marmura (Utah: Brigham Young University Press, 2005), 27.

14 فإنه إذا كذبا معا فى شىء، كان ذلك الشىء ليس بإنسان مثلا، وليس أيضا بلا إنسان. فيكون قد اجتمع الشىء الذى هو اللا إنسان وسالبه الذى هو لا لا إنسان.

2 Arguments against the LNC

Rāzī gives four arguments on behalf of those who deny self-evident assents while defending sensible propositions. The most self-evident assent is defined as "negation and affirmation do not combine and cannot be removed" (*al-nafy wa-l-iṯbāt lā yaǧtamiʿān wa-lā yartafiʾān*).[15] Before discussing the arguments, a note on their method: they are *ad hominem*. Each of them shows that holding the LNC is self-refuting; that is, that accepting the LNC leads to its refutation. Thus, these arguments do not have any presuppositions: they begin with the opponent's stance and then show that it refutes itself. Thus, in order to pose such arguments, one does not have to accept the LNC.

Now let us briefly consider the relationship between these arguments and the theory of states. As we will see, the first argument concerns intentionality. Asserting the law of non-contradiction implies referring to contradiction. Whatever is intendable is real, and thus, the LNC is self-refuting. The second argument concerns existence, suggesting the ontological explanation of existence itself posed by the theory of states. It can be considered as an argument for the claim that existence itself is neither existent nor non-existent. As already mentioned, this claim is an important part of the theory of states. The third argument concerns the problem of universals, and is nothing more than an argument in favour of the theory of states concerning this problem. Again, universals are one of the five categories of states. The fourth argument concerns impossibility and the process of becoming. In these two latter arguments, the middle between existence and non-existence, which is the core of the theory of states, is explicitly advocated. Thus, it is not only the middle between existence and non-existence which makes the connection between these arguments and the theory of states, but also the targets of these arguments are some of the most important aspects of this theory.

Rāzī explains the first argument as follows:

> **[T2]** This assent is based on conceptualising non-existence, and people have been perplexed by this. Because that which is conceptualised must be distinguished from other things, and whatever is distinguished from others must have an essential determination, and whatever is essentially determined is an object in itself. Since the affirmation [of the law of non-contradiction] is

---

15 In this passage, Rāzī sometimes defines the most self-evident assent as "affirmation and negation do not combine and cannot be removed," sometimes as "an object either is or is not," and sometimes as "negation and affirmation do not combine." This may have been influenced by Avicenna's argument that every violation of the LEM is also a violation of the LNC. As we will see, the arguments against the LNC do not always directly target it; sometimes, they reject it via their rejection of the LEM. This makes sense if we accept Avicenna's argument. Thus, for simplicity, I will not explain on every occasion that such a method is being used, but I will ask the reader to bear it in mind while reading this section of the paper.

> based on this conceptualisation and this conceptualisation is impossible, therefore, this affirmation is also impossible (Rāzī, *Muḥaṣṣal*, 30).[16]

Rāzī does not explicitly explain how the claim that "affirmation and negation do not combine and cannot be removed" is based on conceptualising the non-existence. Non-existence, in this affirmation, occurs in two places: first in negation and then regarding truth-value gaps or gluts. Negating something of something is to claim the non-existence of the former in the latter. For instance, asserting that "snow is not black" is to affirm the non-existence of the accident of blackness in snow. Moreover, the claim that affirmation and negation are neither combined nor removed also means that there are no truth-value gaps—that is, removal of both affirmation and negation—or gluts—that is, combinations of affirmation and negation. Thus, it affirms the non-existence of truth-value gaps and gluts. I will consider both cases in order to make the discussion sufficiently exclusive.

Since, following the LNC and the LEM, truth-value gaps and truth-value gluts are non-existent, we must first be able to conceptualise non-existence. It is the same for negation: for example, negating the blackness of snow, which is the non-existence of blackness in snow. However, having a conceptualisation of non-existence is already a problem, though not for the followers of Abū Hāšim. In fact, besides some of the Muʿtazilites, no other Islamic theologians or Muslim peripatetic philosophers held the view that some things do not exist. For this much larger group, every object is existent. Thus, for them, non-existent is not an object and since an object is that which is knowable, whatever does not exist is not an object and cannot be conceptualised.

As for truth-value gaps or truth-value gluts, talking about them implies their objecthood, because in order to be able to refer to them, or in Rāzī's words, to have a conceptualisation of them, they must be determined. Whatever is determined in itself is an object, and if every object is existent, truth-value gaps and truth-value gluts become existent and thus one cannot claim their non-existence. In other words, affirming the LNC and the LEM implies the conceptualisation of truth-value gaps and gluts. Thus, anyone one who accepts the LNC or the LEM cannot hold them to be true.

One possible answer on behalf of the advocates of the view that every object is existent is to make a distinction between mental existence and external existence. Rāzī mentions this possible reply. In fact, Muslim peripatetic philosophers, such as Avicenna, made this distinction. According to Rāzī, one cannot reply to the objection by embracing such a distinction between mental and external existence, because for those who advocate the LNC and the LEM, truth-value gaps and gluts are impossible, not only in the external world, but also in the mind; in their view,

---

16 أن هذا التصديق موقوف على تصور أصل عدم والناس قد تحيروا فيه. لأن المتصور لا بد وأن يتميز عن غيره والتميز عن غيره متعين فى نفسه، وكل متعين فى النفسه فهو ثابت فى نفسه، فكل متصور ثابت فى نفسه، فما ليس بثابت فغير متصور فالمعدوم غير ثابت فلا يكون متصورا وإذا كان ذلك التصديق متفرعا على هذا التصور وكان هذا التصور ممتنعا كان ذلك التصديق ممتنعا.

they are unperceivable. Moreover, when a person negates the blackness of snow, it does not follow that she affirms the existence of blackness in snow in her mind. Thus, the mental–external distinction of existence is of no help here.

It should be noted that if this argument works, it works against those who hold every object to be existent. One might well be an advocate of the LNC and the LEM and yet hold the view that some objects do not exist.

Rāzī, then, explains the second argument:

> **[T3]** Let us accept that non-existence can be conceptualised. However, the assertion that negation and affirmation do not combine implies a distinction between existence and non-existence, and this distinction implies that the referent of non-existence has a distinct essence from that of existence, but this is impossible, because every entity which the mind intends can also be removed by the mind; otherwise, it does not have an opposite. Therefore, non-existence does not have an opposite and this implies the denial of existence, and it is false. Thus, the removal of that which is the referent of non-existence is coherent. However, this removal is a specific one and is counted as a case of absolute non-existence, and it means that the complement of non-existence is a part of it, which is impossible (Rāzī, *Muḥaṣṣal*, 30).[17]

Since negation and affirmation do not combine,[18] there is a distinction between existence and non-existence. In other words, there is nothing which is both existent and non-existent, because affirmation implies the existence of an entity, such as an accident or object, and negation implies the non-existence of an entity, such as an accident or object, and these two—that is, affirmation and negation—do not combine. Consequently, the referents of existence and non-existence are distinct. Rāzī claims that this is not the case. Here is the reason why: on the assumption that existence can be removed by the mind, the removal of non-existence is the opposite of non-existence; however, it is still a sub-case of non-existence, because all privations are cases of non-existence. Hence, a contradiction: the removal of non-existence is non-existence and not-non-existence at the same time. Therefore, the aforementioned distinction between existence and non-existence is impossible. To sum up, the LNC is based on a distinction between existence and non-existence, which is impossible. Therefore, the LNC is not true.

Rāzī then proceeds to the third argument. This is the longest and the most complicated of the four. Thus, I will first attempt to split the argument and then we will discuss the details.

---

17 لو سلمنا إمكان تصور العدم لكان قولنا النفى والإثبات لايجتمعان يستدعى امتياز العدم عن الوجود، وامتياز العدم عن الوجود يستدعى أن يكون لمسمى العدم هوية متميزة عن الوجود لكن ذلك محال، لأن كل هوية يشير العقل إليها والعقل يمكنه رفعها، وإلا لم يكن له مقابل، وكان يلزم أن لا يكون للعدم مقابل وكان يلزم نفى الوجود وهو باطل، فثبت أن ارتفاع الهوية المسماة بالعدم معقول، لكن ارتفاع تلك الهوية ارتفاع خاص فيكون داخلا تحت العدم المطلق، فيكون قسيم العدم قسما منه هذا خلف.

18 Initially, Rāzī defines the conjunction of the LNC and the LEM as the most self-evident. However, on some occasions, instead of the conjunction of the LNC and the LEM, he is only talking about the LNC. One example is this passage of *Muḥaṣṣal*. Thus, the reader should be aware of the way Rāzī uses the conjunction of the LNC and the LEM interchangeably with the LNC.

Rāzī begins by making a distinction between two cases in which affirmation and negation occur:[19] first, affirming and negating the reality (*ṯubūt*) of the object itself, such as "blackness is either existent or non-existent" (*al-sawādu immā an yakūnu mawǧūdān wa-immā an lā yakūnu mawǧūdān*), and second, affirming or negating the reality (*ṯubūt*) of an object in another object, such as "the body is either black or not" (*al-ǧismu wa-immā an yakūnu aswad wa-immā an lā yakūn*).[20]

Consider the first case:

(1) Blackness is either existent or non-existent.

As Rāzī says, (1) can be asserted only after conceiving the meaning of "blackness is existent" and "blackness is non-existent," but this is not going to happen.

(1.1) Blackness is existent.

In this statement, either blackness is identical with existent or blackness is distinct from existent. Rāzī demonstrates that both are false.

Let us take the first. As Rāzī puts it:

> **[T4]** Our statement that blackness is existent [would] be like the statement "blackness is blackness" and the statement "existent is existent," but it is not like these, for the [two] latter [statements] are trivial; however, the former is informative (Rāzī, *Muḥaṣṣal*, 31).[21]

Thus, since "blackness is existent" is informative, it is not a simple identity statement.

Then, Rāzī explains why the second understanding of (1.1)—that is, that blackness and existent are distinct—is also false. He gives two reasons for this:

> **[T5]** One of them: that is, if existence is grounded in blackness, then blackness is not, as such, existent, because otherwise the question arises again, but the same object cannot be existent for a second time. And if so, existence would be grounded in that which is not existent. However, existence is the property of existent. Otherwise, we should accept the middle between existence and non-existence, but you deny that (Rāzī, *Muḥaṣṣal*, 31).[22]

If existence is grounded in blackness, then blackness is metaphysically prior. In other words, blackness is blackness without being existent and then existence is grounded in it. It is obvious that blackness could not be existent before being existent because, as Rāzī explains, this would mean that it would be existent for a second time. Moreover, the question about the latter existence would also arise for the

---

19 Rāzī, *Muḥaṣṣal*, 31.

20 The reader should be aware that, as already mentioned, the theory of *aḥwāl* is based on the idea that some objects do not exist. Thus, the predication relation does not have existential import in itself. My thanks to an anonymous referee for mentioning this point.

21 كان قولنا السواد موجود جاريا مجرى قولنا السواد سواد وقولنا الموجود موجود ومعلوم أنه ليس كذلك لأن هذا الأخير هذر، والأول مفيد.

22 أحدهما: أنه إذا كان الوجود قائما بالسواد فالسواد فى نفسه ليس بموجود وإلا لعاد بحث فيه، ولكان الشىء الواحد موجودا مرتين وإذا كان كذلك كان الوجود قائما بما ليس بموجود، لكان الوجود صفة موجودة وإلا لثبت الواسطة بين الموجود والمعدوم وأنتم أنكرتموه.

former. Thus, blackness in itself is non-existent. Since existence is the property of an existent, it cannot be the property of blackness, because blackness is non-existent. Therefore, Rāzī concludes, existence and blackness, in the statement "blackness is existent," cannot be distinct.[23] Here, Rāzī mentions the middle between existence and non-existence as a solution to this issue. We saw in part 1 that this is the main idea of the theory of states. In fact, this is the first explicit reference to the theory of states in these arguments.

Let us move on to the second argument for the impossibility of the distinction between existence and blackness:

> **[T6]** If existence is distinct from quiddity, the referent of our saying "blackness" is different from our saying "existent." Then, if we say that blackness is existent and by this we mean that blackness is that which is existent, this would be asserting the oneness of the two [non-identicals], which is impossible (Rāzī, *Muḥaṣṣal*, 31).[24]

We also cannot say that that which is existent is blackness, for this implies the identity of the two. This is unacceptable because we have already presupposed that blackness and existent are non-identical. Moreover, we have already seen that blackness and existent cannot be identical.

One might reply that the identity sentence is asserting the identity relationship between "blackness" and "described by being existent" and not that between "blackness" and "existent." Rāzī explains to us that this reading faces similar problems and does not work. As he put it:

> **[T7]** And if you say that this is not what we intend from our saying that blackness is existent—that is, that the referent of "blackness" is the referent of "existent"—but what we intend is that blackness is described by being existent, you have just moved the issue to the referent of "described." Then, if the referent of blackness is the referent of "described by being existent," then this is also impossible, because our saying that blackness is described by being existent becomes the same as our saying that blackness is blackness, and that is impossible. And if [the referent of blackness] is distinct from [the referent of "described by being existent"], then asserting that blackness is described by existence is asserting the oneness of the two (Rāzī, *Muḥaṣṣal*, 31).[25]

Thus, replacing "existent" with "being described by existence" does not solve the problem. Again, we will face the same problems as in the case of "existent."

---

23 However, if we consider that existence is grounded in blackness, this presupposition seems reasonable if we analyse the structure of the sentence "blackness is existent" in a subject–predicate form.

24 إذا كان الوجود مغايرا للماهية كان مسمى قولنا السواد غير مسمى قولنا موجود فإذا قلنا السواد موجود بمعنى أن السواد هو موجود كان ذلك حكما بوحدة الاثنين وهو محال.

25 فإن قلت ليس المراد من قولنا السواد موجود وهو أن مسمى السواد مسمى الوجود، بل المراد أن السواد موصوف بالموجودية قلت فحينئذ ينقل الكلام إلى مسمى الموصوفية، فإنه إما أن يكون مسمى الموصوفية بالوجود محال فيكون قولنا السواد موصوف بالوجود جاريا مجرى قولنا السواد سواد وهو محال، وإما أن يكون مغايرا له فيكون الحكم على السواد بأنه موصوف بالوجود حكم بوحدة الاثنين.

So far, Rāzī has shown that sentence (1.1) is false, or, to be more precise, impossible (*muḥāl*). Obviously, it is impossible for one who endorses the LNC. Now, he proceeds to the second part of (1) and demonstrates that this is also false:

(1.2) Blackness is non-existent.

It is obvious that being blackness cannot be the same as the existence of blackness. Otherwise, as Rāzī explains, (1.2) entails that blackness is not blackness, which is a contradiction.[26] Thus, existence is not identical with blackness. Then, Rāzī argues that blackness and its existence cannot also be distinct, if (1.2) is to be true.

The gist of his argument is as follows. Negating the existence of a quiddity implies the quiddity to be determined and this in itself entails that the quiddity is real (*ṯābit*). To be real is to be existent. Thus, in order to negate the existence of an entity, it should be existent, which is a contradictory. Therefore, (1.2) cannot be asserted.

This is the well-known problem of negative existentials. It is obvious that this argument is only against those who hold every object to be existent, or, in other words, those who hold that all that is real (*ṯābit*) is existent. As already mentioned, this was the dominant view among the *falāsifah* and *mutakallimūn*. However, some Muʿtazilites held the view that some objects do not exist, or, in other words, that there are real entities (*ṯābitāt*) which are non-existent. Among these Muʿtazilites are those who advocated the theory of states; that is, Abū Hāšim and his followers. However, since one can well be an advocate of the theory that some objects do not exist and also be an advocate of the LNC, like some of the early Muʿtazilites, this argument loses its generality.

Then, Rāzī proceeds to the second form of the sentence expressing the LNC:

(2) A body is either black or not.

First, we should grasp the meanings of both "the body is black" and "the body is not black." Rāzī argues that this is impossible. He begins with "the body is black," giving us two reasons for the "impossibility of this statement." First, it cannot be an identity statement; otherwise, it is asserting the oneness of two distinct entities, which is impossible. Second, since it is not an identity statement, it has the form of a subject–predicate sentence in which the subject is described by the predicate. Now, as Rāzī explains, "that the body is described by blackness is either a non-existential property or an existential one."[27]

Being described (*mawṣūf*) is itself a property. It is not a non-existential property, because it is the negation of not being described, which is itself a non-existential property. On the other hand, Rāzī also argues that being described is not an existential property:

26 These arguments are against those who defend the LNC. Thus, any premise which entails a contradiction is not acceptable. See Rāzī, *Muḥaṣṣal*, 32.
27 Rāzī, 33.

> **[T8]** If [being described] is existential, then it is either the same as the existence of the body and blackness or distinct from them. The former is impossible, because it is not the case that whenever one conceives the existence of the body and the existence of blackness, one also conceives that the body is described by blackness. The latter is also impossible, because if being described by blackness were superadded to the body, then being described by that property [i.e., the property of being described] would also be superadded to the body. But this [leads to regress and thus] is impossible (Rāzī, *Muḥaṣṣal*, 33).[28]

Here, Rāzī is asking about the ontological status of a property. Is there a reality which is correspondent to the property of an object? He replies that the answer is neither yes nor no. The property of being described is not existential. In other words, there is no reality (*ṯābit*) which is the property of being described. In order to be real, being described by blackness for a body should either be the same as the body and blackness itself or distinct from these two. It is not the same as the body and blackness, for one can conceive these two without conceiving the body being described by blackness. In other words, the existence of a body and the existence of blackness do not entail that the body is black, or that the body is described by blackness. Thus, there should be another reality which is described (*mawṣūf*) and is distinct from the body and blackness. But this is also impossible, because then there would have to be another reality which is being described by being described, and this leads to regress. Therefore, neither of these possibilities is true.

This argument exposes one of the problems which the theory of states answers: the problem of universals. As already mentioned in section 1, Abū Hāšim's theory of states, among other things, explains the ontological status of properties. According to other *mutakallimūn*, the ontological categories of the world consisted of God, atoms, and accidents. However, these cannot explain what it is to have a property. Relying on the grammarians' analysis of language, Abū Hāšim appealed to states (*aḥwāl*) to explain the ontological status of properties. This argument demonstrates the shortcoming of the theories which Abū Hāšim criticised in order to give a full-fledged ontological explanation of the world.

Let us move on to the fourth argument against the LNC, which is also an explicit defense of the theory of states. This argument concerns impossibility (*imtināʿ*) and the process of becoming (*ḥudūṯ*). First, consider impossibility. Rāzī says that there are three possibilities for that which is called impossibility:[29] either it is existent, or it is non-existent, or it is neither existent nor non-existent. He demonstrates that the first two possibilities are false:

---

28 أن يكون أمرا ثبوتيا لأنه على هذا التقدير إما أن يكون نفس وجود الجسم والسواد وإما أن يكون مغايرا لهما. والأول محال لأنه ليس كل من عقل وجود الجسم ووجود السواد عقل كون الجسم موصوفا بالسواد. والثانى أيضا محال لأن موصوفية الجسم بالسواد لو كانت صفة زائدة لكانت موصوفية الجسم بتلك الصفة زائدة عليها وهو محال.

29 He does not tell us why it cannot be both existent and non-existent. Even if he thinks that the truth-value gap here is the same as the truth-value glut, he does not explain it. However, there is a fourth possibility; i.e., both existent and non-existent.

[T9] It is not existent, because otherwise that which is described by it would also be existent. For existent cannot be placed at non-existent and if that which is described [by impossible] were existent, then the impossible would not be existent; however, it would either be necessary or possible (Rāzī, *Muḥaṣṣal*, 34).[30]

The instances of impossibility are non-existent because otherwise, they would not be impossible. Since those which are described by impossibility are not existent, impossibility is not existent because existent cannot be grounded (*qā'im*) in non-existent.

Impossibility cannot be non-existent either,

[T10] because it is the negation of not-impossibility which can be predicated to non-existent. Thus, since not-impossibility is non-existential, impossibility is not non-existential. Moreover, impossibility is a determined quiddity in itself, which is distinct from other quiddities. Otherwise, it could not be intended by the mind. Thus, it cannot be absolute non-entity (Rāzī, *Muḥaṣṣal*, 34).[31]

There are two arguments in this paragraph in favour of the view that impossibility is not non-existent. It is dubious that the result follows in the first one. For one thing, impossibility is the negation of possibility and a similar argument can be applied to demonstrate that impossibility is non-existential. According to the second argument, what is impossible is supposed not to be an entity simply because it is impossible. However, since it can be intended by the mind, it must be something, and thus, by understanding non-existence in its broadest sense, it is not absolutely non-existent. Rāzī also considers a possible reply according to which the impossible has reality in the mind and not in the external world. Rāzī rejects this reply by emphasising the nature of the impossible which is impossible in itself, whether in the mind or in the external world.[32]

Now consider the process of becoming existent from non-existence. As Rāzī explains, *becoming* means coming out of non-existence into existence. Thus, the referent of *becoming* is distinct from both the referent of *non-existence* and the referent of *existence*, because otherwise, "the referent of *non-existence* or the referent of *existence* will be the referent of *coming out of non-existence into existence*, and that is impossible." He considers three possibilities for that which is the quiddity of the referent of coming out of non-existence into existence. It is either existent, non-existent, or neither existent nor non-existent. Then, he demonstrates that it cannot be existent:

30 لا جائز أن يكون موجودا وإلا لكان الموصوف به موجودا لاستحالة قيام الموجود بالمعدوم ولو كان الموصوف به موجودا لم يكن الممتنع ممتنعا، بل إما واجبا أو ممكنا.

31 لأنه نقيض اللا امتناع الذى يمكن حمله على المعدوم فيكون اللا امتناع عدميا فلا يكون الامتناع عدميا. ولأن الامتناع ماهية متعينة فى نفسها متميزة عن سائر الماهيات إذ لو لم يكن كذلك لاستحال إشارة العقل إليها، وإذا كان كذلك استحال أن يكون نفيا محضا.

32 Rāzī, *Muḥaṣṣal*, 34.

[T11] If it was existent, it would be true of the existent that it is coming out of non-existence into existence, and if this is so, it would be like saying that the existent is coming out towards existence and thus the object is existent twice, and that is impossible (Rāzī, *Muḥaṣṣal*, 34).[33]

It is also not non-existent, because otherwise, there would be no change from non-existence to existence. As Rāzī put it:

[T12] When it is non-existent, the original non-existence endures, and as long as the original non-existence endures, no change from non-existence occurs (Rāzī, *Muḥaṣṣal*, 34).[34]

Since it cannot be either existent or non-existent, Rāzī concludes that when something is in the process of becoming, it is neither existent nor non-existent. As already mentioned, the same is true of the impossible. Therefore, the LNC is not true.

So far, we have discussed the four arguments against the LNC which, according to Rāzī, were posed by those thinkers who accept sensible assents, but altogether deny self-evident assents. We also saw that these arguments are based on the theory of states, or, to be more precise, the middle between existence and non-existence, which is the core idea of the theory of states, as well as the idea that some objects do not exist.

As already mentioned, they also argue that other well-known self-evident assents are based on the LNC, and since the LNC has been rejected, those assents are also unacceptable. These self-evident assents are the subject of the next section.

3 Other Self-Evident Assents

There are three other well-known self-evident assents: (i) The whole is bigger than its parts; (ii) identity is transitive; (iii) an object cannot be in two different places at the same time. The philosophers in question argue that all three of these assents are based on the LNC. We will see their arguments one by one.

33 فإن كانت موجودة فقد صدق على الموجود أنه يخرج من العدم إلى الوجود فيكون ذلك كأنه يقال الموجود يخرج إلى الوجود فيكون الشيء موجودا مرتين وهو محال.

34 متى كانت معدومة كان العدم الأصلى باقيا ومتى كان العدم الأصلى باقيا لم يكن النقل فى التغير من العدم حاصلا.

3.1 The whole is bigger than its parts

As Rāzī put it:

> **[T13]** The whole is bigger than its parts because otherwise, the existence of the other part and its non-existence would be the same. Thus, being existent and being non-existent is combined, at the same time, in the other part (Rāzī, *Muḥaṣṣal*, 29).[35]

It is obvious that here, parthood is proper parthood. Thus, consider an object, namely *a*, which has two proper parts, namely *b* and *c*. By using mereological formalisation, we have:

$$b < a \quad , c < a$$

Since *b* and *c* are two distinct parts of *a*, they do not overlap. Thus:

$$(c \circ a) \;\wedge\; \neg(c \circ b)$$

It follows that:

$$(1)\ \exists x((x \circ a) \wedge \neg(x \circ b))$$

On the other hand, if *a* was not bigger than *b*, we would have $a = b$, which means that parthood is improper. Then, it follows:

$$(2) \neg\exists x((x \circ a) \wedge \neg(x \circ b))$$

Therefore, (1) and (2) are contradictories. They express that there both is and is not an object—namely, *c*—which is the other part of *a*, while *b* is a proper part of *a* and is not smaller than *a*.

The rejection of this assent leads to contradiction. Thus, this is true as long as the LNC is true. In other words, the statement that the whole is bigger than its parts is based on the LNC.

3.2 Identity is transitive

Identity is transitive. In other words, two objects which are identical with a third object are themselves identical with each other:

> **[T14]** [Two] objects that are identical with another object are identical, because otherwise, [an object, namely] *a* which is identical with blackness is blackness. And since it is identical with that which is not blackness, it is not blackness. Hence, if *a* is identical with two different things,

35 الكل أعظم من الجزء لأنه لو لم يكن ذلك لكان وجود الجزء الآخر وعدمه بمثابة واحدة، فحينئذ يجتمع فى ذلك الجزء الآخر كونه موجودا معدوما معا.

> it follows that a is in itself blackness and also that a is not in itself blackness. Thus, the affirmation and the negation are combined (Rāzī, *Muḥaṣṣal*, 29).[36]

Let us simplify the example. Suppose there are three objects—a, b, and c,—and suppose, for the sake of simplicity, that there are only two properties, P_1 and P_2. a has both properties and b only has the property P_2. c is identical with a, and thus it has both properties. Therefore, P_1c and P_2c are both true. On the other hand, c is identical with b. Thus, c has the property P_2, but it does not have the property P_1. This means that $\neg P_1c$ and P_2c are true. Therefore, $P_1c \wedge \neg P_1c$, which is a contradiction. The table below shows the details:

	P_1	P_2
a	+	+
b	−	+
c	±	+

If two distinct objects, a and b, are identical with a third object, c, then the third object will have contradictory properties.[37] Thus, the rejection of the transitivity of identity entails contradictions. If the LNC is not true, then at least some contradictions are true, and therefore, it is not necessarily true that identity is transitive.

3.3 An object cannot be in two different places at the same time

> **[T15]** A single object cannot be in two different places at the same time, because if this was possible, then there would be no difference between an object being in two places or two objects being so, and meanwhile, the existence of the other object would be the same as its non-existence, and it would be true to say that it is both existent and non-existent (Rāzī, *Muḥaṣṣal*, 29).[38]

Suppose that a is an object which is located at l_1 and suppose that at the same time a is also located in another place, namely, l_2. This scenario is similar to a scenario in

---

36 الأشياء المساوية لشيء واحد متساوية لأنه لو لم يكن كذلك لكان الألف المحكوم عليه بأنه يساوى السواد سوادا لا محالة، ومن حيث إنه محكوم عليه بأنه يساوى ما ليس بسواد يجب أن لا يكون سوادا فلو كان الألف مساويا للأمرين لزم أن يكون الألف فى نفسه سوادا وأن لا يكون فى نفسه سوادا فيجتمع النفى والإثبات.

37 Here, the Leibnizian definition of identity is used. According to this definition, two objects are identical if they share the same properties. For more on the non-transitivity of identity and its resulting contradictions, see Graham Priest, *One: Being an Investigation into the Unity of Reality and of Its Parts, Including the Singular Object which Is Nothingness* (Oxford: Oxford University Press, 2014), chapter 2.

38 إن الجسم الواحد فى الآن الواحد لا يكون فى مكانين معا، لأنه لو جاز ذلك لما تميز الجسم الواحد الحاصل فى مكانين متباينين عن الجسمين اللذين حصلا كذلك، وحينئذ لا يتميز وجود الجسم الآخر عن عدمه فيصدق عليه كونه موجودا معدوما معا.

which there are two objects, *a* and *b*, which both have exactly the same properties with only one exception: that is *a* is located in l_1 and *b* is located in l_2. According to this argument, since these two scenarios are not distinguishable, *b* can be considered as both existent and non-existent. Thus, the rejection of this assent entails a contradiction. Since the LNC is already rejected, at least some contradictions might be true. Therefore, if the LNC is not true, this assent is not true either.

There is also another more straightforward argument that this assent is based on the LNC. Let us write the property of being located in l_1 as *P*. Thus, if *a* is located in l_1 and is not located in l_1, we have $P_a \wedge \neg P_a$, which is a contradiction.

4 Dialetheism and Paraconsistency

In this part, we will see how the violation of the LNC co-exists with the theory of states without demolishing the coherency of the theory of states and the ability to reason within this theory.[39] Let us accept that the aforementioned arguments against the LNC work. Since this rejection of the LNC logically follows from the theory of states, the theory of states is a dialetheic theory. However, we already saw that some contradictions are true in the theory of states; for example, that states are and are not objects. Now, we should discuss how the theory can be logically modelled.

The logic of a dialetheic theory should be a paraconsistent one. A paraconsistent logic is a logic whose consequence relation is not explosive;[40] that is, explosion, or *ex contradiction quodlibet* (ECQ), is not a valid argument in that logic. Otherwise, the theory will be trivial; that is, everything, and consequently every contradiction, would be true according to this theory.[41] There are many paraconsistent logics on the market. Before choosing one, we should take another look at the theory of states and focus on the claims which are critical to the LNC and the LEM. These are two: (1) "states are neither existent nor non-existent" and (2) "states are and are not ob-

---

39 Marwan Rashed gives another interpretation of the theory of states by focusing only on truth-value gaps and applying intuitionistic logic. For more, see Marwan Rashed, "Abū Hāshim al-Ǧubbā'ī sur le langage de l'art," *Histoire Épistémologie Langage* 36, no. 2 (2014): 85–96, and Rashed, "Abū Hāshim al-Ǧubbā'ī et l'infini," in *Miroir de l'amitié: Mélanges offerts à Joël Biard*, ed. Cristophe Grellard (Paris: Vrin, 2017).

40 In classical logic, explosion—i.e., the argument $A \wedge \neg A \vdash B$—is valid. According to explosion, or *ex contradiction quodlibet*, from an arbitrary contradiction, everything follows. However, in his defence of the principle of non-contradiction, Aristotle does not propose this argument, although Avicenna later defended a primary form of explosion in *Metaphysics of the Healing*, 42. Finally, in the twelfth century, Willian of Soissons formalised the argument for explosion in a way which is valid in classical logic. For more, see Benham Zolghadr, "Avicenna on the Law of Non-Contradiction," *History and Philosophy of Logic* 40, no. 2 (2019): 105–15.

41 It is obvious that the theory of states is not trivial: "nothing exists in the world" is not true in this theory.

jects." The former seems to state a truth-value gap, while the latter is apparently a truth-value glut. If we accept both truth-value gaps and gluts in the theory of states, we need a four-valued logic. A good candidate might be the logic of FDE.[42] However, there are other ways to choose the proper logic for the theory of states while keeping the theory as simple as possible. Perhaps the best choice is the three-valued logic LP.[43] In LP, propositions can be only true, only false, and both true and false. By applying LP, not only do we have a simpler logic with more straightforward interpretations, but we can also have the truth of (1) without the need for a fourth value; that is, neither true nor false. We will see the details after an explanation of LP. Since, for the purpose of this paper, the propositional logic of LP suffices, I will restrict the discussion to it and skip the first-order LP.

A logic is defined by the structure $< V, \ D, \ \{f_c : c \in C\} >$. V is the set of truth values, D, which is a subset of V, is the set of designated values, C is the set of connectives, and f_c is the corresponding truth function. For LP, we have:

$$\mathrm{V} = \{0, \ 0.5, \ 1\}$$

$$\mathrm{D} = \{0.5, \ 1\}$$

$$\mathrm{C} = \{\neg, \ \wedge, \ \vee\}$$

Thus, the truth values are true, both true and false, and false, which are conventionally represented by 0, 0.5, and 1 respectively. The designated values are true and both true and false. The connectives are negation, conjunction, and disjunction. The material conditional can be defined as follows:

$$A \supset B \equiv \neg A \vee B$$

An interpretation for the language is a map, v, from the set of propositional parameters, P, into V. This is extendable to all the formulas of the language by applying truth functions recursively. This extension is done through the following conditions:

$v(\neg A) = 1 - \ v(A)$

$v(A \wedge B) = \min\{v(A), \ v(B)\}$

$v(A \vee B) = \max\{v(A), v(B)\}$

The notions of logical truth and semantic consequence are defined as follows:

$\Sigma \models A$ *iff* there is no interpretation, v, such that for all $B \in \Sigma$, $v(B) \in D$, but $v(A) \notin D$.

A is a logical truth *iff* for every interpretation $v(A) \in D$.

42 For more about FDE, see Graham Priest, *Non-Classical Logic, An Introduction to Non-Classical Logic: From If to Is*, 2nd ed. (Cambridge: Cambridge University Press, 2008), chapter 8.
43 For more about LP, see Priest, chapter 7.

It can easily be checked that LP is not explosive; that is, ECQ is not a valid argument in LP.

$$P \wedge \neg P \nvDash Q$$

For a counter-model, suppose:

$$v(P) = 0.5$$

$$v(Q) = 0$$

Thus, contradictions can be true in LP without making the system trivial.

Two comments are in order here. LP has the simplest semantics among all the paraconsistent logics. In fact, LP is a minimal revision to classical logic and classical logic is a proper extension of LP. These make LP a good choice for the desired paraconsistent logic.

Now let us return to the theory of states. As already seen, states both are and are not objects. Hence, for every state, we have the following sentence: $P \wedge \neg P$. LP handles such contradictions without falling into triviality. Moreover, one important aspect of the theory of states is the middle between existence and non-existence. We saw that this is usually seemingly uttered as a truth-value gap. There are three values in LP, none of which is neither true nor false. Therefore, we should explain how to understand "a state is neither existent nor non-existent" without appealing to the truth value of "neither true nor false." This can easily be done via the following inference from double negation:

$$\neg P \wedge \neg\neg P \vDash \neg P \wedge P$$

One can easily check that this argument is valid in LP. In section 1, we saw the same argument from Avicenna, demonstrating that any gappy sentence results in a contradiction. Thus, to say that states are neither existent nor non-existent is another form of saying that states are both existent and non-existent. Therefore, we can preserve the truths of the theory of states in LP, especially about the claims critical to the LNC and the LEM. This is also true of *the impossible* and *becoming*, which we met in the discussion of arguments against the LNC.

As already mentioned, at the beginning of this section, we chose LP over FDE. Now we know the reasons for this choice. In addition to the merits of LP due to its simplicity and its relation to classical logic, seeming truth-value gaps in the theory of states can be paraphrased as truth-value gluts by applying LP.

Before finishing this section, I should say more about the use of contemporary logical apparatus which we saw in this paper. It is obvious that the theory of states and the context in which it was stated are very far away from modern logic. Thus, I do not claim that Abū Hāšim and his followers would accept this methodology or that they would confirm the brief formalisation of the theory of states which we have seen here. The aim of applying LP to some aspects of the theory of states is nothing more to show that it is possible to have a model of the theory—or to be more precise, a dialetheic reading of the theory—without losing its coherency. It is also obvious that a model of a theory is merely an abstraction which only grasps some aspects of the theory in question. Here, we have only provided a sketch of the idea, which I think suffices for the purposes of this paper.

5 Conclusion

After discussing the arguments against the LNC, Rāzī says:

> These objections are only a drop of the sea of objections which have been raised against our saying that an object either is or is not. And if this is the state of the strongest self-evident [assent], what do you think about the weaker ones?[44]

Rāzī seems to be exaggerating about the number of objections to the LNC, but his discussion of the LNC shows that there were ongoing debates about it which were driven by the proponents of the theory of states. Unfortunately, none of Abū Hāšim's works has survived and most of the works of his followers are no longer extant. Hence, it is not possible to trace back the arguments discussed above to their advocates. As already mentioned, Rāzī himself is not an opponent of the LNC. He is well known to have enjoyed rejecting different views only for the sake of argument. However, it seems unlikely that the arguments that we discussed in this paper are his. For one thing, he explicitly talks about *others* who hold this view, at various points in the text, though with all that said, we do not know who posed those arguments first. Let us leave it to historians and scholars to decide. Moreover, we saw that the theory of states, understood as a theory which violates the LNC, can, by applying modern logical apparatus, be coherently uttered within the contemporary metaphysical scene.

Bibliography

Allāmeh-Hellī. *Kašfu l-Murād*. Qom: Mu'assese Našr-e Eslami, 1995.

Alwishah, Ahmed, and David Sanson. "The Early Arabic Liar: The Liar Paradox in the Islamic World from the Mid-Ninth to the Mid-Thirteenth Centuries CE." *Vivarium* 47, no. 1 (2009): 97–127.

Avicenna. *The Metaphysics of the Healing*. Translated by Michael E. Marmura. Utah: Brigham Young University Press, 2005.

Dhanani, Alnoor. *The Physical Theory of* Kalām: *Atoms, Space, and Void in Basrian Mu'tazilī Cosmology*. Leiden: Brill, 1994.

El-Rouayheb, Khaled. "Theology and Logic." In *The Oxford Handbook of Islamic Theology*, edited by Sabine Schmidtke, 408–32. Oxford: Oxford University Press, 2016.

Frank, Richard. *Beings and Their Attributes: The Teaching of the Basrian School of the Mu'tazila in the Classical Period*. Albany, NY: SUNY Press, 1978.

44 Rāzī, *Muḥaṣṣal*, 35.

Priest, Graham. *An Introduction to Non-Classical Logic: From If to Is*. 2nd ed. Cambridge: Cambridge University Press, 2008.
Priest, Graham. *Doubt Truth to Be a Liar*. Oxford: Oxford University Press, 2006.
Priest, Graham. *One: Being an Investigation into the Unity of Reality and of Its Parts, Including the Singular Object which Is Nothingness*. Oxford: Oxford University Press, 2014.
Qaramaleki, Ahad. "Khafri and the Liar Paradox." [Persian.] *Journal of Religious Thought of Shiraz University* 11 (2004): 33–44.
Rashed, Marwan. "Abū Hāshim al-Ǧubbāʾī et l'Infini." In *Miroir de l'amitié: Mélanges offerts à Joël Biard*, edited by Cristophe Grellard. Paris: Vrin, 2017.
Rashed, Marwan. "Abū Hāshim al-Ǧubbāʾī sur le langage de l'art." *Histoire Épistémologie Langage* 36, no. 2 (2014): 85–96.
Rāzī, Faḫr al-Dīn al-. *Muḥaṣṣal afkār al-mutaqaddimīn wa-l-mutaʾaḫḫirīn*. Cairo: Dār al-Turaṯ, 1995.
Šahrastānī, Muḥammad. *Nihāyah*. Beirut: Dār al-Kutub al-Ilmiyyah, 1956.
Schmidtke, Sabine. "The Muʿtazilite Movement III: The Scholastic Phase." In *The Oxford Handbook of Islamic Theology*, edited by Sabine Schmidtke, 159–80. Oxford: Oxford University Press, 2016.
Thiele, Jan. "Abū Hāšim al-Ǧubbāʾī's (d. 321/933) Theory of 'States' (*aḥwāl*) and Its Adaption by Ashʿarite Theologians." In *The Oxford Handbook of Islamic Theology*, edited by Sabine Schmidtke, 364–83. Oxford: Oxford University Press, 2016.
Van Ess, Josef. "The Logical Structure of Islamic Theology." In *Logic in Classical Islamic Culture*, edited by Gustave E. von Grunebaum, 21–50. Wiesbaden: Otto Harrassowitz, 1970.
Zolghadr, Behnam. "Avicenna on the Law of Non-Contradiction." *History and Philosophy of Logic* 40, no. 2 (2019): 105–15.

Máté Veres

The Causes of *Epochē*: A Note on *Stromateis* 8.22.1–4[1]

The majority of the excerpts traditionally taken to derive from a planned book 8 of Clement of Alexandria's *Stromateis* concern the theory of demonstration (*apodeixis*) and related matters of logic. The suspension of judgement (*epochē*), a recognisably sceptical response to disagreement and a lack of demonstrative certainty, receives two brief treatments in this context. Apart from an attempted refutation of scepticism which points to the allegedly self-refuting character of universal *epochē* (5.15.2–16.3), the text also includes an account of the causes that lead one to suspend judgement (7.22.1–4).

The source of this treatment of scepticism is unknown.[2] In his recent edition, translation, and commentary, Matyáš Havrda gives a judicious overview of the scholarly proposals and argues that a *liber logicus*, perhaps Galen's lost work *On Demonstration* (*De Demonstratione* or *Peri Apodeixeōs*), could have been the direct or indirect source of parts or the whole of the alleged book 8.[3] At the same time, Havrda is not convinced that this material ever made it into a Clementine book. In his view, it is

1 This paper was presented at a workshop entitled "'Liber Logicus' by Clement of Alexandria: Proof, Inquiry, Scepticism, Causation in an Early Christian Text" held at the Czech Academy of Sciences, Prague, April 20–22, 2017. I thank the DFG-2311 for their financial support and Matyáš Havrda and my anonymous reviewers for comments on an earlier draft.

2 Scholars either cautiously refer to an otherwise unknown *liber scepticus*, treat the anti-sceptical strategy of the first excerpt as reflecting a Stoic position, or suggest that Clement's source had something to do with Aenesidemus, the most notable sceptical figurehead of his native Alexandria. See Hans von Arnim, *De octavo Clementis Stromateorum libro* (Rostock: Adler, 1894), 11–15; Christiane von Wedel, *Symbola ad Clementis Alexandrini Stromatum l. VIII interpretandum* (Berlin: s.n., 1905), 5–8; Eric F. Osborn, *The Philosophy of Clement of Alexandria* (Cambridge: Cambridge University Press, 1957), 149–51; Karel Janáček, "Ainesidemos und Sextos Empeirikos," in *Studien zu Sextus Empiricus, Diogenes Laertius und zur Pyrrhonischen Skepsis*, ed. Filip Karfík and Jan Janda (Berlin and New York: De Gruyter, 2008), 251–64 (originally published in *Eirene* 17 [1980]: 5–16); Karel Janáček, "Zur Interpretation des Photiosabschnittes über Ainesidemos," in *Studien zu Sextus Empiricus, Diogenes Laertius und zur Pyrrhonischen Skepsis*, 216–24 (originally published in *Eirene* 14 [1976]: 93–100). For the suggestion that the so-called book 8 transmits Antiochean material, perhaps with Peripatetic mediation, see Reginald E. Witt, *Albinus and the History of Middle Platonism* (Cambridge: Cambridge University Press, 1937), 31–41.

3 Matyáš Havrda, *The So-Called Eighth* Stromateus *by Clement of Alexandria: Early Christian Reception of Greek Scientific Methodology* (Leiden and Boston: Brill, 2016), 34–50 (with reservations as to whether the Galenic source would have covered the excerpts on suspension); see also Matyáš Havrda, "*Galenus Christianus?* The Doctrine of Demonstration in *Stromata* VIII and the Question of Its Source," *Vigiliae Christianae* 65, no. 4 (2011): 368–72.

https://doi.org/10.1515/9783110618839-006

unlikely that these excerpts were intended, "at least in the form preserved for us," as another book of the *Stromateis*.[4]

It goes without saying that theories concerning the text's provenance depend to a large extent on the appreciation of its philosophical content. The excerpts on scepticism are interesting both with regard to the broader question of the text's origin and in their own right. In this paper, I aim to contribute to the analysis of the sceptical ideas contained in the alleged book 8. I offer considerations in favour of the scholarly consensus that the version of scepticism discussed therein is different from, and earlier than, the variant of Pyrrhonism formulated by Sextus Empiricus.[5] In doing so, I revisit and partially correct some of the received arguments in support of this claim,[6] so as to place the position on a more solid footing.

First, I will give a brief overview of both the argument from self-refutation employed in the first Clementine excerpt and the Sextan rejection of a related argument. I will argue that the comparison does not support definite claims about the origin of the Clementine material (section 1). Then I will discuss the second excerpt, which offers a series of remarks on the causal origin of *epochē* (section 2). I will then contrast it to Sextus's account of the origin of suspensive disposition in terms of originative principles (*archai*) rather than causes (*aitiai*), arguing that Sextus's formulation allows for a more nuanced sceptical position (section 3). In conclusion, I will briefly consider possible views on the relevance of this material, or the lack thereof, for Clement's Christian project (section 4).

1

The first excerpt, which appears under the inserted heading *Pros tous Pyrrōn[e]ious* (using the Byzantine spelling), criticises the sceptical position on two counts. The first objection is that one could hold the statement "nothing is certain" (*bebaion einai mēden*) to be true only on pain of self-refutation (15.2–7; cf. 16.1). This is then followed by the claim that one cannot even begin to inquire without a previous apprehension of some states of affairs, which in turn implies that something certain can be apprehended (15.7–9).[7]

---

4 Havrda, *So-Called Eighth* Stromateus, 10, cf. 14, 25, 50. Nevertheless, he entertains the possibility that while composing book 7, Clement made use of something *like* the so-called book 8 (70, cf. 74). For his summary statement, see 76–77.

5 See Janáček, "Ainesidemos und Sextos Empeirikos," 252–54; Havrda, *The So-Called Eighth* Stromateus, 244 n. 299.

6 Janáček's main argument for a non-Sextan variety of scepticism is puzzling at best; see my discussion of T4 below.

7 For Sextus's claim that the sceptic can act and investigate without dogmatic apprehension, see his distinctions between two senses of *dogma* (*PH* 1.13) and two senses of *katalēpsis* (*PH* 2.1–11).

More precisely, the charge of self-refutation is introduced in book 8 as one horn of a dilemma. In response to the question of whether anything is true, one might either give an affirmative answer, which amounts to a rejection of the claim that "nothing is certain," or hold that nothing is true, which turns out to be a self-refuting position. Similar objections were routinely raised about a variety of sceptical positions.[8] A version of the objection is addressed by Sextus Empiricus, who states that utterances such as "everything is false" (*panta esti pseudē*) or "nothing is true" (*ouden estin alēthes*) apply to themselves and cancel themselves out (*symperigraphei*, *PH* 1.14). Sextus gives a variety of responses in the *Outlines* (*PH* 1.14, 206; 2.188, all three passages featuring the term *symperigraphein*; cf. also *PH* 2.84), as well as in the second book of *Against the Logicians* (*M* 8.55, 480–81).

According to the received view, Sextus's main strategy is to embrace the self-refuting nature of self-referring sceptical utterances.[9] According to a recently proposed alternative view, however, the term *symperigraphein* invokes a specifically Pyrrhonian strategy which was designed to avoid self-refutation. On this proposal, self-referring utterances are self-refuting only if and when they are advanced with a dogmatic mindset; i. e., when the premises and the conclusion are endorsed as true by the one who utters or entertains them. By contrast, when uttered or entertained with a suspensive disposition already in place, they "bracket themselves" (this being the proposed translation of *symperigraphein*), as a result of which no contradiction arises.[10]

On either reading, Sextus is ready with an answer to the sort of objection from self-refutation that we find endorsed in the first of the two excerpts from book 8. On the novel reading, however, it becomes possible to claim that the sceptical response that Sextus preserved is reported under Clement's name as well. One of two occurrences of the key term in Clement comes "seemingly with the same meaning as in Sextus and in a sceptical context."[11] One could suggest on this basis that the two texts show little difference in terms of philosophical sophistication. Furthermore, even if we discount this possibility and admit that the source informing the Clementine passages is philosophically less advanced than Sextus, it might still not suffice to establish a chronological difference. After all, individuals working at the same time and place might achieve different levels of philosophical acuity.[12]

---

8 For similar objections, see Cicero, *Acad.* 2.29, 109; Aristocles *apud* Eusebius, *Praep. evang.* 14.18.5–24; Photius, *Bibl.* 212.169b.18–31; Sextus Empiricus, *M* 7.399, 440; 8.55; cf. already Aristotle, *Metaph.* 4.8.1012b13–22.

9 See especially Mark L. McPherran, "Skeptical Homeopathy and Self-Refutation," *Phronesis* 32 (1987): 290–328.

10 Luca Castagnoli, *Ancient Self-Refutation: The Logic and History of the Self-Refutation Argument from Democritus to Augustine* (Cambridge: Cambridge University Press, 2010), 251–307, especially 266, 269, 277.

11 Castagnoli, 351 n. 137, pointing to 8.22.2 (T4 below). The other Clementine passage, *Strom.* 6.15.119.4, is irrelevant for, and impossible to square with, Castagnoli's interpretation.

12 It is worth mentioning that if these passages had indeed been derived from Galen, then one could even come to imagine—though not in any way to establish—that Sextus was responding to the sort of

Nor does it follow that the second excerpt, a causal account of the origins of *epochē* which does not even mention the argument from self-refutation, represents a less developed form of scepticism. With these reservations in mind, I will now turn to the causal account developed in the second excerpt.

2

Even though the first excerpt rejects universal *epochē*, it does allow for the legitimacy of suspending one's judgement on particular occasions:

> **[T1]** Not only the "suspenders," but every "dogmatist" too is used to suspending judgement in some matters, either from weakness of thought, [or from] the obscurity of things, [or from] the equipollence of arguments (16.3).[13]

The second excerpt arguably picks up on the conciliatory tone of this remark and goes on to explain the reasons for widespread disagreement in the domains of inquiry and ordinary life. It is, of course, questionable whether these two discussions derive from the same source or even express the same generic outlook concerning scepticism. At any rate, even if it is possible to detect a transition from a distinctly anti-sceptical perspective to another which is mildly sympathetic to sceptical considerations,[14] the crucial link seems to be provided by the observation that localised *epochē* is a common resort of both sceptics and doctrinaire philosophers.

On the account at hand, there seem to be three reasons why one might suspend one's judgement. The first has to do with a subjective condition; namely, the weakness or infirmity of one's own cognitive capacities. By contrast, the second reason has to do with the objects of inquiry: they are unclear or obscure, and hence they do not deliver immediate knowledge of themselves. Finally, the arguments one might consider for and against accepting a given proposition are said to be equipollent. Going only by this brief sentence, it is unclear how these three sources of suspension are supposed to relate to each other.

The second excerpt is almost entirely devoted to the discussion of the causes (*aitiai*) of the suspension of judgement. To begin with, it states:

---

anti-sceptical considerations that Galen had put forward. Unfortunately, the limits of historical fancy are set by the available evidence, and nothing seems to support such a connection.

13 Οὐ μόνον οἱ ἐφεκτικοί, ἀλλὰ καὶ πᾶς δογματικὸς ἔν τισιν ἐπέχειν εἴωθεν ἤτοι παρὰ γνώμης ἀσθένειαν ἢ παρὰ πραγμάτων ἀσάφειαν ἢ παρὰ τὴν τῶν λόγων ἰσοσθένειαν. Translations taken from Havrda, *So-Called Eighth* Stromateus.

14 "This excerpt or paraphrase [i.e., the second] was probably made by the same author who accepts the limited validity of *epochē* in 16,3, not in order to embrace scepticism, but rather to represent the sceptic's position with his [i.e., the sceptic's] own words" (Havrda, 241).

[T2] The most general causes of suspension of judgement are two. One is the fickleness and instability of the human mind; it naturally generates disagreement, either that of one with another, or that of one with oneself. Second is the disagreement in things, which also, with good reason, comes down to inducing suspension of judgement (22.1 = 93.19–23).[15]

This passage differs from **T1** in a number of respects. First, it leaves out the equipollence of arguments mentioned above. Second, it explicitly casts the other two sources of *epochē* in causal terms. In addition, it seems that equipollence is here replaced by disagreement (*diaphōnia*), which figures twice in **T2.** On the one hand, it is a consequence of the fallibility of the human mind, which, by its nature (note the term *gennētikon*, which will reappear in **T3** below), brings about a variety of differing opinions on both the interpersonal and the intrapersonal level. On the other hand, disagreement applies to the nature of things themselves (*hē en tois ousi diaphōnia*), perhaps in the sense that they give rise to anomalous appearances.[16]

The final, lengthiest discussion of the causes of *epochē* gives yet another variation on the same theme. According to the passage below, which includes a vivid description of the dogmatic predicament, disagreement is the proximate cause of suspension of judgement (*hē de diaphōnia proseches aition tēs epochēs*), and it is ultimately due to the condition of the human mind, with no explicit mention of the nature of things themselves:

[T3] Out of these most principal [causes] of suspension, the infirmity of the mind is the one that generates disagreement, and disagreement is the proximate cause of the suspension of judgement. As a consequence, life is full of courts, councils, assemblies, and, generally speaking, choice <and avoidance> regarding so-called good and evil things. These are tokens [showing] that the mind is puzzled and that it wavers vis-à-vis *the equipollence* of opposing things. And libraries are full of books of people disagreeing with one another in their beliefs, each having convinced himself that he knows the truth in things (22.3–4 = 93.27–94.4).[17]

---

15 Τὰ ποιητικὰ τῆς ἐποχῆς αἴτια δύο ἐστὶν τὰ ἀνωτάτω, ἓν μὲν τὸ πολύτροπον καὶ ἄστατον τῆς ἀνθρωπίνης γνώμης, ὅπερ γεννητικὸν εἶναι πέφυκεν τῆς διαφωνίας ἤτοι τῆς ἀλλήλων πρὸς ἀλλήλους ἢ τῆς ἑαυτῶν πρὸς ἑαυτούς, δεύτερον δὲ ἡ ἐν τοῖς οὖσι διαφωνία, ἣ καὶ εἰκότως ἐμποιητικὴ καθέστηκε τῆς ἐποχῆς.

16 But cf. von Wedel's alternative classification: "Et eodem modo bipartitae sunt causae ἐποχῆς in Clementis § 16 (I) γνώμης ἀσθένεια (II) 1. πραγμάτων ἀσάφεια 2. λόγων ἰσοσθένεια" (von Wedel, *Symbola ab Clementis Alexandrini*, 8). My understanding is closer to the contrast in Osborn, *The Philosophy of Clement of Alexandria*, 151 (on the one hand, people change their minds and disagree with each other; on the other hand, the world as such presents us with conflicting appearances).

17 Τούτων δὲ αὐτῶν τῶν ἀρχικωτάτων <αἰτίων> τῆς ἐποχῆς τὸ μὲν ἀβέβαιον τῆς διανοίας γεννητικόν ἐστι διαφωνίας, ἡ δὲ διαφωνία προσεχὲς αἴτιον τῆς ἐποχῆς, ὅθεν πλήρης μὲν ὁ βίος δικαστηρίων τε καὶ βουλευτηρίων καὶ ἐκκλησιῶν καὶ καθόλου τῆς περὶ τὰ λεγόμενα ἀγαθὰ καὶ κακὰ αἱρέσεως <καὶ φυγῆς>, ἅπερ ἠπορημένης ἐστὶ διανοίας καὶ πρὸς τὴν τῶν ἀντικειμένων πραγμάτων *ἰσοσθένειαν* μετοκλαζούσης τεκμήρια. πλήρεις δ' αἱ θῆκαι τῶν βιβλίων [καὶ αἱ συντάξεις καὶ αἱ πραγματεῖαι] τῶν διαφωνούντων ἐν τοῖς δόγμασι καὶ πεπεικότων ἑαυτοὺς τὴν ἐν τοῖς οὖσιν ἀλήθειαν γινώσκειν. Havrda added <αἰτίων> to this passage; Stählin added <καὶ φυγῆς>; Heyse and Weydel modified *εἰς ἀσθένειαν* to *ἰσοσθένειαν*; and Havrda removed [καὶ αἱ συντάξεις καὶ αἱ πραγματεῖαι].

Whether or not one is sympathetic to the assumption that all three passages (**T1–T3**) belong to one and the same account concerning the causes of suspension, one might find the last version to be the closest to Sextus Empiricus's Pyrrhonian stance. After all, in **T3**, the major consideration in favour of *epochē* appears to be equipollence. On the one hand, there is widespread disagreement in matters of theory and practice; on the other, contrary arguments turn out to be equally persuasive, and this is in fact the main feature that defines Sextan scepticism (*PH* 1.8). Insofar as this passage downplays or drops the idea that disagreement is due to the nature of the things investigated, it might after all manage to escape a lurking dogmatic commitment concerning the objective limitations of our capacity for knowledge.[18]

Nevertheless, the emphasis on our subjective limitations might still hold the danger of latent dogmatism. In this respect, it is not so much the characterisation of the human mind as feeble or changeable (*polytropos* and *astatos* in **T2**; *abebaios* in **T3**) which seems problematic, but rather the claim that it is a congenital condition.[19] If equipollence is merely a consequence of the infirmity of the human mind, the question once again becomes whether this infirmity is due to our limited cognitive capacities or rather to the nature of things themselves. In order to read **T3** as obeying the spirit or perhaps even the letter of Sextan Pyrrhonism, this aspect of the account should be downplayed.

The following two considerations indicate a harmonious rapport between Sextan Pyrrhonism and the sceptical stance reported by Clement. Firstly, Sextus's classification of the Ten Modes for *epochē* shows affinity with the causes specified by Clement: as Sextus puts it, some of the modes have to do with the one who judges, some with the thing being judged, and some with a combination of both (*PH* 1.38).[20] The same

---

18 On the other hand, this might be nothing more than a change of terminology. Perhaps the disagreement *en tois ousi* amounts to what Sextus calls "anomalies" or conflicting appearances. See, e.g., Sextus's references to *anōmalia* in his narratives of sceptical conversion (*PH* 1.12, 29, 214; *M* 1.6), as well as throughout his description of the Ten Modes of Aenesidemus (*PH* 1.112, 114, 132, 163, 198, 218, 220; cf. 3.233–35, *M* 9.191, and *M* 1.154, 236, 240). See also DL 9.76: "Whenever things are at odds with each other and arguments have equal strength, ignorance of the truth follows suit" (τῶν μὲν γὰρ πραγμάτων διαφωνούντων τῶν δὲ λόγων ἰσοσθενούντων ἀγνωσία τῆς ἀληθείας ἐπακολουθεῖ, trans. Scharffenberger and Vogt).

19 Interestingly, a related remark from Sextus appears in the context of defending Pyrrhonism against self-refutation by invoking the idea that the human mind will not put lasting trust in the argument against the possibility of demonstration: "However, if the skeptics have to answer for themselves, they will answer in a safe way. For they will say that the argument against demonstration is merely persuasive, and that for the moment it persuades them and induces assent, but that they do not know whether it will also be like this in the future given the fickle character of human thought" (ὅμως δὲ καὶ τοὺς σκεπτικοὺς ἂν δέῃ ὑπὲρ αὐτῶν ἀποκρίνασθαι, ἀσφαλῶς ἀποκρινοῦνται. φήσουσι γὰρ τὸν κατὰ τῆς ἀποδείξεως λόγον πιθανὸν εἶναι μόνον καὶ πρὸς τὸ παρὸν πείθειν αὐτοὺς καὶ ἐπάγεσθαι συγκατάθεσιν, ἀγνοεῖν δέ, εἰ καὶ αὖθις ἔσται τοιοῦτος διὰ τὸ πολύτροπον τῆς ἀνθρωπίνης διανοίας; *M* 8.473, trans. Bett).

20 "Superordinate to these [i.e., to the Ten Modes] are three modes: that deriving from the subject judging; that deriving from the object judged; that combined from both" (τούτων δὲ ἐπαναβεβηκότες

classificatory concern explains his presentation not only of the Ten Modes, but also that of the Five Modes, the Two Modes, and the Eight Modes (*PH* 1.31–186). Secondly, an additional argument offered in-between **T2** and **T3** sounds especially close to a statement of Sextus's official position:

> **[T4]** We can neither believe all presentations, because of their conflict; nor disbelieve them all, because the one saying that all [presentations] are untrustworthy, is bracketed along with the "all," being one of them; nor can we believe some and disbelieve others, because of their equality; thus we are brought to suspension of judgement (22.2 = 93.23–27).[21]

Surprisingly, Janáček invokes this passage as the main argument in favour of a pre-Sextan variety of scepticism informing the so-called book 8's treatment of scepticism.[22] Admittedly, the argument itself has a pre-Sextan pedigree, but there is no reason to believe that it might not form part of a position sufficiently similar to that of Sextus. He himself found it congenial to his presentation of the Pyrrhonist position, as evidenced primarily by his discussion of the criterion of truth, which, at *M* 7.389, starts out with exactly the same sort of argumentation. As things stand, therefore, we find little in book 8's presentation of scepticism which cannot, at the end of the day, be made to square with Sextus's flexible Pyrrhonism.

3

I have argued thus far that there is a more or less straightforward, if somewhat overzealous, way of reading Clement's account as pointing away from a sceptical view based on dogmatic commitments concerning the nature of reality and towards one that lives up to the Sextan gold standard. My purpose in pushing for such a reading was not to argue that the Clementine excerpts do in fact understand scepticism in the vein of Sextus Empiricus; rather, my intention was to prepare the ground for the next step, which is to draw attention to a more significant difference between these presentations. Quite importantly, while Clement or his source gives a causal account of the suspension of judgement, Sextus embeds his own account in a biographical narrative concerning the conversion to scepticism—not least because a causal account, by his standards, would qualify as dogmatic.

In his idealised biography of an individual who sets out to discover the truth but ends up suspending judgement instead (*PH* 1.12, 26, 29), Sextus speaks not of causes

---

εἰσὶ τρόποι τρεῖς, ὁ ἀπὸ τοῦ κρίνοντος, ὁ ἀπὸ τοῦ κρινομένου, ὁ ἐξ ἀμφοῖν, trans. Annas and Barnes). See Havrda, *So-Called Eighth* Stromateus, 241–42.

21 μήτε γὰρ πάσαις ταῖς φαντασίαις πιστεύειν δυνηθέντες διὰ τὴν μάχην μήτε πάσαις ἀπιστεῖν διὰ τὸ καὶ τὴν λέγουσαν πάσας ἀπίστους ὑπάρχειν ἐξ ἁπασῶν οὖσαν συμπεριγράφεσθαι πάσαις μήτε τισὶ μὲν πιστεύειν, τισὶ δὲ ἀπιστεῖν διὰ τὴν ἰσότητα, κατήχθημεν εἰς ἐποχήν.

22 Janáček, "Ainesidemos und Sextos Empeirikos," 253: "Das Hauptargument aber für eine ältere skeptische Quelle als Sextos im Klemensabschnitt ist die dreifache Begründung des Skeptizismus."

(*aitiai*), but rather of originating principles (*archai*) which motivate the adoption of a suspensive disposition. Sextus's story describes the interplay of two such principles. The first, originative principle (*archē aitiōdēs*) is the hope of tranquillity: individuals sensitive to philosophical concerns and troubled by their inability to decide between conflicting appearances set out to discover what is true and what is false, in the hope of achieving intellectual tranquillity. The second sustaining or constitutive principle (*archē systaseōs*) appeals to their eventual recognition that to every account an equally persuasive account is opposed, which leads them to suspend their judgement and, in turn, to a fortuitous state of tranquillity. Since the first principle applies to all philosophers (labeled by Sextus *hoi megalophyeis*) while the second only applies to those who have become Pyrrhonians, Sextus's narrative links together these two groups in a way that enables him to claim that sceptics are, as their chosen denomination suggests, proper inquirers.[23]

Sextus not only distances himself from the all-too-formal accounts of what brings about *epochē*; he also manages to establish a continuity between what he takes to be the original motivation for inquiry and the aim of the full-fledged Pyrrhonist. By contrast, book 8 does not engage in conjectures concerning the background of a philosophical sceptic—it does not mention tranquillity or, for that matter, any other possible motivation for adopting a sceptical position.[24] Nor does it present sceptics as genuine inquirers; rather, the opening passage of book 8 suggests that the sceptical predilection for an aporetic outcome is nothing but an obstacle to genuine inquiry. Insofar as neither of these characteristic features of Sextan Pyrrhonism make an appearance in it, book 8 can hardly be taken to expound a form of scepticism sufficiently similar to that of Sextus.

At this point, one might pose the question why, given such an unflattering view of scepticism, Clement would show any interest in it. If book 8's discussion of scepticism is in fact compatible with Clement's Christian agenda, then the guiding idea must perhaps be that genuine inquiry takes its departure from divine authority, and the alternative to accepting that authority is to engage in the fruitless, sophistical bickering of scepticism. Whether or not this was in fact Clement's position, as we shall see in the concluding section, is ultimately unclear.

4

Scholars have understandably found it puzzling that a Christian theologian of Clement's standing would express interest in sceptical arguments, let alone recognise

---

23 For a detailed defence of this reading, see Máté Veres, "Keep Calm and Carry On: Sextus Empiricus on the Origins of Pyrrhonism," *History of Philosophy and Logical Analysis* 23, no. 1 (2020), special issue on *Ancient Modes of Philosophical Inquiry*, ed. Jens Kristian Larsen and Philipp Steinkrüger.
24 It has been suggested to me that this might indicate an Aenesideman source, but it is difficult to square this suggestion with *DL* 9.108; cf. also Aristocles *apud* Eusebius, *Praep. evang.* 14.18.2–4.

their potential legitimacy.[25] On the basis of the discussion in the previous sections, two possibilities seem to emerge: either to find a place for these excerpts in Clement's Christian thought or to discard them as inauthentic, meta-Stromatic material that bears no relation to it.

As to the former option, one might say that it is not unheard of that a Christian thinker would employ scepticism in support of their religious stance. In such a view, the sceptical predicament shows that human reason naturally generates disagreement, which, if unaided by divine revelation, it cannot ever hope to overcome.[26] It is conceivable that sceptical argumentation could play an inadvertent role in the formulation of such a quasi-fideistic position: by calling into question our ability to acquire demonstrative knowledge, scepticism can be seen as giving support to revealed starting points for inquiry.[27]

A general context of this sort may bring us closer to appreciating Clement's interest in the book 8 material. Such a reading, then, is at the very least possible; the question is whether there is a textual basis to support it. One might point to the opening paragraphs of book 8, where Clement contrasts more recent Greek philosophers, who reduced themselves to elenctic and eristic argumentation, with both the more ancient Greek philosophers and the true philosophy of those who are led to

---

25 See Janáček, "Ainesidemos und Sextos Empeirikos," 252: "It [i.e., the second excerpt] is clearly an excerpt from a sceptical author, just as the polemic in III 89–90 [i.e., the first excerpt] betrays an anti-sceptical author. One could hardly claim that Clement the Christian, an admirer of Plato, Pythagoras, the Stoics, was a sceptic." ("Es ist natürlich ein Exzerpt aus einem skeptischen Autor, gerade wie das Pamphlet III 89–90 einen antiskeptischen Autor verrät. Den Christen Klemens, einen Bewunderer von Platon, Pythagoras, den Stoikern, könnte man mit Mühe einen Skeptiker nennen.")

26 On the Christian use of sceptical arguments, see Gábor Kendeffy, *Az egyházatyák és a szkepticizmus* (Budapest: Áron, 1999), especially 48–60. On Clement, see George Boys-Stones, *Post-Hellenistic Philosophy: A Study of Its Development from the Stoics to Origen* (Oxford: Oxford University Press, 2001), 188–94, and George Karamanolis, *The Philosophy of Early Christianity* (Durham: Acumen, 2013), 120–27. For a classic statement of such a reading, see Eugène De Faye, "The Influence of Greek Scepticism on Greek and Christian Thought in the First and Second Centuries," *Hibbert Journal* 22 (1924): 717–20, who argues that the structure and style of the entire *Stromateis* is ultimately due to Clement being shaken by the force of sceptical arguments. On Clement's project in general, see also Pierre Nautin, "La fin des *Stromates* et les *Hypotyposes* de Clément d'Alexandrie," *Vigiliae Christianae* 30, no. 4 (1976): 268–302, Silke-Petra Bergjan, "Logic and Theology in Clement of Alexandria. The Purpose of the 8th Book of the *Stromata*," *Zeitschrift für Antikes Christentum* 12, no. 3 (2008): 394–411, and Matyáš Havrda, "Clement's Exegetical Interests in *Stromateis* VIII," in *Clement's Biblical Exegesis*, ed. Veronika V. Černušková, Judith L. Kovacs, and Jana Plátová (Leiden and Boston: Brill, 2016), 162–78.

27 For a recent argument that fideism as a distinct philosophical stance appeared in the Alexandrian context as a combination of Jewish monotheism and Aenesideman scepticism in the works of Philo of Alexandria, see Carlos Lévy, "De l'*epochè* sceptique à l'*epochè* transcendantale: Philon d'Alexandrie fondateur du fidéisme," in *Scepticisme et religion: Constantes et évolutions, de la philosophie hellénistique à la philosophie médiévale*, ed. Anne-Isabelle Bouton-Touboulic and Carlos Lévy (Turnhout: Brepols, 2016), 57–73.

investigation by Scripture (8.1.1).[28] This might also fit well with Clement's project of carving out a niche for a Christian stance which does not reject philosophical inquiry, but ties its possibility and eventual success to the prior acceptance of Christian revelation. After all, book 8 opens with the interpretation of a biblical passage, Matthew 7.7: "Seek and you will find, knock and it will open, ask and it will be given for you."[29]

The other option, that scepticism is of little relevance to Clement's stated purposes in the *Stromateis*, nevertheless remains equally plausible. The closest Clement comes to engaging with sceptical strategies is probably his discussion of the argument from disagreement in book 7 (7.89–90).[30] In that context, he appears adamant that his method of scriptural interpretation leads to a sort of demonstrative knowledge which is not vulnerable to the challenge of *diaphōnia*. This, of course, could still be squared with the quasi-fideistic suggestion that, if it were not for Scripture, one would never be able to reason one's way out of disagreement; the emphasis, however, is very much on the message of success, not on the appeal of scepticism.[31]

In sum, the questions concerning the origin of these excerpts and their connection to the main text of the *Stromateis* are at the very least contentious, and suggestions concerning the Clementine appropriation of scepticism for Christian purposes will inevitably stand on shaky grounds.[32] Hence the modest aim of this paper was to provide an analysis of the causal account of suspension offered in the alleged book 8 and to compare it with the mature Pyrrhonist position of Sextus Empiricus. A more ambitious discussion could venture to situate the Clementine material in its supposedly Alexandrian context and to analyse its possible ties not only to Clement's epistemology and theory of faith, but also to the pre-Sextan form of scepticism reported by Philo of Alexandria and championed by Aenesidemus.[33] This, however, remains a task for another day.

---

28 Cf. *Stromateis* 7.4 and Nautin, "La fin des *Stromates*," 291. Bergjan, "Logic and Theology in Clement of Alexandria," 406–9, suggests that the proponents of universal suspension of judgement that Clement has in mind are not so much sceptical philosophers, but rather the followers of Tertullian, who reject the religious relevance of rational argumentation.

29 Αἰτεῖτε, καὶ δοθήσεται ὑμῖν· ζητεῖτε, καὶ εὑρήσετε· κρούετε, καὶ ἀνοιγήσεται ὑμῖν.

30 For his rather limited awareness of sceptics, see also references to Pyrrho in *Stromateis* 1.14, 64.4 and 7.16, 101.4.

31 See Matyáš Havrda, "Demonstrative Method in *Stromateis* VII: Context, Principles, and Purpose." In *The Seventh Book of the* Stromateis: *Proceedings of the Colloquium on Clement of Alexandria (Olomouc, 21–23, 2010)*, edited by Matyáš Havrda, V. Hušek, and J. Plátová (Vigiliae Christianae Supplements 117) (Leiden: Brill, 2012), 261–75. Havrda argues that "the material in the eighth book [...] seems to reflect Clement's theological goals to a limited extent only" (263; cf. 270–71).

32 See Havrda, *So-Called Eighth* Stromateus, 54–55; cf. 27–28 for the claim that only the first chapter of book 8 is of an undoubtedly Christian character.

33 Havrda, 241, points out that other than Clement, only Philo, *Ebr.* 170–71 speaks of the causes of disagreement and suspension.

Bibliography

Arnim, Hans von. *De octavo Clementis Stromateorum libro.* Rostock: Adler, 1894.

Bergjan, Silke-Petra. "Logic and Theology in Clement of Alexandria. The Purpose of the 8th Book of the *Stromata.*" *Zeitschrift für Antikes Christentum* 12, no. 3 (2008): 394–411.

Boys-Stones, George. *Post-Hellenistic Philosophy: A Study of Its Development from the Stoics to Origen.* Oxford: Oxford University Press, 2001.

Castagnoli, Luca. *Ancient Self-Refutation: The Logic and History of the Self-Refutation Argument from Democritus to Augustine.* Cambridge: Cambridge University Press, 2010.

De Faye, Eugène. "The Influence of Greek Scepticism on Greek and Christian Thought in the First and Second Centuries." *Hibbert Journal* 22 (1924): 702–21.

Diogenes Laertius. "Lives of Pyrrho and Timon (9.61–116)." Translated by Elisabeth Scharffenberger and Katja Maria Vogt. In *Pyrrhonian Skepticism in Diogenes Laertius: Introduction, Text, Translation and Interpretive Essays*, edited by Katja Maria Vogt, 16–51. Tübingen: Mohr Siebeck, 2015.

Havrda, Matyáš. "Clement's Exegetical Interests in *Stromateis* VIII." In *Clement's Biblical Exegesis*, edited by Veronika V. Černušková, Judith L. Kovacs, and Jana Plátová, 162–78. Leiden and Boston: Brill, 2016.

Havrda, Matyáš. "Demonstrative Method in *Stromateis* VII: Context, Principles, and Purpose." In *The Seventh Book of the* Stromateis: *Proceedings of the Colloquium on Clement of Alexandria (Olomouc, 21–23, 2010)*, edited by Matyáš Havrda, V. Hušek, and J. Plátová (Vigiliae Christianae Supplements 117), 261–75. Leiden: Brill, 2012.

Havrda, Matyáš. "*Galenus Christianus?* The Doctrine of Demonstration in *Stromata* VIII and the Question of Its Source." *Vigiliae Christianae* 65, no. 4 (2011): 343–75.

Havrda, Matyáš. *The So-Called Eighth* Stromateus *by Clement of Alexandria: Early Christian Reception of Greek Scientific Methodology.* Leiden and Boston: Brill, 2016.

Janáček, Karel. "Ainesidemos und Sextos Empeirikos." In *Studien zu Sextus Empiricus, Diogenes Laertius und zur Pyrrhonischen Skepsis*, edited by Filip Karfík and Jan Janda, 251–64. Berlin and New York: De Gruyter, 2008. Originally published in *Eirene* 17 (1980): 5–16.

Janáček, Karel. "Zur Interpretation des Photiosabschnittes über Ainesidemos." In *Studien zu Sextus Empiricus, Diogenes Laertius und zur Pyrrhonischen Skepsis*, edited by Filip Karfík and Jan Janda, 216–24. Berlin and New York: De Gruyter, 2008. Originally published in *Eirene* 14 (1976): 93–100.

Karamanolis, George. *The Philosophy of Early Christianity.* Durham: Acumen, 2013.

Kendeffy, Gábor. *Az egyházatyák és a szkepticizmus.* Budapest: Áron, 1999.

Lévy, Carlos. "De l'*epochè* sceptique à l'*epochè* transcendantale: Philon d'Alexandrie fondateur du fidéisme." In *Scepticisme et religion: Constantes et évolutions, de la philosophie hellénistique à la philosophie médiévale*, edited by Anne-Isabelle Bouton-Touboulic and Carlos Lévy, 57–73. Turnhout: Brepols, 2016.

McPherran, Mark L. "Skeptical Homeopathy and Self-Refutation." *Phronesis* 32 (1987): 290–328.

Nautin, Pierre. "La fin des *Stromates* et les *Hypotyposes* de Clément d'Alexandrie." *Vigiliae Christianae* 30, no. 4 (1976): 268–302.

Osborn, Eric F. *The Philosophy of Clement of Alexandria.* Cambridge: Cambridge University Press, 1957.

Sextus Empiricus, *Against the Logicians.* Translated by Richard Bett. Cambridge: Cambridge University Press, 2005.

Sextus Empiricus, *Outlines of Scepticism.* Translated by Julia Annas and Jonathan Barnes. Cambridge: Cambridge University Press, 2000.

Veres, Máté. “Keep Calm and Carry On: Sextus Empiricus on the Origins of Pyrrhonism.” *History of Philosophy and Logical Analysis* 23, no. 1 (2020). Special issue on *Ancient Modes of Philosophical Inquiry*, edited by Jens Kristian Larsen and Philipp Steinkrüger.
Wedel, Christiane von. *Symbola ad Clementis Alexandrini Stromatum l. VIII interpretandum*. Berlin: s.n., 1905.
Witt, Reginald E. *Albinus and the History of Middle Platonism*. Cambridge: Cambridge University Press, 1937.

Ze’ev Strauss

Meister Eckhart Reading Ibn Gabirol’s *Fons vitae*[1]

Introduction

The High Scholastic German Dominican Meister Eckhart (1260–1328) is renowned for his innovative philosophy that dared to break the mould of the accepted Scholastic worldview of his days.[2] Some of the heterodox elements of his thought also seem to have originated from his favourable view of Jewish philosophical sources. The celebrated Jewish philosopher Maimonides (1135/38–1204) immediately comes to mind since his Judaeo-Arabic Aristotelianism exerted a formative impact on Eckhart.[3] However, Maimonides is not the only Jewish philosopher to have shaped Eckhart’s body of thought. A further philosophical source requiring scholarly attention is *Fons vitae* by the eleventh-century Jewish Andalusian Solomon Ibn Gabirol (1020/21–58), generally denoted in Scholasticism as *Avicebron* or as *Avencebrol*. Eckhart, who was not aware of Avicebron’s Jewish faith, approvingly invokes his philosophical authority on several occasions. His treatment of *Fons vitae*’s metaphysical assertions represents a constructive way of approaching Gabirolean thought which deviates from his Dominican predecessors, the ardent Aristotelians Albertus Magnus (c. 1200–80) and Thomas Aquinas (1225–74). These two thirteenth-century Scholastics are rather notorious for their critical attitude towards Avicebronian hylomorphism.[4]

1 The title of my article alludes to Yossef Schwartz’s *“To Thee Is Silence Praise”: Meister Eckhart’s Reading in Maimonides’ Guide of the Perplexed* [Hebrew] (Tel Aviv: Am Oved Publishers, 2002), in which he methodically reveals Meister Eckhart’s dependency upon Maimonides, showing that he regarded “Rabbi Moyses” as a philosophical authority of the highest order.

2 See Kurt Flasch, *Meister Eckhart: Philosopher of Christianity*, trans. Anne Schindel and Aaron Vanides (New Haven: Yale University Press, 2015), 234–73.

3 For a further article by Schwartz on this subject matter, see Yossef Schwartz, “Meister Eckhart and Moses Maimonides: From Judaeo-Arabic Rationalism to Christian Mysticism,” in *A Companion to Meister Eckhart*, ed. Jeremiah M. Hackett (Leiden: Brill, 2013), 389–414.

4 Cf. Alessandro Palazzo, “Eckhart’s Islamic and Jewish Sources: Avicenna, Avicebron, and Averroes,” in *A Companion to Meister Eckhart*, ed. Jeremiah M. Hackett (Leiden: Brill, 2013), 274; Fernand Brunner, “Maître Eckhart et Avicébron,” in *Lectionum Varietates. Hommage à Paul Vignaux (1904–1987)*, ed. Jean Jolivet, Zenon Kaluza, and Alain de Libera (Paris: Vrin, 1991), 150; Bernard McGinn, “Ibn Gabirol: The Sage among the Schoolmen,” in *Neoplatonism and Jewish Thought*, ed. Lenn E. Goodman (Albany, NY: SUNY Press, 1992), 96–97. As Alessandro Palazzo correctly notes, despite Eckhart’s positive approach to Ibn Gabirol, his treatment of his system of thought is still not on the same level of complexity and of similar elaboration as those of his fellow Dominicans: Palazzo, “Eckhart’s Islamic and Jewish Sources,” 274.

https://doi.org/10.1515/9783110618839-007

In this study, I set out to provide a brief overview of Eckhart's explicit and implicit references to Avicebron's *Fons vitae.* The main argument I will put forward is that Avicebron's metaphysics influenced Eckhart's Christian philosophy at significant points. I will aim to uncover this extensive usage and appropriation by bringing six central themes of Gabirolean thought to the fore: (1) divine creation; (2) the ultimate epistemological goal of the human soul; (3) emanation and its fourfold *ordo essendi*; (4) negative theology; (5) God's absolute transcendence and pure unity; and (6) universal hylomorphism. Our approach will not confine itself solely to Eckhart's explicit references to *Fons vitae*, but will also have recourse to implicit ones, especially when they appear to echo ideas present in the former. I wish to argue that Eckhart does not affirm particular thought patterns from *Fons vitae* while dismissing others. Rather, he appears to wholeheartedly endorse doctrines that are central to Avicebronian philosophy. Furthermore, I will attempt to establish that Eckhart was intimately familiar with *Fons vitae*, aptly evaluating it as advancing a metaphysics of the One.

I will show that through Ibn Gabirol's philosophy, Eckhart not only explicates the philosophical meaning of key verses from the Pentateuch, such as Gen 1:1, but also that of verses from the New Testament, a prime example being the prologue to the Gospel of John. In fact, contrary to his use of Maimonides, which is almost entirely limited to his exegesis of Jewish Scripture,[5] Eckhart mainly invokes Avicebron in his reading of the New Testament. Nearly half of his mentions of *Fons vitae* are found in his *Commentary on the Gospel of John.*[6] It is also noteworthy that seven out of Eckhart's overall twelve allusions to Ibn Gabirol refer specifically to topics found in the fifth chapter of *Fons vitae.* This particular chapter seems to bear the strongest resemblance to Christian theology, denoting, for example, God's Will (*voluntas*) in terms of the all-pervading divine Word (*verbum*) in section 36.[7] By emphasising these points throughout this survey, I will prompt the conclusion that Eckhart assumed Avicebron to be a Christian thinker, through whom he can philosophically

---

5 See Bernard McGinn, *The Mystical Thought of Meister Eckhart: The Man from Whom God Hid Nothing* (New York: Crossroad, 2001), 271 n. 47.

6 See Palazzo, "Eckhart's Islamic and Jewish Sources," 275.

7 The writings of Ibn Gabirol will be cited in brackets according to Baeumker's critical edition: *Avencebrolis (Ibn Gabirol) Fons Vitae, ex Arabico in Latinum Translatus ab Johanne Hispano et Dominico Gundissalino*, ed. Clemens Baeumker, in *Beiträge zur Geschichte der Philosophie des Mittelalters, Texte und Untersuchungen*, vol. 1.2, ed. Clemens Baeumker and Georg von Hertling, 2nd ed. (Münster: Aschendorff, 1995), 323.17 (5:36) (henceforth FV). For the possible influence of the *Longer Theology of Aristotle* on Ibn Gabirol's theory of the divine Will, see Jacques Schlanger, *The Philosophy of Solomon Ibn Gabirol* [Hebrew] (Jerusalem: Magnes Press, 1980), 65–66; Sarah Pessin, "Solomon Ibn Gabirol [Avicebron]," in *The Stanford Encyclopedia of Philosophy* (Winter 2016 Edition), ed. Edward N. Zalta, https://plato.stanford.edu/archives/win2016/entries/ibn-gabirol/ (under section 4.2: "Will, Wisdom, Word, Intellect and the Cosmos"); Pessin, *Ibn Gabirol's Theology of Desire: Matter and Method in Jewish Medieval Neoplatonism* (Cambridge: Cambridge University Press, 2013), 177.

uncover core tenets of Christian faith.[8] However, the ultimate claim I will bring forth is that it was not Avicebron's supposed religious affiliation that was the decisive factor in attracting Eckhart to his philosophy, but rather its speculative and universal approach, metaphysically accounting for theological notions concerning God, creation, and man.

Although several scholarly works have been produced that concentrate on Avicebron's general reception history within High Scholasticism,[9] there has been scarcely any attention devoted to Meister Eckhart's extensive usage of *Fons vitae*. Two scholarly works in which Eckhart's reception of Ibn Gabirol takes centre stage merit special attention, the first being the article entitled "Maître Eckhart et Avicébron" by the Swiss historian of philosophy Fernand Brunner published in 1991.[10] In fact, this is the only article of its kind, inasmuch as it deals exclusively with this subject matter. Brunner offers a concise and profound overview of Eckhart's employment of *Fons vitae* in his Latin works (henceforth LW) as well as in his German ones (henceforth DW). The second is the article by Alessandro Palazzo entitled "Eckhart's Islamic and Jewish Sources: Avicenna, Avicebron, and Averroes" published in 2013. This article is the first major contribution to this topic in recent scholarship. Palazzo predicates his exploration of Eckhart's Jewish sources almost exclusively on Ibn Gabirol's *Fons vitae*, which entails an 11-page section that systematically tackles the theme at hand.[11] His study offers us a more detailed account of Eckhart's approach to *Fons vitae* throughout his writings than Brunner's. Through a statistical evaluation of Eckhart's quotations of Avicebron, Palazzo outlines, in a succinct and carefully thought-out manner, the explicit mentions of *Fons vitae* in the corpus of Eckhart's works.

---

8 It was not unprecedented in the scholastic tradition to consider Avicebron to be a Christian thinker. See James A. Weisheipl, "Albertus Magnus and Universal Hylomorphism: Avicebron. A Note on Thirteenth-Century Augustinianism," in *Albert the Great: Commemorative Essays*, ed. Francis J. Kovach and Robert W. Shahan (Norman: University of Oklahoma Press, 1980), 244.

9 To mention a few: McGinn, "Ibn Gabirol: The Sage among the Schoolmen," 77–109; Jacob Guttmann, *Die Scholastik des dreizehnten Jahrhunderts in ihren Beziehungen zum Judenthum und zur jüdischen Literatur* (Breslau: M. & H. Marcus, 1902), 1–46, 60–85, 154–67; Guttmann, *Die Philosophie des Salomon ibn Gabirol* (Göttingen: Vandenhoeck & Ruprecht's Verlag, 1889), 39–65; Weisheipl, "Albertus Magnus and Universal Hylomorphism," 239–60; Michael Wittmann, *Die Stellung des hl. Thomas von Aquin zu Avencebrol (Ibn Gebirol) im Entwicklungsgang der arabischen Philosophie*, in *Beiträge zur Geschichte der Philosophie des Mittelalters*, vol. 3.3, ed. Clemens Baeumker and Georg von Hertling (Münster: Aschendorff, 1900); John A. Laumakis, "Avicebron (Solomon ibn Gabirol) and Aquinas on Primary and Secondary Causality" (PhD diss., Marquette University, 2001); Mária Mičaninová, "Was Aquinas Right When He Claimed That Avicebron Departed from Aristotle?" *Archive of the History of Philosophy and Social Thought* 62 (2017): 281–305.

10 Brunner, "Maître Eckhart et Avicébron," 133–39. Another noteworthy article, which I unfortunately came across too late, is Alessandra Beccarisi's "Zwischen Averroes, Avicenna und Avicebron. Meister Eckhart und die Noetik im Islam und Judentum," in Christine Büchner, Markus Enders and Dietmar Mieth, eds., *Meister Eckhart – interreligiös* (Stuttgart: Kohlhammer, 2016), 223–40.

11 Palazzo, "Eckhart's Islamic and Jewish Sources," 271–81.

In Eckhart's writings, one encounters explicit references to *Fons vitae*, but also several passages which implicitly exhibit similarities to Gabirolean notions. In his Latin works, Eckhart directly cites Avicebron on *thirteen* different occasions:[12]

- *once* in his *Expositio libri Genesis* (henceforth *In Gen.I*): (1) n. 27 [LW 1, 205.8–15]: FV 5:25 [301.20–302.1, 10–14];
- *twice* in his *Liber parabolarum Genesis*[13] (henceforth *In Gen.II*): (2) n. 113 [LW 1, 580.2–4]: FV 1:3 [5.13–16], (3) n. 209 [LW 1, 684.5–685.4]: FV 5:24 [301.16–302.15];
- *four* times in his *Expositio libri Exodi* (henceforth *In Ex.*): (4) n. 58 [LW 2, 64.12–13]: FV 5:23 [300.16], (5) n. 90 [LW 2, 93.11–13]: FV 5:23 [300.16–17], (6) n. 271 [LW 2, 218.13–219.2]: FV 5:43 [338.22–27], (7) n. 281 [LW 2, 225.5–12]: FV 1:4 [6.16–22, 24];
- *once* in *Sermones et Lectiones super Ecclesiastici C. 24:23–31* (henceforth *In Eccli.*): (8) n. 9 [LW 2, 238.9–12]: FV 5:22 [298.18–21];
- on *five* separate occasions in *Expositio sancti evangelii secundum Johannem* (henceforth *In Ioh.*): (9) n. 263 [LW 3, 218.4–11]: FV 3:14 [108.19–20, 108.23–109.3], (10) n. 389 [LW 3, 332.10–11]: FV 5:23 [300.16], (11) n. 463 [LW 3, 396.10–397.3]: FV 3:14 [108.19–20, 108.22–109.1], (12) n. 467 [LW 3, 399.9–10]: FV 3:14 [108.25–26], (13) n. 554: [LW 3, 483.13–484.3]: FV 2:9 [40.6–27], 2:10 [41.11–16], 3:4 [82.9–83.10].

Moreover—as Brunner revealed—two further parts of Eckhart's German sermons, rendering a philosophical conversation between a teacher (*meister*) and his pupil (*jünger*), echo specific sections from *Fons vitae*: DW 2, 612.3–10 [sermon 58]: FV 3:56 [204.10–13] and DW 3 302.3–14 [sermon 75]: FV 3:56–57 [204.13–205.18].[14] As in his treatment of Maimonides, Eckhart also entirely refrains from *openly* mentioning Ibn Gabirol in his Middle High German works.[15] The critical edition of Eckhart's writings suggests the presence of numerous passages bearing resemblance to formulations and ideas found in Ibn Gabirol's *Fons vitae*. As Palazzo notes, some of these parallels come across as being rather coincidental, whereas others—such as LW 1,

12 Both Palazzo and Schwartz enumerate twelve explicit references by means of direct citations of Ibn Gabirol's *Fons vitae* in Eckhart's Latin writings: Palazzo, "Eckhart's Islamic and Jewish Sources," 272; Schwartz, *"To Thee Is Silence Praise,"* 80; Schwartz, "Meister Eckhart and Moses Maimonides," 392. Eckhart's writings will be cited in brackets according to the critical edition: *Meister Eckhart, Die deutschen und die lateinischen Werke*, commissioned by the DFG, comprising *Die deutschen Werke*, ed. Josef Quint and Georg Steer, 5 vols. (Stuttgart: W. Kohlhammer Verlag, 1936–2016) (henceforth DW) and *Die lateinischen Werke*, ed. Konrad Weiss, Loris Sturlese, Albert Zimmermann, Joseph Koch, Heribert Fischer, Bruno Decker, Ernst Benz, Karl Christ, Bernhard Geyer, and Erich Seeberg, 5 vols. (Stuttgart: W. Kohlhammer Verlag, 1936–) (henceforth LW). McGinn alludes to the fact that Eckhart mentions Ibn Gabirol by name seventeen times.

13 Cf. Schwartz, *"To Thee Is Silence Praise,"* 80.

14 Brunner, "Maître Eckhart et Avicébron," 133–39, 149.

15 McGinn, *The Mystical Thought of Meister Eckhart*, 267–68 n. 21.

196.4–5 and LW 2, 64.6–7—are quite persuasive.[16] In total, the editors of the critical edition of the DW and the LW suggest that there are *nine* sections in which Eckhart makes recourse to sections from *Fons vitae* in order to implicitly support his arguments:

- *once* in his Middle High German sermon 102 *Ubi est, qui natus est rex Iudaeorum?*: (1) [DW 4, 420.3 n. 31]: FV 1:3 [5.22–23];
- *four* times in his *In Gen.I*: (2) n. 12 [LW 1, 196.4–5 n. 3]: FV 3:22 [131.3], (3) n. 32 [LW 1, 210.3–4 n. 1]: FV 3:3 [81.17], FV 5:42 [335.9], (4) n. 36 [LW 1, 213.9–10 n. 3]: FV 5:11 [277.5–9], (5) n. 88 [LW 1, 246.6–7 n. 3]: FV 4:11 [236.11–237.9];
- *twice* in his *In Gen.II*: (6) n. 11 [LW 1, 483.1–3 n. 2]: FV 5:23 [300.16], (7) n. 197 [LW 1, 669.12–13 n. 2]: FV 1:3 [5.13–16.22];
- *twice* in his *In Ex.*: (8) n. 18 [LW 2, 24.2–4 n. 2]: FV 5:24 [301.16–302.15], (9) n. 58 [LW 2, 64.6–7 n. 2]: FV 4:13 [239.12–17].

1 Emanation of the One or the Two? Avicebron's Fourfold *Ordo Essendi*

On examining Eckhart's *Opus tripartitum*, one immediately encounters Ibn Gabirol's impact on it, seeing that the Dominican already reverts to his thought for his metaphysical reading of Gen 1:1:

> Finally, it must be observed that it says "God" created "heaven and earth in the principle," two things, not more, such as three or four and so forth. It does not say that he created one thing. The reason is because by the fact that anything is or has been created it falls away from unity and simplicity. Unity and simplicity are proper to God and are his property, as I have written at length on the text "God is one" (Dt. 6:4; Gal. 3:20). Again, everything that falls away from the One, the First of all things, immediately falls into two and unto the other numbers by means of duality. And so Avicebron says that "the question 'whether a thing is' is asked in relation to the One because it is pure existence," "Under him the questions 'whether a thing is' and 'what a thing is' in the manner of duality belong to the intelligence, which is the first thing below God. In a threefold way the questions 'whether a thing is,' 'what it is,' and 'what sort it is' belong to the soul. In a fourfold way, the questions 'whether,' 'what,' 'what sort,' and 'why' a thing is belong to nature or the realm of generation that is below the soul. The fourth question refers to the first three." The first equal number, namely two, is the root of all division, plurality and number, just as the unequal or the One is the root and reason of lack of division (*In Gen.I* n. 26–28 [LW 1, 205.1–206.4]).[17]

16 Palazzo, "Eckhart's Islamic and Jewish Sources," 272 n. 48.

17 Meister Eckhart, *Meister Eckhart: The Essential Sermons, Commentaries, Treatises, and Defense*, trans. Edmund Colledge and Bernard McGinn (New York: Paulist Press, 1981), 91. The printed translation uses the name "Ibn Gabirol," which I chaged to "Avicebron" since Eckhart did not know his true identity. For Eckhart as an interpreter of the book of Genesis, see Flasch, *Meister Eckhart: Philosopher of Christianity*, 115–30.

In the quoted passage, Eckhart tackles the exegetical difficulty of providing a reason as to why God in His singularity initially calls into being two creations as opposed to one or any other number. He explains this problem away by employing the Avicebronian notion of the *binarium famosissimum.*[18] The precept at hand, positing that from the *first* unified principle, *two* further principles emerge—those of form and matter—is critical for the understanding of Ibn Gabirol's metaphysical undertaking. It entails Gabirolean hylomorphism, inferring a composite of a spiritual *Materialprinzip* (*materia spiritualis/universalis*) with a universal form (*forma universalis*) underlying all levels of the created realm.[19] As Jacques Schlanger contends, this notion addresses the chief systematic objective of Gabirolean philosophy, reconciling the oneness of the Creator and the multiplicity of His created beings.[20] Prior to quoting from *Fons vitae*, Eckhart seems to reiterate this Gabirolean notion, asserting that everything emanating from the One, it being the origin of the universe, must directly fall into the rank of the Two. If one keeps in mind Albertus Magnus's explicit Aristotelian rebuttal of this particular Avicebronian proposition, then Eckhart's acceptance thereof becomes even clearer. In the first book of his commentary on *Liber de causis*, a pseudo-Aristotelian paraphrase of Proclus's *Elementatio theologica*,[21] Albert singles out this one tenet as the most problematic element underlying Avicebron's philosophy.[22]

Following his endorsement of this metaphysical view, Eckhart then quotes a passage from the fifth chapter of *Fons vitae* (section 24), a section of which he seems particularly fond, citing it on three different occasions. In this passage, which resembles a *Präpositionsmetaphysik*,[23] Ibn Gabirol lays out his Neoplatonic outlook in an

---

18 See Paul V. Spade, "Binarium Famosissimum," in *The Stanford Encyclopedia of Philosophy* (Fall 2008 Edition), ed. Edward N. Zalta, https://plato.stanford.edu/archives/fall2008/entries/binarium/; Pessin, *Ibn Gabirol's Theology of Desire*, 186–88; McGinn, "Ibn Gabirol: The Sage among the Schoolmen," 86–87, 96–97.

19 For detailed explanations of Gabirolean hylomorphism, see Pessin, *Ibn Gabirol's Theology of Desire*, 66–90, 106–12, 165–88; Schlanger, *The Philosophy of Solomon Ibn Gabirol*, 177–218; Nicola Polloni, "Toledan Ontologies: Gundissalinus, Ibn Daud, and the Problems of Gabirolian Hylomorphism," in *Appropriation, Interpretation and Criticism: Philosophical and Theological Exchanges Between the Arabic, Hebrew and Latin Intellectual Traditions*, ed. Alexander Fidora and Nicola Polloni (Barcelona: FIDEM, 2017), 19–49. For an analysis of Albertus Magnus's critique of this specific metaphysical tenet of Avicebronian philosophy, see McGinn, "Ibn Gabirol: The Sage among the Schoolmen," 97.

20 Schlanger, *The Philosophy of Solomon Ibn Gabirol*, 135.

21 Cf. Yossef Schwartz, "Die Rezeption philosophischer Schriften aus dem Judentum," in *Grundriss der Geschichte der Philosophie. Die Philosophie des Mittelalters IV: 13. Jahrhundert*, ed. Peter Schulthess and Christoph Flüeler (Basel: Schwabe Verlag 2017), 1:206–16.

22 Albertus Magnus, *Buch über die Ursachen und den Hervorgang von allem aus der ersten Ursache: Liber de causis et processu universitatis a prima causa*, ed. and trans. Henryk Anzulewicz, Maria Burger, Silvia Donati, Ruth Meyer, and Hannes Möhle (Hamburg: Felix Meiner Verlag, 2006), 46. See Kurt Flasch, *Meister Eckhart. Die Geburt der "Deutschen Mystik" aus dem Geist der arabischen Philosophie*, 2nd ed. (Munich: Beck, 2008), 82–83.

23 See, for example, Philo's *Cher.* 125–27. For a further examination of *Präpositionsmetaphysik* in relation to Philo and Jewish and Christian sources in antiquity, see Gregory E. Sterling, "Prepositional

interrogative fashion, supplying the answers through the order of the first four natural numbers. These in turn epitomise the four respective metaphysical principles and ranks of being (God, intellect, soul, and nature).[24]

It is still not quite clear how Ibn Gabirol's explanation of the fourfold structure of being aligns with his specific reading of Gen 1:1 here. However, the previous paragraphs provide an indication. Actually, *In Gen.I* n. 24–25 is the point of departure for Eckhart's drawing on Avicebron. There, he symbolically equates the creation of heaven and earth with that of form and matter, arguing that the latter pair also came into being *simultaneously*. Eckhart then goes on to ascribe activity to form and, accordingly, passivity to matter, the first inhabiting the higher realm, the latter the lower one.[25] This may be the point that Eckhart aims to highlight: Ibn Gabirol assigns the universal intellect, representing the category of the Two, this specific ontological rank, inasmuch as its innate structure comprises of *genus* and *differentia*, both fundamental categories denoting form and matter,[26] which in turn make up its being.[27] Moreover, one need look no further than the concluding remark of section 23 in chapter 5 of *Fons vitae* in order to observe that Ibn Gabirol's framework is almost identical to Eckhart's. In it, the *magister* sets out to answer the basic question as to why matter and form come into being, the answer supplying the groundwork for the subsequent discussion in the following section (24).[28] The passage at hand depicting God's productive essence as the One from which the twofold reality comprised of matter and form emerges is quite revealing, inasmuch as it seems to

---

Metaphysics in Jewish Wisdom Speculation and Early Christological Hymns," *Studia Philonica Annual* 9 (1997): 219–38. As Altmann and Stern note, a similar presentation is already present in Isaac Israeli's *Book of Definitions*, in *Isaac Israeli. A Neoplatonic Philosopher of the Early Tenth Century*, trans. Alexander Altmann and Salomon M. Stern (Chicago: University of Chicago Press, 2009), 23. See also Pseudo-Aristotle, *The Theology of Aristotle*, 5:30–52, trans. Geoffrey Lewis, in, *Plotini Opera. Tomus II: Enneades IV – V*, ed. Hans-Rudolf Schwyzer and Paul Henry (Oxford: Oxford University Press, 1977), 437–39 (henceforth *ThA*). Cf. Schlanger, *The Philosophy of Solomon Ibn Gabirol*, 87–89, 98.

24 FV 5:24 [301.20–302.1, 302.10–14]. For the English translation, see Solomon ibn Gabirol (Avicebron), *The Font of Life (Fons Vitae)*, trans. John A. Laumakis (Milwaukee: Marquette University Press, 2014), 270 (henceforth FL). In chapter 1 of *Fons vitae* (section 16 [20.15–17]), Ibn Gabirol also discusses matter and form in relation to both the heavenly realm and the earthly one, though he rather contends that their mergence takes place in both ontological realms. For further similar depictions in *Fons vitae*, see FV 4:11 and 4:13. For Ibn Gabirol's Neoplatonic theory of creation, see Yossef Schwartz, "Salomon Ibn Gabirol: Zwischen Schöpfungsmythologie und Geheimnis der Einheit," in *Platonismus im Orient und Okzident: Neuplatonische Denkstrukturen im Judentum, Christentum und Islam*, ed. Raif G. Khoury, Jens Halfwassen, and Frederek Musall (Heidelberg: Universitätsverlag Winter, 2005), 144–59; Schlanger, *The Philosophy of Solomon Ibn Gabirol*, 229–42.

25 LW 1, 204.1–16. See Burkhard Mojsisch, *Meister Eckhart: Analogy, Univocity and Unity*, trans. Orrin F. Summerell (Amsterdam: B.R. Grüner, 2001), 70–71, 105.

26 For the definitions of universal matter and form, see FV 1:10 [13.23–14.5], 1:13 [16.9–13].

27 See, for example, FV 3:41 [172.10–20]; 4:6 [223.17–224.1] 5:12 [279.6–19]. For the unifying and dynamic feature of the divine Will in relation to universal matter and form, see FV 2:13 [46.2–12].

28 FV 5:23 [300.14–26].

constitute a great deal of Eckhart's metaphysical understanding and approach to Gen 1:1 in *In Gen.I* n. 24–28. Ibn Gabirol's passage reads as follows:

> S. Why did matter and form exist?
> T. Because of the first essence and its property. Moreover, two was placed under the One, and they are the expressions of it, and if what was created were one, there would be no difference here, because difference is found only under the One. Moreover, because non-being does not have form, it is necessary that being has form. Moreover, since being needed to be finite in itself, it is necessary that it should be made finite by form, because a form is what encompasses a thing. Moreover, since the first acting unity does not have *hyle*, it is necessary that the unity that follows this one should consist of *hyle*. Hence, it receives unity and is made two, that is, subjected *hyle* and sustained unity. Moreover, since the first unity is a self-sufficient agent, it is necessary that the unity that follows it needs what sustains it. Hence, there must by *hyle* here that sustains it, and consequently, two come to be (FV 5:23 [300.14–26]).[29]

The way in which Eckhart characterises matter and form, the two building blocks of reality, in this context also bears a resemblance to Ibn Gabirol's conceptualisation thereof: (1) Like Ibn Gabirol,[30] Eckhart posits the passivity of matter[31] and the activity of form. (2) Eckhart infers the ontological priority of form over matter in a manner similar to descriptions found in *Fons vitae*.[32] While Ibn Gabirol usually portrays form as occupying the higher noetic realm, in his empirical diversification, he often relegates matter to the lowest ontological level.[33] (3) Eckhart's emphasis laid on the *creatio simultanea* of matter and form is a feature that conforms well to Avicebronian thought. On numerous occasions, Ibn Gabirol underpins the fact that God is above time and eternity and therefore that His creative activity is not constrained to the realm of time. (4) Eckhart's specific terminology used to account for worldly divisibility in light of the indivisibility of God's unity by means of numerological argumentation is reminiscent of Gabirolean thought. Following his quotation of Avicebron, Eckhart links the "root of division of all things" (*radix omnis divisionis*) to the

29 FL 239 (emphasis in original).
30 See, for example, FV 2:9 [40.6–27].
31 See Brunner, "Maître Eckhart et Avicébron," 149.
32 See, for example, FV 2:9 [40.6–27], 4:11 [236.5–237.17]. For an analysis of this Gabirolean theme, see Schlanger, *The Philosophy of Solomon Ibn Gabirol*, 161; McGinn, "Ibn Gabirol: The Sage among the Schoolmen," 87–89.
33 The critical edition of Eckhart's LW alludes to the overlap of features (1) and (2) in *In Gen.I* n. 32 [LW 1, 210.3–4 n. 1] ("Inferiora autem passiva sunt, informia et nuda"). Two passages from *Fons vitae* are quoted in this respect: FV 3:3 [81.17] ("Omne quod est, quanto superius est, tanto formae similius est") and 5:42 [335.9–11] ("Forma unenit a superiore, et materia eam recipit ab inferiore, quia materia est subiecta, per hoc quod habe esse sub forma, et forma est sustentata super eam"). The additional notion here that is similar to Ibn Gabirol's *Fons vitae* is that the ideas *perfect* matter: "*perficere essentiam* illius in quo est" (FV 1:13 [16.10–11]; emphasis in original; cf. also FV 1:13 [16, 17], 2:10 [60.28–61.3], 3:36 [161.20–162.1], 4:10 [233.17–19]). For a detailed presentation of the ontological difference between matter and form in Gabirolean philosophy, see Schlanger, *The Philosophy of Solomon Ibn Gabirol*, 141–46.

concept of the first even natural number, the Two,[34] and the notion of strict indivisibility to the first uneven natural number, the One.[35] In *Fons vitae*, one encounters various such lines of argumentation using similar jargon.[36] Eckhart's wording in this respect, speaking of the Two—that is, matter and form—in terms of *radix omnis divisionis* also stands out. In a key section (11) of chapter 4 of *Fons vitae*, Ibn Gabirol specifically describes matter as epitomising the nature of the Two and as being, by virtue thereof, the root of all divisibility. Accordingly, he renders matter together with form—with both of them together conforming to the nature of the uneven number three—the ultimate *radix omnium.*[37] Interestingly, the critical edition of Eckhart's works does not allude to the possible correlation here, but to a similar correlation in *In Gen.I* n. 88 [LW 1, 246.6–7]. In this section, Eckhart attempts to explain why in Gen 1:8b, the second day of creation does not end with God's otherwise usual concluding proclamation that it was good. In order to account for this exegetical difficulty, Eckhart once again reverts to the metaphysical idea that the Two is the root of all division ("binaries radix est et origo omnis divisionis"), claiming that in this instance, division embodies evil.[38] Eckhart concludes his ontological view on the subject with the assertion that the Two and division are always a falling from being.[39]

At this thematic conjuncture and against this backdrop, another remark in the critical edition of Eckhart's LW concerning his reading of Gen 1:2 ("tenebrae errant super faciem abyssi") attracts our attention. In *In Gen.I* n. 36, Eckhart explicates the ontological potentiality (*potentia*) of matter in terms of being lacking in form ("privatio formae").[40] As Konrad Weiss suggests,[41] this depiction closely resembles a section from chapter 5 of *Fons vitae:*

> Consider the privation of matter to be like the darkness of air, and consider form in it to be like light. And consider dark air as having being in itself and as having the being of light in potency when it lacks light. Similarly, consider that matter has being in itself and has being in potency when it lacks form, and that being is produced from the joining of matter and form.[42]

---

34 Cf. *In Gen.I* n. 88 [LW 1, 246.6–7].
35 *In Gen.I* n. 28 [LW 1, 205.16–206.4].
36 See, for example, FV 4:14 [242.8–24].
37 FL 203: "Moreover, in its first division, matter is divided into two parts according to the nature of two, namely, into the matter of the simple substances and into the matter of composite substances. The property of two, therefore, also belongs to it in this way.–Therefore, it is now clear to you that form is likened to one and matter to two. And since this is so and matter and form are the root of all things, it is, therefore, clear that three is the root of all things [*et materia et forma sunt radix omnium: ergo manifestum est quod tria sunt radix omnium*]" (FV 4:11 [237.9–17]). See McGinn, "Ibn Gabirol: The Sage among the Schoolmen," 79.
38 The editors provide the following passage from *Fons vitae:* FV 5:11 [236.18–237.9].
39 *In Gen.I* n. 90 [LW 1, 248.15–16].
40 *In Gen.I* n. 36 [LW 1, 213.9–10].
41 LW 1, 213 n. 3.
42 FL 226: FV 5:11 [277.5–10]. Weiss specifically refers to *Fons vitae* 5:11 [277.5–9]. Weiss also mentions Eckhart's *In Ioh.* n. 440 [LW 3, 378.5–6] and 492 [LW 3, 424.7–11], which appears to echo Ibn Gabirol's

The similarity becomes even more prominent if one accounts for Eckhart's symbolic exegesis of the darkness hovering over the abyss. The darkness, as he explains, is to be understood in terms of deprivation of light, while the abyss manifests the shapeless *prima materia*.[43] If we develop the logical consequence of this Eckhartian depiction further, then light, accordingly, represents form. If our purported reading holds true here, then this reveals that the LW aligns Augustine's view of matter (in book 12 of his *Confessions*) with the premises of Gabirolean hylomorphism.[44]

In Eckhart's reading of Gen 1:1 in *In Gen.II* n. 11, the Gabirolean *binarium famosissimum* appears to recur. In this section, he aims to account for diversity within the created realm by stating that it is in keeping with the nature of the Two, typified by God's simultaneous creation of heaven and earth.[45] His Neoplatonic line of thought is quite straightforward: while no form of multiplicity can be present within the category of the One, anything that stems from it—hence occupying an ontological realm *outside* of it—conforms to the category of the Two. In this way, Eckhart clarifies Augustine's dichotomous view of God's creation of the heavens and the earth in section 7 of book 12 of his *Confessions*.[46] A brief analysis of section 23 of chapter 5 of *Fons vitae* suffices to discern the parallels it bears with Eckhart's rendition of the thought pattern of *binarium famosissimum*: "And if what was created were one, there would be no difference here, because difference is found only under the One."[47]

The editors of the LW are correct in drawing support for their purposed linkage of *In Gen.II* n. 11 to this specific passage from *Fons vitae* by alluding to *In Ioh.* n. 389.[48] There, within an exegesis of John 4:38,[49] Eckhart explicitly refers to Ibn Gabirol's assertion that each differentiation occurs under the One ("omnis differentia sub uno est") in *Fons vitae* 5:23.[50] Admittedly, the context in which Eckhart employs this Gabirolean idea is entirely different. However, more importantly, it uncovers Eckhart's

---

attitude towards matter as pure potentiality and as a deficiency of form. Eckhart's depiction of matter's lack of form in terms of a throne (*In Ioh.* n. 36 [LW 1, 213.11–12]) in the same paragraph, a metaphor which can be found in *Fons vitae* 5:42 [335, 22–24], is also striking. See Schlanger, *The Philosophy of Solomon Ibn Gabirol*, 53.

43 *In Gen.I* n. 35 [LW 1, 212.4–5].

44 For the correlation of Augustine (and Boethius) with Ibn Gabirol's *Fons vitae* within High Scholasticism, see McGinn, "Ibn Gabirol: The Sage among the Schoolmen," 93, 95.

45 *In Gen.II* n. 11 [LW 1, 482.8–10].

46 Augustine, *St. Augustine's Confessions in Two Volumes*, trans. William Watts (London: William Heinemann; Cambridge, MA: Harvard University Press, 1912), 2:269–71: "For you are almighty and good in making all things good, the great heaven and the small earth: two entities [*duo quaedam*], one near to you, the other near to nothing. You alone would be greater than the one, while nothing would be lesser than the other."

47 FL 239: FV 5:23 [300.14–20].

48 LW 1, 483 n. 2.

49 "Ego misi vos metere quod vos non laborastis: alii laboraverunt, et vos in labores eorum introistis."

50 *In Ioh.* n. 389 [LW 3, 332.11].

view thereof as a true metaphysical precept. With the aid of this Avicebronian axiom, he can reinforce, *ex negativo*, the unification of the faithful man with God by means of his love of Him.[51]

As Konrad Weiss aptly observes, Eckhart appears to resort to an Avicebronian precept at the outset of *In Gen.I* (n. 12), in a further metaphysical rationalisation of Gen 1:1, attempting once more to explain how multiplicity emerges from the One.[52] Conceding that from the One, nothing other than a thing that is one in essence can emerge, Eckhart accounts for the multiplicity of God's creation by putting forward the claim that the universe, notwithstanding God's ostensible plurality of forms, is one in nature.[53] He proceeds to apply this contention to God, arguing in analogous terms that God is also purely one in light of His attributes of being, life, thought, and activity.[54] Eckhart's dependence upon *Fons vitae* at this juncture does not become evident until he makes the metaphysical assertion that the simpler and more perfect an existing thing is, the more diverse it is in relation to its powers:

Debet necessario ut omnis causa sit imprimens figuras suas et formas in suo causato. et nullum impressum est tale quale id a quo imprimitur in fortitudine et *perfectione*. et substantia *simplex* causa est *compositae*. ergo debet necessario ut substantia *simplex* sit plurium figurarum et formarum quam substantia composita (FV 3:22 [130.25–131.4]; my emphasis).	Ubi notandum primo quod, quanto res est *perfectior* in esse et *simplicior*, tanto *copiosior* est secundum rationes et potentias (*In Gen.I* n. 12 [LW 1, 196.4–5]; my emphasis).

Through this argument, Ibn Gabirol endeavours to show that the *substantia simplex*, such as the universal intellect that encompasses all forms,[55] constitutes a structure of totality which entails all of created reality within it.[56] However, not only does Eckhart's claim itself bear a resemblance to the Gabirolean one, but so too does the example he adduces for it. Eckhart moves on to contrast the rational soul with the forms of matter, maintaining that the substance of the soul is more unified than any material form, while its powers are more diverse.[57] An examination of the next section (23) of *Fons vitae's* chapter 3 reveals the fact that Ibn Gabirol alleges almost the exact same example for his contention. In connection with the soul, he also dem-

51 Brunner underpins this fact here: "Maître Eckhart et Avicébron," 146. Cf. Flasch, *Meister Eckhart: Philosopher of Christianity*, 108.
52 LW 1, 196 n. 3. Cf. LW 1, 194.7–8. Palazzo also evaluates this passage's link to *Fons vitae* as being one of "the most convincing ones among those suggested by the editors of Eckhart's *Opera omnia*" (Palazzo, "Eckhart's Islamic and Jewish Sources," 272 n. 48).
53 *In Gen.I* n. 12 [195.10–196.3].
54 *In Gen.I* n. 12 [195.12–196.2].
55 See Schlanger, *The Philosophy of Solomon Ibn Gabirol*, 194–95.
56 See, for example, FV 2:3 [30.6–20], 3:9 [98.24], 3:22 [131.14–17], 3:23 [132.22–133.6], 3:24 [137.26–27], 3:32 [153.1–4, 154.15–155.6].
57 *In Gen.I* 12 [196.5–8].

onstrates that the spiritual elemental substance (*substantia simplex spiritualis*) is "better suited for gathering many forms in itself than is a corporeal composite substance."[58] If our tracing of Eckhart's contention to *Fons vitae* holds true here, then this means that he had aptly grasped Gabirolean thought as ultimately being a metaphysics of the One. That is to say, Eckhart would thereby be indicating that in the final analysis, Avicebron does not subscribe to the precept of the *binarium famosissimum*, but rather to the notion of the primary unity from which a further unity emerges, albeit one that is created and ontologically subordinated to the first.[59]

2 The Soul's Intellectual Ascent to the Noetic Realm

Eckhart is by no means exclusively interested in Ibn Gabirol's abstract metaphysical premises: he is also interested in their anthropological framework; in the final goal of the human soul, which is its reunification with its pure counterpart, the higher world.[60] The following point of interest finds its apt expression in *In Gen.II* n. 113. In his Philonic reading of Gen 2:18 ("Non est bonum hominem esse solum, faciamus ei adiutorium simile sibi")—with man epitomising the faculty of reason (*intellectus*) while his female counterpart epitomises that of sensual perception (*sensitivum*)[61]—Eckhart adduces an idea from the third section of chapter 1 of *Fons vitae* for his exegesis. He specifically quotes Ibn Gabirol's contention "that the perfection of the soul is its knowledge, whereas its imperfection is its ignorance [...]. And [...] from its beginning in this world, the soul is changed from ignorance to knowledge and thus goes forth from potency to act."[62] In fact, this citation seems to constitute an important reference point for Eckhart's claim that sense perception is indispensable for the initial phases of the intellectual actualisation of the human soul.[63] Ibn Gabirol depicts this intellectual process in Aristotelian terms, as a transition from potentiality to actuality, which in turn means the transition from ignorance to knowledge. However, he places this Aristotelian portrayal as part of a broader Neoplatonic theme of the soul's intellectual ascent to the divine realm. As the Andalusian philosopher proclaims from the outset, the subject at hand, providing the answer as to why God created man, actually marks the focal thread running through the entire metaphysical

---

58 FL 141: FV 3:22 [131.23–24]. See FV 3:23 [132.5–17].

59 See, for example, FV 2:20 [60.18–62.9], 4:13–14 [238.23–245.4], 5:23 [300.23–26].

60 See, for example, FV 1:2 [4.24–25]. See Schlanger, *The Philosophy of Solomon Ibn Gabirol*, 2–4.

61 For the influence of Philo's allegorical reading of Adam and Eve on the Church Fathers, see David T. Runia, *Philo in Early Christian Literature: A Survey* (Assen: Van Gorcum; Minneapolis: Fortress Press, 1993), 309, 352.

62 It is noteworthy that in the original text, the passage in question appears in the interrogative form: FL 63 (FV 1:3 [5.3–16]).

63 *In Gen.II* 113 [LW 1, 580.5–6]: "Hoc autem fit ministerio sensuum, sensitivi scilicet."

examination of *Fons vitae*.[64] Thus, the critical edition of the LW is correct in noting[65] that a similar depiction is found in Eckhart's allegorical reading of the expression "duae gentes" (Vg. Gen 25:23) in *In Gen.II* n. 197, by which God alludes to Rebecca's impending delivery of Jacob and Esau.[66]

Drawing upon the invocation of Avicebron in *In Gen.II* n. 113, the critical edition of the DW notes that Eckhart may have intended to echo this Gabirolean idea once again in his German sermon 102,[67] claiming "Got hât den menschen gemachet, daz er wizze."[68] This appears to be Ibn Gabirol's unequivocal answer to the question of why God created man in *Fons vitae* 1:3: "It follows that the end and aim of the creation of man is for his conscious knowing to develop from potentiality to actuality."[69] Yet it could also be the case that Eckhart is simply giving voice to a theological premise prevalent among Dominican circles of his day. However, the clear-cut manner in which Eckhart brings forth this argument would suggest otherwise, since it seems to be distinctively Gabirolean.

Palazzo rightfully points to the thematic correlation between Eckhart's citation of Ibn Gabirol in *In Gen.II* n. 113 and his implicit citations from *Fons vitae* in sermons 58 (FV 3:56 [204.10–13]) and 75 (FV 3:56–57 [204.13–205.18]). All three references to *Fons vitae* pivot around Ibn Gabirol's theme of the soul's intellectual ascent to the upper world. Sermon 75 is particularly interesting in view of Eckhart's invocation of Ibn Gabirol in his interpretation of John 4:38. The theological foundation of this reference was the attainment of a *unio mystica* through divine love. In the German sermon at hand, this is where Eckhart's employment of *Fons vitae* takes its departure: in order to attain the perfect love and subsequently unify with God's being, one must first be endowed with the *ûfklimmenden geist*, through which the individual can spiritually ascend to the Godly realm.[70] Even though Ibn Gabirol does not directly deal with the theme of the soul's *amor Dei* in these specific instances, Eckhart was most likely aware of its significance within the overall framework of the Avicebronian metaphysics of longing. Section 32 of chapter 5 of *Fons vitae* revolves, for example,

---

64 FV 1:1 [2.8–10], 1:2 [3.30–4.9], 1:3 [5.22–25], 2:6 [36.6–14]. On Eckhart's connection of Avicenna with Ibn Gabirol at this juncture, see the insightful remarks by Palazzo: "Eckhart's Islamic and Jewish Sources," 276–77. Cf. also Brunner, "Maître Eckhart et Avicébron," 141; Schlanger, *The Philosophy of Solomon Ibn Gabirol*, 243–46.

65 LW 1, 669 n. 2.

66 After his interpretation of the *duae gentes* in Rebecca's womb as the two respective tendencies innate to the human soul, the sensual (Esau) and the rational (Jacob) (*In Gen.II* 196 [LW 1, 668.8–669.11]), Eckhart proceeds to present them both in terms of knowledge and ignorance ("*duae gente* in nobis sunt [...] scientian et ignorantia").

67 DW 4, 420 n. 31.

68 DW 4, 420.127.

69 FL 6 [5.22–23].

70 Cf. Palazzo's apt description of this passage: "Eckhart's Islamic and Jewish Sources," 277–78. Cf. Brunner, "Maître Eckhart et Avicébron," 135–39.

around the notion of all things ascending towards the goodness of the transcendent One, a drive stemming from their fundamental desire for unity.[71]

However, whereas Eckhart exploits these Avicebronian portrayals to buttress man's viable epistemological prospect of a spiritual union with God's being, Gabirolean philosophy certainly does not wish to go this far. Despite the ultimate striving to be at one with God's primary being that is intrinsic to the human soul, Ibn Gabirol restricts the pinnacle of its attainable intellectual endeavours to its unification with the noetic realm.[72]

Another noteworthy feature of this Eckhartian reference to Ibn Gabirol in the DW is the fact that he also deems it important to disclose the means by which *Fons vitae* conveys these theoretical notions: through an intellectual instruction between a *magister* and his *discipulus*. This may show that Eckhart considered the dialogical form of *Fons vitae* to have a special pedagogical thrust, which proved effective within the framework of the religious instruction in his sermons. Eckhart's German sermon 58 also underpins this hypothesis, for there he quotes passages from *Fons vitae* in their original literary form as part of a lesson between a *meister* and his *jünger*.

3 A Gabirolean Reading of the Gospel of John's Metaphysics of the *Verbum*

Eckhart's insertion of Avicebron's philosophy into a distinctly Christian theological discourse finds its culmination in his reading of the prologue of the Gospel of John (1:1–2).[73] At this thematic juncture, we find a perplexing passage in *In Ioh.* n. 47:

> From a careful investigation of what Augustine said, we can understand what the second book of the *Posterior Analytics* says: "The questions we ask are equal in number to the things we truly know." We can ask about things whether they are, what sort they are, what they are and why they are. The four passages we have here answer these in order. "In the beginning was the Word" gives you the fact that the thing exists, for he says that it "was." What sort of thing the Word is, is pointed out in what follows, "and the Word was with God." This will be made

71 FV 5:32 [317.4–12]. For further analysis of this Gabirolean theme, see Schwartz, "Salomon Ibn Gabirol," 157–59; Warren Zev Harvey, "Filosofía y poesía en Ibn Gabirol," *Anuario Filosófico* 33, no. 2 (2000): 491–504 (I would like to thank Harvey for giving me the English translation of this article). For a systematic treatment of the subject of desire in Gabirolean metaphysics, see Pessin, *Ibn Gabirol's Theology of Desire*, especially 15–52, 66–90; Schlanger, *The Philosophy of Solomon Ibn Gabirol*, 213–19.

72 Brunner points to *ThA*, specifically *ThA* 1, §§ 21–42 [*Enn.* 4.8.1], as the possible source at the root of these Gabirolean depictions: "Maître Eckhart et Avicébron," 138–39. Cf. Schlanger, *The Philosophy of Solomon Ibn Gabirol*, 64.

73 On Eckhart's intention in his philosophical reading of the Gospel of John as a whole, see Flasch, *Meister Eckhart: Philosopher of Christianity*, 166–89.

clear below. What the Word is follows in the text, "the Word was God." The reason why is shown when it says, "this Word was in the beginning with God" (LW 3, 39.1–8).[74]

On examining the passage in question, one might justifiably ask how this even relates to Ibn Gabirol. On the face of it, Eckhart is merely citing the opening sentence of book 2 of Aristotle's *Posterior Analytics*. Yet a closer look at it exposes—as the critical edition of the LW notes[75]—that in fact, it partially has its origin in *Fons vitae* 5:24. The Dominican master begins by simply citing the opening assertion of *An. post.* 2.89b23, which reads: "Quaestiones sunt aequales numero his, quaecumque vere scimus."[76] However, Eckhart proceeds to predicate his detailed presentation of the questions themselves on *Fons vitae* 5:24 [301.18]:

Est autem quaestio de rebus: *an sint, quales sint, quid sint, et quare* (*In Ioh.* n. 47 [LW 3, 39.3]).	Dico quod esse ab extremo supremo usque ad extremum infimum distinctum est quatuor ordinibus, scilicet *an est, quid est, quale est, quare est* (FV 5:24 [301.16–18]).	Quaerimus autem quatuor: quia, propter quid, si est, quid est (*An. Post.* 2.89b23).

This unexpected use of section 24 of chapter 5 of *Fons vitae*, a section with which Eckhart was very familiar, appears to be anything but coincidental. This passage prompts the following three inferences: (1) The fact that Eckhart draws upon Ibn Gabirol without explicitly mentioning him does not diminish Ibn Gabirol's significance within his body of thought. On the contrary, it would rather suggest that the impact of Gabirolean philosophy on Eckhart was so profound that he—possibly even unknowingly—uses it to complete Aristotelian ideas. Bearing such non-explicit invocations in mind, one may be tempted to put forward the claim that Eckhart not only read *Fons vitae* firsthand—as Palazzo convincingly contends[77]—but that he even knew several sections of it by heart. (2) One must also account for the particular context in which Eckhart makes use of *Fons vitae:* an interpretation of the two opening verses of the prologue of the Gospel of John, whose sense he aims to uncover by allusion to the sixth chapter of book 8 of Augustine's *On the Trinity* (section 10). He then, in turn, goes on to elucidate Augustine's portrayal of the soul's intuitive inner

74 *Meister Eckhart: The Essential Sermons, Commentaries, Treatises, and Defense*, 138.

75 See the remark by the editors of the LW: "*Quattuor hae quaestiones non sunt eaedem quas Aristoteles I. c. enumerate* (quaerimus autem quattuor: quia, propter quid, si est, quid est), *sed quas Avencebrol point, Fons vitae V. c. 24*" [emphasis in original]. Cf. Niklaus Largier's remark, seemingly based on the verdict of the critical edition: "532,8 *an* < ... > *quare]* In den vier Fragen folgt Eckhart, obwohl er sie Aristoteles zuschreibt, nicht diesem, sondern Avencebrol, *Fons vitae* V c.24" [emphasis in original] (*Meister Eckhart: Werke II*, ed. Niklaus Largier, trans. Josef Quint [Frankfurt am Main: Deutscher Klassiker Verlag, 1993], 866).

76 Aristotle, *Posterior Analytics. Topica*, trans. Hugh Tredennick and E.S. Forster (Cambridge, MA: Harvard University Press, 1960), 174–75.

77 Palazzo, "Eckhart's Islamic and Jewish Sources," 273.

knowledge of its own being[78] by making recourse to Aristotle's *An. post.* 2.89b23. Eckhart employs Ibn Gabirol at a point intersecting with the most crucial sources of his Christian philosophy: the prologue of the Gospel of John and the two most cited thinkers in his oeuvre, Augustine and Aristotle. Against this background, Avicebron's centrality to Eckhart's thought becomes all the more evident. (3) More importantly, the Dominican appears to apply the four Aristotelian inquiries, in their Gabirolean incarnation, to John 1:1–2 in a similar fashion to *Fons vitae* 5:24. That is to say, Eckhart adopts this method for the purpose of outlining his speculative worldview. Taking his cue from Avicebronian metaphysics, Eckhart implements the Aristotelian questions in his Neoplatonic reading of Vg. John 1:1–2: "In principio erat Verbum" provides an answer to the initial question of whether the Word even exists. The subsequent statement "et Verbum erat apud Deum" yields the Word's essence. The third assertion "et Deus erat Verbum" discloses the inner makeup of the divine Word. The concluding formulation "hoc erat in principio apud Deum" reveals the reason for its existence.[79]

Of course, one may dismiss this proposed comparison with Ibn Gabirol's philosophy, arguing that Eckhart's exegesis here does not exhibit any distinct Neoplatonic features. Given, however, the broader scope of his understanding of the prologue of the Gospel of John, one may raise the following three objections: (1) Eckhart describes the Word's coming into being in emanatistic terms, as perpetually overflowing from God the Father.[80] From this perspective, one may liken this employment of *Fons vitae* 5:23 to the one in *In Gen.I* n. 26–28, inasmuch as both deal with basic questions emerging from Scripture pertaining to the creation of the universe.[81] (2) Like Ibn Gabirol in *Fons vitae's* chapter 5, Eckhart, in sections 1 to 51 of *In Ioh.*, quite often equates the divine Word with the universal intellect.[82] (3) Although the Dominican applies the four questions exclusively to the Word, he still appears to want to depict a different ontological stage of its unfolding with each respective an-

78 See Augustine, *On the Trinity: Books 8–15*, ed. Gareth B. Matthews, trans. Stephen McKenna (Cambridge: Cambridge University Press, 2002), 16. See Flasch, *Meister Eckhart: Philosopher of Christianity*, 181.

79 Cf. Jens Halfwassen, "Ein Meister der Philosophie. Neue Literatur zu Meister Eckhart," *Philosophische Rundschau* 58 (2011): 172–73.

80 *In Ioh.* nn. 8 [LW 3, 8.10–13], 25 [LW 3, 20.5–6], 27 [LW 3, 21.28], 35 [LW 3, 29.6–30.9] 56 [LW 3, 46.13–47.1], 82 [LW 3, 70.6–8]. See also *In Ioh.* nn. 185 [LW 3, 154.9–14], 342 [LW 3, 291.6–9], 489 [LW 3, 489.10–12], 546 [LW 3, 492.9–12]. See Brunner, "Maître Eckhart et Avicébron," 145; Flasch, *Meister Eckhart: Philosopher of Christianity*, 185.

81 For Eckhart's strong correlation of the account of the creation with the prologue of the Gospel of John, see Flasch, *Meister Eckhart: Philosopher of Christianity*, 41.

82 *In Ioh.* n. 34 [LW 3, 27.12–28.3]. For the Neoplatonic thought pattern of the universal intellect in Eckhart's philosophy, see Jens Halfwassen, "Gibt es eine Philosophie der Subjektivität im Mittelalter? Zur Theorie des Intellekts bei Meister Eckhart und Dietrich von Freiberg," *Theologie und Philosophie* 72 (1997): 338–60; Halfwassen, "Gott als Intellekt: Eckhart als Denker der Subjektivität," *Meister-Eckhart-Jahrbuch* 5 (2011): 13–25.

swer. Taking this view, one may further develop this claim by aligning Eckhart's reading of John 1:1–2 with Ibn Gabirol's application of the four questions to reality's fourfold order of being in *Fons vitae* 5:24:[83]

- *Existence* (*ei esti*/*an est*): John 1:1a, signifying the Word in its original coexistence with God.
- [Existence,] *quiddity* (*ti esti*/*quid est*): John 1:1b, signifying the Word's manifestation as intellect.
- [Existence, quiddity,] *quality* (*to hoti* [*hopoion ti esti*]/*quale est*): John 1:1c, signifying the Word's embodiment as the soul.[84]
- [Existence, quiddity, quality,] *quality* (*to dioti* [*ti dia esti*]/*quare est*): John 1:2, signifying the Word's epitome of the physical world.[85]

Even in light of this comparison, the difference between Eckhart's ontological scheme and that of Ibn Gabirol is still substantial and in some aspects simply irreconcilable. In stark contrast to Ibn Gabirol, the sequence of emanation does not actually seem to entail a *fundamental* ontological difference for Eckhart, but has at its core God's absolute being.[86] As Schwartz aptly observes, the Eckhartian "God is simultaneously the most transcendent and most immanent entity."[87] In *In Ioh.* nn. 34–35, Eckhart highlights his view of John 1:1–2 as philosophically mirroring the tenet of Holy Trinity.[88] Perhaps the fact that Ibn Gabirol's *Fons vitae* appears to coincide with notions present in the prologue of the Gospel of John on several central points might have stimulated Eckhart to draw support from it.[89]

83 See Brunner, "Maître Eckhart et Avicébron," 140, 142.

84 See Flasch, *Meister Eckhart: Philosopher of Christianity*, 246, 259.

85 For Eckhart's equation of God and the world, see Flasch, *Meister Eckhart: Philosopher of Christianity*, 175, 189, 243–44. For a detailed discussion of the development of these four Aristotelian inquiries within Islamic and Jewish philosophy, see Altmann and Stern, *Isaac Israeli*, 13–23.

86 For Eckhart's "metaphysics of the verbum," see Flasch, *Meister Eckhart: Philosopher of Christianity*, 36–37, 70, 89–90, 117, 169, 177–81, 185–87, 244, 246.

87 Schwartz, "Meister Eckhart and Moses Maimonides," 399.

88 LW 3, 20–21, 28–30. Cf. *In Ioh.* nn. 38 [LW 3, 32–33], 45 [LW 3, 37.13–38.2], 54 [LW 3, 45.2–7]. See Flasch, *Meister Eckhart: Philosopher of Christianity*, 178, 201–2, 239.

89 In section 16 of chapter 3 of *Fons vitae*, for example, we read: "This is why the primary Creator is said to be *present in all existing things, and without him nothing can exist*" [et ideo dictum est quod factor primus est *in omnibus quae sunt et nihil sine eo esse potest*] (FV 3:16 [114.4–6]; my emphasis). This is quite reminiscent of the Vulgate's Latin translation of John 1:3, which reads: "Omnia per ipsum facta sunt et sine ipso factum est nihil." See also FV 5:39 [327.15–17].

4 Avicebron's Chain of Being and Jacob's Ontological Ladder

Concluding their reference to *Fons vitae* 5:24, the editors of the LW call further attention to *In Gen.II* n. 209,[90] since Eckhart explicitly invokes this specific Avicebronian paragraph there. This exegetical reference is particularly interesting considering the fact—based on Abraham ibn Ezra's fragmentary account—that Ibn Gabirol actually advanced a philosophical reading of Jacob's vision of the heavenly ladder.[91] In this section, Eckhart offers a philosophical analysis of Jacob's ladder in Vg. Gen 28:12–13.[92] At the outset of his interpretation of this biblical depiction and prior to his mention of *Fons vitae*, he draws on chapter 10 of the second part of Maimonides's *Guide of the Perplexed*, using it as the main reference point for his entire exegetical discussion.[93] Within a general metaphysical analysis of the number four, Maimonides then clarifies Jacob's dream of the heavenly ladder. The category of four is relevant for Maimonides's understanding of Jacob's vision, inasmuch as he assumes that four angels made use of the ladder, two ascending and two descending. According to Maimonides, the four angels on the ladder allegorically correspond to the four elements of the sublunar universe: fire and air match the two ascending angels, while water and earth match the two descending ones.[94] The manner in which Eckhart goes on to explicate Maimonides's explanation of Jacob's ladder is quite surprising. He does not simply subscribe to Maimonides's specific alignment of the four angels to the four empirical elements, but perceives it as an ontological claim markedly wider in scope, reproducing the fourfold structure of the cosmos.[95]

---

90 LW 3, 39 n. 2.

91 See Schlanger, *The Philosophy of Solomon Ibn Gabirol*, 20. For further analysis of the Gabirolean exegesis of Genesis, see Julius Guttmann, "Zu Gabirols allegorischer Deutung der Erzählung vom Paradies," *Monatsschrift für Geschichte und Wissenschaft des Judentums* 80 (1936): 180–84; Sara Klein-Braslavy, "The Philosophical Exegesis," in *Hebrew Bible/Old Testament: The History of Its Interpretation*, ed. Magne Sæbø (Göttingen: Vandenhoeck & Ruprecht, 2000), vol. 1, part 2: 304–6.

92 "Vidit quoque in somnis scalam stantem super terram et cacumen eius tangens caelum et angelos dei ascendentes et descendentes per eam et dominum innixum scalae" (LW 1, 667.2–5).

93 LW 1, 677–79. For a comprehensive and insightful interpretation of Eckhart's usage of Maimonides's exegesis of Jacob's ladder, see Schwartz, *"To Thee Is Silence Praise,"* 256–65.

94 See Sara Klein-Braslavy, "Bible Commentary," in *The Cambridge Companion to Maimonides*, ed. Kenneth Seeskin (Cambridge: Cambridge University Press, 2005), 260–61; Schwartz, *"To Thee Is Silence Praise,"* 257; Gad Freudenthal, "Maimonides on the Scope of Metaphysics *alias* Ma'aseh Merkavah: The Evolution of His Views," in *Maimónides y su época*, ed. Carlos del Valle, Santiago Garcia-Jalón, and Juan Pedro Monferrer (Madrid: Sociedad Estatal de Conmemoraciones Culturales, 2007), 224–30. Eckhart is alluding, of course, to this reading: LW 1, 678.14–15 ("quarto elementa mundi quattuor"), 681.13–683.11. For the other two instances in which Maimonides interprets Jacob's ladder in *The Guide of the Perplexed*, see *The Guide of the Perplexed*, trans. Shlomo Pines, 2 vols. (Chicago: University of Chicago Press, 1963), 1:12–13 [part 1, introduction], 1:41 [1:15] (henceforth *Guide*).

95 *In Gen.II* 209 [LW 1, 684.1–2].

This becomes apparent in Eckhart's recourse to Proclus's *Elementatio theologica*, by which he reframes the Maimonidean reading of Gen 28:12–13 in distinctly Neoplatonic terms.[96] Drawing on proposition 20 of *Elementatio theologica*—which states that "beyond all bodies is the soul's essence, beyond all souls, the intellective principle; and beyond all intellective substances, the One"[97]—Eckhart makes Maimonides's Aristotelean explication of Jacob's ladder conform to Proclus's outlined structure of reality in its fourfold hierarchical structure of emanation: (1) the One, (2) the universal intellect, (3) the soul, and (4) the corporeal realm.[98] The merging of Ibn Gabirol with Proclus's *Elementatio theologica* has a provocative edge, if one keeps in mind Albertus Magnus's critical treatment of Avicebronian philosophy specifically within the context of his commentary on this work in its Latin rendition as *Liber de causis*.[99]

At this point, seeking to reinforce his own Neoplatonic and dynamic take on the heavenly ladder, Eckhart proceeds to cite from what appears to be his favourite part of *Fons vitae*; namely, section 24 of chapter 5. This is the specific section through which he supports his reading of John 1:1–2 as well as that of Gen 1:1. Citing almost three entire paragraphs, constituting two successive answers from the *magister* (!), in which Ibn Gabirol provides a succinct account of his ontological *Weltanschauung* (301.16–302.7, 9–15), Eckhart brings his Neoplatonic view of Gen 28:12–13 to its conceptual peak.[100]

While one cannot simply ignore the central role of Maimonides here, it does appear that Eckhart adduces him particularly for his basic exegetical intuition, through which he is able to discern the ladder's basic symbolic meaning as signifying the fourfold order of reality. However, to address the question of what ultimately constitutes each of the respective ontological ranks, he takes an Avicebronian perspective. The three different layers of philosophical views on the structure of the universe—which, on the face of it, Eckhart seems to conflate[101]—might entail a certain hierarchical order: having Jewish Aristotelianism (Maimonides) as his starting point, then turning to its logical continuation in the form of pagan Neoplatonism (Proclus), and culminating in what Eckhart thought to be Christian Neoplatonism (Avicebron).

---

96 Cf. Schwartz, *"To Thee Is Silence Praise,"* 259.

97 Proclus, *The Elements of Theology*, trans. Eric R. Dodds, 2nd ed. (Oxford: Oxford University Press, 1971), 23.

98 For Eckhart's dynamic understanding of the ladder as a symbol of the fourfold structure of the universe, see Schwartz, *"To Thee Is Silence Praise,"* 258, 260.

99 Nevertheless, as McGinn observes, Eckhart's specific unifying approach seems to have become more prevalent in his time, with thinkers such as Berthold Moosburg (?–after 1361): McGinn, *The Mystical Thought of Meister Eckhart*, 173. Cf. Brunner, "Maître Eckhart et Avicébron," 139. On the possible impact that Proclus's *Elementatio theologica* could have exerted on *Fons vitae*, see Schlanger, *The Philosophy of Solomon Ibn Gabirol*, 68–71; McGinn, "Ibn Gabirol: The Sage among the Schoolmen," 79–80; Brunner, "Maître Eckhart et Avicébron," 142–43, 145.

100 For an accurate account of this reference to Ibn Gabirol, see also Palazzo, "Eckhart's Islamic and Jewish Sources," 278–79.

101 Schwartz lays out this insightful perspective: Schwartz, *"To Thee Is Silence Praise,"* 259.

5 Seeing God's Hidden Face through Avicebron's Negative Theology

As McGinn hints, another, even more explicit conflation of Maimonides with Ibn Gabirol in Eckhart's thought finds its expression in his negative theological reading of Exod 33:23b ("faciem meam videre non poteris").[102] This time, however, Eckhart's exegesis takes its departure not from *The Guide of the Perplexed*, but rather from *Fons vitae:*

> In Book 1, Chapter 4, of the *Fountain of Life* Avicebron asks, "Is there any way to understand the first essence?" His response is that "this is not impossible, but it is not possible in any way whatever. It is impossible to know the first of essences apart from the created things that come from it; it can be known from its works. [...] He adds that [full] knowledge of the essence of the First, that is, God's essence, is impossible, because it is above all things, because it is infinite, because it is not like the intellect, and because the First Essence and the intellect do not agree and have a mutual bond." In Book 1, Chapter 58, Rabbi Moses says, "The sages agree that the sciences do not apprehend the Creator; only God himself can grasp his essence." Nevertheless, Moses knew everything that we can know about him to the limit of our apprehension. What he knew no one before or after him knew, as Rabbi Moses says in the same chapter (*In Ex.* n. 281 [LW 2, 225.5–226.7]).[103]

In spite of the fact that Eckhart appears to be following Maimonides's metaphysical illumination of Exod 33:23, he commences his interpretation by presenting the chief epistemological and ontological premises at the core of Ibn Gabirol's apophatic theology in its original interrogative form: that directly apprehending God's absolute transcendent being as the first essence (*essentiae primae*) is, strictly speaking, *impossible.* The sole *possible* way of knowing Him—that is, merely knowing that He exists—is by means of mediation; namely, through His created works.[104] Eckhart proceeds to bring forth the four reasons that Ibn Gabirol subsequently provides which completely rule out human soul's potential to conceive of God's primary essence: (1) God "is above all things";[105] (2) He is infinite;[106] (3) God's purely unified being is dissimilar in every sense from the universal intellect and its dual structure compounded of matter and form,[107] which is rather analogous to the human intellect;[108] and finally,

---

102 McGinn, *The Mystical Thought of Meister Eckhart*, 173. For Eckhart's programmatic exegesis of the book of Exodus, see Flasch, *Meister Eckhart: Philosopher of Christianity*, 142–54.

103 I took this citation from the English translation by McGinn: *Meister Eckhart: Teacher and Preacher*, ed. and trans. Bernard McGinn (New York: Paulist Press, 1986), 129 (henceforth LWE).

104 FV 1:4 [6.16–22].

105 FL 64 [FV 1:5 (6.24)].

106 FV 1:5 [6.24].

107 See, for example, FV 1:7 [10.20]. For further analysis, see Pessin, *Ibn Gabirol's Theology of Desire*, 82–84.

108 FV 1:5 [7.1–7].

(4) the primary essence does not have any ontological point of commonality with the universal intellect.[109] It is important to note that at the core of all these arguments lies Ibn Gabirol's hylomorphic hypothesis, from which a fundamental division emerges between the pure One and the twofold being consisting of universal matter and form.

Intending to reaffirm his Avicebronian reading, Eckhart cites excerpts from chapters 54[110] and 59[111] from the first part of *The Guide of the Perplexed*, which are broadly congruent with Ibn Gabirol's negative theology.[112] The basic ideas Eckhart employs here are already, for the most part, present in Ibn Gabirol's thought:[113] the human intellect's absolute incapability of apprehending God's essence, thereby restricting man's knowledge of God to His attributes of action and the mere fact of His existence.[114] Maimonides's relevance for Eckhart's exegesis of Exod 33:23 actually appears to lie not so much in a tangible philosophical contribution, but rather in an exegetical one: drawing support for his negative theological understanding of Jewish Scripture. Interestingly, at the same time, Eckhart exploits two elements from Maimonides's explanations—Moses's exceptional knowledge of God and God's apprehension of Himself—to buttress his own affirmative theology, which, by supernatural means, permits the human intellect to access God's being.[115] Thus, while Eckhart puts forward the tenets of Gabirolean apophatic theology in strict terms, he pinpoints the "weak spots" in the Maimonidean ones.

Ibn Gabirol's speculative philosophy also proves significant for Eckhart's initial reading of the analogous depiction of Vg. Exod 33:13b ("ostende mihi faciem tuam") in its peculiar Latin rendition,[116] in which Moses asks God to reveal His face. Follow-

109 FV 1:5 [7.2–3]. Palazzo provides the most comprehensive interpretation of this passage: "Eckhart's Islamic and Jewish Sources," 280–81. Cf. also Schwartz, *"To Thee Is Silence Praise,"* 178–79, and Udo Kern, *"Gottes Sein ist mein Leben." Philosophische Brocken bei Meister Eckhart* (Berlin: De Gruyter, 2003), 231–32.

110 *Guide* 1:123 (given as chapter 53 in the Latin edition).

111 *Guide* 1:139 (given as chapter 58 in the Latin edition).

112 For a detailed and profound analysis of Eckhart's use of Maimonidean apophatic theology here, see Schwartz, *"To Thee Is Silence Praise,"* 177–79; Flasch, *Meister Eckhart: Philosopher of Christianity*, 247.

113 For a more thorough examination of Ibn Gabirol's apophatic theology, see Schlanger, *The Philosophy of Solomon Ibn Gabirol*, 98, 220–22; Pessin, *Ibn Gabirol's Theology of Desire*, 118–39. For Ibn Gabirol's negative theology, see, for example, FV 2:20 [61.3–8], 3:3 [78.24–79.1]: 3:5 [88.18–25], 3:6 [91.7–29], 3:7 [92.9–10], 3:8 [93.19–28], esp. 3:11 [103.21–104.17], 3:15 [109.22–27], 3:56 [204.13–205.7].

114 This is the limited knowledge of God that Maimonides ascribes to Moses in the context of the passage cited from chapter 54 of book 1 of the *Guide:* "The answer to the two requests that He, may He be exalted, gave him consisted in His promising him to let him know all His attributes, making it known to him that they are His actions, and teaching him that His essence cannot be grasped as it really is" (*Guide* 1:123).

115 LW 2, 225.13–226.7. Cf. Kern, *"Gottes Sein ist mein Leben,"* 232–33; Schwartz, *"To Thee Is Silence Praise,"* 179.

116 The original Hebrew reads: *hodiʿeni naʾ et derakhekha.*

ing citations from Augustine's *On the Trinity* and *Confessions*, he goes on to quote the concluding section of *Fons vitae:*[117] "S: What will help attain this noble hope [sc. the avoidance of death and union with the origin of life]? T: To be separated first from sensible things, to be poured out in intelligible things by the mind, and to be completely suspended from the giver of goodness."[118]

In the cited passage, Ibn Gabirol—with an allusion to Ps 36:10—delineates his conception of man's aspiration to apprehend God in Neoplatonic terms: as the ascendance of the human soul to the transcendent Good and the source of life for the purpose of unification, thus triumphing over death and attaining salvation. This invocation of *Fons vitae* implies that Eckhart—similar to his understanding of Maimonides—ultimately construes Ibn Gabirol's metaphysics as resorting to cataphatic theology. As Palazzo concisely points out, this specific employment shares a thematic feature common to several of the other Eckhartian references to Avicebron: "The teaching on intellectual perfection is one of the main focuses of Eckhart's reading of the *Fons vitae*."[119] In his employment of *Fons vitae*, Eckhart, as Palazzo claims, "is inclined to emphasize [...] the existential dimension in the process of achieving intellectual perfection."[120]

The aligning of Avicebronian and Maimonidean negative theology is, however, even more far-reaching. In fact, already in Eckhart's *In Ex.* n. 58, as part of a lengthy philosophical elucidation of one of the opening lines of the Song of the Sea (Vg. Exod 15:3b: "Omnipotens nomen eius"; *YHWH iš milḥamah YHWH šemo* in the original Hebrew), he resorts to Ibn Gabirol and Maimonides.[121] The discussion at hand revolves around his understanding of the Vulgate formulation "Omnipotens nomen eius" ("Almighty is his name"), which he explicates by recourse to thought patterns that are pivotal to negative theology. The main thread of his exegesis lies in the assumption that the omnipotence of God's name insinuates that His essence is inexpressible in terms of discursive thought.[122] This, in turn, implies God's absolute transcendence and pure unity, which, strictly speaking, does not entail multiplicity in any form:

> The distinction which the term "all" implies is indeed guilt, fault, and defect in existence and unity. Everything that exists is either above all and above number, or is numbered among all things. But above all and outside number there is only the One. No difference at all is or can be in the One, but "All difference is below the One," as it says in the *Fountain of Life*, Book

117 Eckhart actually states that the source of this quotation is book 3 of *Fons vitae*. Cf. Palazzo, "Eckhart's Islamic and Jewish Sources," 273 n. 51.

118 FL 261 (FV 5:43 [338.22–27]).

119 Palazzo, "Eckhart's Islamic and Jewish Sources," 280.

120 Palazzo, 280. Cf. Schwartz, "Meister Eckhart and Moses Maimonides," 409–12; Brunner, "Maître Eckhart et Avicébron," 150–51.

121 Brunner, 143.

122 See, for example, *In Ex.* n. 35 [LW 2, 41.6–15, 42.1–5] and 57 [LW 2, 63.7–13]. For a detailed and profound analysis of the influence of Maimonides on Eckhart in relation to speaking or naming God, see Schwartz, *"To Thee Is Silence Praise,"* 201–37.

5. "That is truly one in which there is no number," as Boethius says. And Rabbi Moses, as mentioned above, says that God is one "in all ways and according to every respect," so that any "multiplicity either in intellect or in reality," is not found in him. Anyone who beholds [the number] two or who beholds distinction does not behold God, for God is one, outside and beyond number, and is not counted with anything (*In Ex.* n. 58 [LW 2, 64.10–65.6]).[123]

In this citation, Eckhart addresses the question of whether ascribing positive attributes to God is merely a medium through which the human intellect is able to construe God's entity in terms that are more concrete.[124] Eckhart's response is decisive: God's essence is inconceivable by means of His primary determinations—such as wisdom, power, and goodness[125]—for His pure unified being transcends all these conceptual differentiations. It is at this thematic juncture that Eckhart invokes Ibn Gabirol, Maimonides, and Boethius in order to strengthen his negative theological viewpoint. As in *In Ioh.* n. 389,[126] here too, Eckhart refers to the Gabirolean premise from section 23 of *Fons vitae's* chapter 5 [300.16–17], which situates the category of "difference [...] below the One" ("differentia non est nisi sub uno").[127] He then proceeds to quote chapter 51 of the first book of *The Guide of the Perplexed*, in which Maimonides claims that in God's strict oneness, no "multiplicity either in the thing as it is outside of the mind or as it is in the mind" can be present.[128] Eckhart's correlation of Ibn Gabirol and Maimonides, both in general and in this specific passage, caught the attention of Bernard McGinn. In his monograph on Meister Eckhart entitled *The Mystical Thought: The Man from Whom God Hid Nothing*, he states that "Eckhart cited Gabirol in his Latin works mostly for the Jewish philosopher's emphasis on God as Absolute Unity," while stressing that "it is interesting to see him combining Maimonides and Gabirol with the Christian witness of Boethius to proof-text his teaching on God's oneness."[129]

A further examination of Eckhart's own line of reasoning—laid out prior to his invocation of these three authoritative sources—prompts the conjecture that it echoes distinct Gabirolean views. Even the editors of the LW have partially recognised this fact, inasmuch as they suggest that Eckhart's assertion regarding the numerical order

---

123 LWE 63 (emphasis in original). Cf. *In Ex.* n. 90 ("In uno autem non est maius aut minus. Omnis enim 'differentia sub uno' est, ut supra, dictum est"). For further analysis of this Eckhartian theme, see Mojsisch, *Meister Eckhart: Analogy, Univocity and Unity*, 98–99; Flasch, *Meister Eckhart: Philosopher of Christianity*, 138–39.
124 *In Ex.* n. 58 [LW 2, 63.14–64.1].
125 For a detailed account of Eckhart's primary determinations (*Erstbestimmungen*), see Flasch, *Meister Eckhart: Philosopher of Christianity*, 77–85.
126 Cf. LW 2, 64 n. 3.
127 Brunner, "Maître Eckhart et Avicébron," 142–43.
128 *Guide* 1:113.
129 McGinn, *The Mystical Thought of Meister Eckhart*, 172–73. Cf. Flasch, *Meister Eckhart: Philosopher of Christianity*, 200–202, 244–46, 268–69, 276.

of being falling under the ultimate One stems from *Fons vitae* 4:13 (239.12–17). By comparing both passages, their similarities come effectively to the fore:

Creatures, by the fact that they are from the One but below the One, necessarily fall into number, plurality, distinction, guilt, and fault, a condition by which they are numbered among all the things that are. That which commits an offense in the One and against the One incurs the guilt of distinction and *falls into the all* [et per creaturas, ubi necessario hoc ipso quod ab uno quidem, sed sub uno sunt, incidunt in numerum multitudinem et distinctionem et reatum seu maculam, quo inter omnia numerantur. Quod enim in uno et *unum offendit cadens ab uno,* incidit reatum distinctionis et cadit inter omnia] (*In Ex.* n. 58 [LW 2, 64.3–7]; my emphasis).[130]	And it seems to me that the substances that are ordered below the intelligence are ordered because of the order of numbers, *since they fall under one.* [...] In general, when you consider all the things that exist, you will find them ordered and constituted according to the nature of number, *and you will find all things falling under the form of the intelligence, which is unity, for the reason that every number falls under one.* Because of this, the form of the intelligence and its essence gather everything and comprehend everything [et omnimo cum consideraueris omnia ea quae sunt, invenies ea ordinata et constituta secundum naturam numeri, et invenies omnia cadentia infra formam intelligentiae, quae est unitas, ideo *quia omnis numerus cadit sub uno*; et propter hoc forma intelligentiae et eius essentia est collectiva omnis rei et comprehendens omne] (FV 4:13 [239.12–17]; my emphasis).[131]

In fact, Eckhart's identification of the category of the true One as being above the concept of number is not what renders this parallel so special.[132] Rather, it is the logical development of this view, which claims that created reality rests on a fixed numerical structure. Similarly, Eckhart's subsequent exegetical analysis of the concept of *omnia* from Vg. Jas 2:10[133] appears to be illuminated in terms of this Avicebronian division, differentiating between a purely unified totality that transcends number altogether and a totality confined to numerical order: "Everything that exists is either above all and above number, or is numbered among all things. But above all and outside number there is only the One."[134]

---

130 LWE 63.

131 FL 204.

132 Cf. *In Eccli.* n. 149 [LW 2, 487.2–5].

133 "Quicumque autem totam legem servaverit offendat autem in uno factus est *omnium* reus" (my emphasis).

134 LW 2, 64.10–11 [LWE 63]. Another thematic juncture in Eckhart's *Commentary on Exodus*, in which he philosophically expounds on one of the divine names, is *In Ex.* n. 18. In this section, Eckhart seemingly applies the four Aristotelian forms of inquiries from book 2 of *An. post.* to God's revealed name in Vg. Exod 3:14a ("Ego sum qui sum"). However, as the editors of the LW point out (LW 2, 24 n. 2), Eckhart—who sees a direct conceptual link between this biblical verse and John 1:1–3 (*In Ex.* n. 15 [LW 2, 20.11–12], 16 [LW 2, 22.7–9])—applies these questions to God as the absolute being in a similar fashion to *Fons vitae* 5:24. Like Ibn Gabirol, he too essentially asserts that it only makes sense to apply the question "whether" to God: "But in God where 'that-it-is' is the 'what-it-is' itself, a fitting answer to someone asking who or what is God is 'God is,' for God's existence is his 'what-it-

6 The Absolute Transcendence of God's Universal Form

Eckhart's invocation of Avicebron in section 9 of his commentary on the Wisdom of Sirach is also revealing. In his speculative reading of Sir 24:23, he exhibits his particular appreciation of Ibn Gabirol's insight into concepts relating to the transcendent realm. Here, Eckhart brings up Avicebron within a context that is pivotal to his understanding of what constitutes Godly wisdom:

> To the realm of the purity of substance belongs the meaning of the [personal] pronoun "I." This is the initial thing which is sought after in the beloved; that is, purity without any admixture. This [purity] truly pertains to Divine wisdom. For this reason, the author of *Fons vitae* contends the following about God, the transcendent and Holy, in book 5: "[He is the] substance constituting the existence of all forms; the perfect wisdom and purest illumination" (*In Eccli.* n. 9 [LW 2, 238.7–12]; my translation).

As Joseph Koch notes in the critical edition of the LW, in the original passage in *Fons vitae* 5:22 [298.17–21], Ibn Gabirol is in fact not depicting God, but rather the universal form.[135] In spite of this blatant misrepresentation of Ibn Gabirol's philosophical intent, it assists us in unearthing Eckhart's programmatic picture of Avicebron: probably regarding him as neither a Jewish nor a Muslim thinker, but as a Christian Platonist writing in Arabic. The identification of Avicebron's Neoplatonic conceptualisation of wisdom as the totality of forms with his own notion of intellect entailing the world of ideas[136] could have enticed Eckhart to misconstrue Ibn Gabirol as a Christian Platonist positing a consubstantial worldview, which puts the intellect as the divine Word and wisdom on the same ontological level as God.[137] Following Brunner,[138] Palazzo is right to suggest that "the very combination of the notions of

---

is.' He says, 'I am who am'" (LW 2, 24.5–7 [LWE 47]). Drawing on these Aristotelian types of questions, Ibn Gabirol prompts the conclusion that God's mere existence is the only fact that human knowledge can infer regarding Him (FV 5:24 [301.18–302.1]). For Schwartz's extensive treatment of Eckhart's drawing on the Maimonidean explication of the divine name in Exod 3:14, see Schwartz, *"To Thee Is Silence Praise,"* 229–37; Schwartz, "Meister Eckhart and Moses Maimonides," 401–8. Cf. also Brunner, "Maître Eckhart et Avicébron," 144. On Eckhart's equation of God with the category of being, see Mojsisch, *Meister Eckhart: Analogy, Univocity and Unity*, 51–65; Roberto Vinco, "Meister Eckhart's Nonstandard Natural Theology," *Neue Zeitschrift für Systematische Theologie und Religionsphilosophie* 58, no. 4 (2016): 485–89; Flasch, *Meister Eckhart: Philosopher of Christianity*, 138, 143–51.

135 LW 2, 238 n. 6. Cf. Brunner, "Maître Eckhart et Avicébron," 144, and Palazzo, "Eckhart's Islamic and Jewish Sources," 275.

136 See, for example, *In Eccli.* n. 10 [LW 2, 240.3–7].

137 Cf. Brunner, "Maître Eckhart et Avicébron," 144–45.

138 Brunner, 144.

purity and wisdom in the passages of the *Fons vitae* would have led Eckhart to single out and apply them, in contrast with their original meaning, to God."[139]

While this assertion is in itself true, one must also account for the wider framework in which Eckhart puts this Gabirolean notion to use. The specific paragraph takes its thematic and exegetical departure from John 1:1, to which Eckhart alludes for the sake of underpinning the human aspiration to grasp God the Father as the ultimate origin.[140] Here, Eckhart explicates the opening verse of the prologue of the Gospel of John through the complementary relationship between thought and its discursive expression by means of uttered words.[141] Eckhart also reads this Avicebronian passage in analogous terms: as a creative divine wisdom, which is purely one and from which the noetic world emanates. He seemingly conflates the universal form with the universal intellect as the totality of forms[142] and then moves on to equate the universal intellect with God. Examining subsequent sections from chapter 5 of *Fons vitae*, one actually encounters similar depictions of the universal intellect. In *Fons vitae* 5:30, for example, Ibn Gabirol portrays the universal intellect as the divine *verbum*, which he identifies with the noetic light which establishes forms.[143] At the outset of the concluding section of *Fons vitae*, Ibn Gabirol likens God's divine creation to the utterance of a word.[144] Furthermore, his distinct tendency to portray the universal intellect in analogous terms to his description of God in chapter 5 may also have facilitated Eckhart's misinterpretation of *Fons vitae* 5:22 [298.17–21]. In chapters 38 and 39 of *Fons vitae*, Ibn Gabirol asserts that the divine Will is indescribable,[145] above the categories of movement and time,[146] omnipresent, and that "nothing exists apart from it inasmuch as the existence and constitution of all things depends on it."[147]

139 Palazzo, "Eckhart's Islamic and Jewish Sources," 275.

140 Eckhart proceeds to cite John 14:8: "Domine, ostende nobis Patrem, et sufficit nobis."

141 LW 2, 237.11: "Verbum enim ad notitiam et intellectum pertinent." Cf. *In Eccli.* n. 12 [LW 2, 241.8–242.8].

142 See, for example, FV 3:32 [153.1–4]. For the Gabirolean conflation of the divine Will with the universal form, see Schlanger, *The Philosophy of Solomon Ibn Gabirol*, 195–97.

143 313.15–17. Cf. FV 4:20 [254.19–24], 3:53 [196.24–197.15].

144 For further analysis of this Gabirolean depiction, see Schlanger, *The Philosophy of Solomon Ibn Gabirol*, 32, 48, 53, 230–32; McGinn, "Ibn Gabirol: The Sage among the Schoolmen," 90–92.

145 FV 5:38 [326.3].

146 FV 5:39 [328.11–13]; also FV 3:57 [205.23–206.5]. See Schlanger, *The Philosophy of Solomon Ibn Gabirol*, 166.

147 FV 5:39 [327.14–17]. For Ibn Gabirol's portrayals of the divine Will in transcendent terms, see Schlanger, *The Philosophy of Solomon Ibn Gabirol*, 172, 223–28.

7 Platonising Avicebronian Hylomorphism

One of the unifying features of Eckhart's employment of *Fons vitae* in his *Commentary on the Gospel of John* is Ibn Gabirol's universal hylomorphism. In his reading of three different verses—Vg. John 1:48, 8:34, and 14:8—this Avicebronian doctrine, which claims that all levels of the cosmos consist of the universal principles of matter and form, is brought to the fore.

Aiming to establish that the divine omniscience and providence allegedly purported in Vg. John 1:48[148] match up with accepted philosophical tenets,[149] Eckhart turns to draw support from this Avicebronian worldview. The primary notion that Eckhart has in mind here appears to be a Platonic one; namely, the demiurge's goodness as the impetus for his creation of the cosmos (*Ti.* 29e1).[150] Through Plato's cosmologic concept of the demiurge, Eckhart is able to substantiate one of his initial assertions regarding Vg. John 1:48: by virtue of his artistic skill, the artificer sees the piece of art prior to his production of it.[151] After two citations from Avicenna,[152] he moves on to quote two interrelated passages from chapter 3 of *Fons vitae*, towards which he—as Palazzo notes—particularly gravitated:[153]

> Hence, it is necessary that form give itself and its own form to what is prepared to receive it. This is a more evident sign that form proceeded from the first maker and is obedient to him, because it is compelled in its own nature to give itself and to confer its own form, when it finds matter receptive of it. Moreover, because there was a first making and, similarly, a first action, it was necessary that this making and this action penetrated through everything until there was a lack of something able to receive them (FV 3:14 [108.19–20, 108.23–109.3]).[154]

It might not immediately stand to reason why Eckhart continues his citation of Avicenna's *Metaphysics* with these specific passages from chapter 3 of *Fons vitae*. Eckhart seems to place both the Avicennian and the Avicebronian descriptions within

148 "Priusquam te Philippus vocaret cum esses sub ficu vidi te."
149 *In Ioh.* n. 251 [LW 3, 209.11–12].
150 *In Ioh.* n. 262 [LW 3, 217.6–7].
151 *In Ioh.* n. 255 [LW 3, 212.3–4]. For further analysis, see Flasch, *Meister Eckhart: Philosopher of Christianity*, 182–83.
152 For Eckhart's linkage of Avicenna and Avicebron in the context of the passage at hand, see Palazzo, "Eckhart's Islamic and Jewish Sources," 276. Cf. Yossef Schwartz, "Thirteenth Century Hebrew Psychological Discussion: The Role of Latin Sources in the Formation of Hebrew Aristotelianism," in *The Letter Before the Spirit: The Importance of Text Editions for the Study of the Reception of Aristotle*, ed. Aafke M.I. van Oppenraay and Resianne Fontaine (Leiden: Brill, 2012), 175–78, 190 ("As claimed above, a close reading of Gundissalinus' *Tractatus de anima* reveals a unique Latin mixture of Avicenna with Gabirol, creating a form of 'Gabirolian Avicennism.' The 'Jewish,' Gabirolian part of this synthesis does not disturb the early translator of *Tractatus de anima* but is absolutely neglected by Gershom and Hillel, who systematically ignore the Gabirolian themes and terminology").
153 Schwartz, "Thirteenth Century Hebrew Psychological Discussion," 276 n. 60.
154 FL 128.

the cosmologic account of Plato's *Timaeus*.[155] Both philosophers advance a similar Neoplatonic outlook towards God and His creation: in His overflowing goodness and pure actuality as the first cause, God brings the world into existence and makes it comply with His divine will.[156] While Eckhart, like Avicenna, stresses the goodness of creation resulting from God's own overflowing goodness, he analogously shows through Avicebron how the universal form, by directly imitating God's pure active power, enforces itself upon its receptive counterpart, the *materia spiritualis*.[157] The closing remark of this paragraph also provides a further indication as to what Eckhart has in mind when quoting Avicebron here: "Doctores etiam dicunt quod angeli superiores illuminant inferiores naturaliter de omnibus quae noverunt."[158] The overlapping characteristic between Avicebronian and Christian teaching is that God endows His pure intelligible substances (universal form/high angels) with knowledge, which they then convey and apply onwards to lower ontological realms (matter/lower angels).[159]

Eckhart's exegetical observations regarding Vg. John 8:34b ("omnis qui facit peccatum, servus est peccati") in *In Ioh.* n. 463 also centre on Avicebronian hylomorphism. In fact, it serves as an important philosophical reference point for his entire reading of this verse. Through Ibn Gabirol's *Fons vitae* 3:14, Eckhart substantiates

---

155 It is not surprising that Eckhart links Avicebron to cosmologic notions from Platonic philosophy, since Plato is the only philosopher referred to in *Fons vitae* (FV 4:8, 5:17). For a discussion of Gabirolean hylomorphism seen from the perspective of Platonic metaphysics, see Pessin, *Ibn Gabirol's Theology of Desire*, 165–67; McGinn, "Ibn Gabirol: The Sage among the Schoolmen," 89–90.

156 Cf. Palazzo's concise interpretation of this link made by Eckhart in "Eckhart's Islamic and Jewish Sources," 276: "While addressing the prophetical flow from God in the *Commentary on the Gospel of John*, Eckhart connects Avicenna's *Metaphysics* Book 8 Chapter 6 (which, as said, describes the Necessary Existent's ability to overflow and produce reality) with the *Fons vitae* text which argues that it befits form to communicate itself to matter, whenever the latter is apt to receive the former, and that in its process of self-communication form obeys the First Maker, Who is the originator of the universal flow of form" (emphasis in original). Cf. Brunner, "Maître Eckhart et Avicébron," 147.

157 For a detailed analysis of this topic in Ibn Gabirol's theory of creation, see Schlanger, *The Philosophy of Solomon Ibn Gabirol*, 161–64, 166; McGinn, "Ibn Gabirol: The Sage among the Schoolmen," 84–85.

158 *In Ioh.* n. 263 [LW 3, 218.11–12]. Cf. the parallel passage in *In Gen.II* 146 [LW 1, 616.1–3]. Here, Eckhart describes the higher realm's contact with the lower realm in terms of love. Interestingly, Ibn Gabirol explains the relationship between God's creative activity, mediated by form in shaping passive matter, in quite similar terms in FV 3:13 [106.13–17]: "As to the nature of this activity, it is one entity endowing another with its own form, when each of them lends itself to this. As to how, it is by combining, either without or with an intermediary; either with alteration and decrease of the active form or on the contrary with no decrease of the active quality; either by the timeless impress of the active potency on the receptive entity, or by thought and imagination, as by something loved on the one who loves" (FL 100). The idea of this allusion to the Christian teachers in *In Ioh.* n. 263 also aligns well with Eckhart's initial citation of Avicenna relating to the angels' noetic communication with the inferior human souls (*In Ioh.* n. 262 [217.7–15]). For further analysis of Eckhart's use of Avicenna, see Palazzo, "Eckhart's Islamic and Jewish Sources," 257–71 (esp. 264).

159 Cf. Brunner, "Maître Eckhart et Avicébron," 145–46.

two of his four clarifications of John 8:34b. He explicates this apparent moral notion that a person who commits a sin becomes its servant in Aristotelian terms,[160] as the precept that each agent brings about that which ontologically matches itself ("cum omne agens agat sibi simile").[161] It is within this framework that Eckhart lays out the Avicebronian viewpoint, citing the exact same passages as in his reading of Vg. John 1:48. To this end, Eckhart insinuates that the principle of action is the universal form, imprinting its qualities on its receptive counterpart of universal matter. It is tempting to equate the sinful act with the mergence of form with matter, but that would be to miss Eckhart's philosophical point. His claim seems to be more basic. Taking its cue from God's transcendent activity, form endows its own nature on passive matter. The ontological symmetry is not so much related to the substrate of matter, but rather to form's articulation of its own essence.[162] Earlier, Eckhart expresses this idea in conjuncture with Exod 3:14, arguing that God's *modus operandi*, by virtue of Him being being itself, consists of calling things into being.[163] Proceeding to quote—as in his *In Ioh.* n. 263—Ibn Gabirol's subsequent depiction of God's creative productivity mediated by the universal form (108.25–109.3), Eckhart may also have been indicating that he has the same scheme in mind here: God's noetic being establishing the universe through the universal form, which reproduces His creative activity within a lower ontological realm.[164] On examining the next paragraph (464), one finds further reinforcement for this assessment, inasmuch as Eckhart speaks of the Gabirolean concept of the universal form as God's mediator within the natural realm between activity (form) and passivity (matter) ("mediatio inter agens et patiens").[165]

Again, referring to one of the above citations from *Fons vitae* 3:14,[166] Eckhart attempts to show that the motif of servitude in Vg. John 8:34b implies a further metaphysical idea; that is, that the activity of each object *serves* its corresponding form by predisposing matter for its exhaustive absorption therein.[167] Similar to his interpretation of Vg. John 1:48, instead of stressing the ontological symmetry between universal matter and form lying at the core of Avicebronian hylomorphism, Eckhart rather chooses to accentuate the clear ontological priority of creative form over pas-

160 See LW 3, 396 n. 1.
161 *In Ioh.* n. 463 [LW 3, 396.5].
162 Palazzo, "Eckhart's Islamic and Jewish Sources," 276.
163 *In Ioh.* n. 463 [LW 3, 396.5–7].
164 For an elucidation of the participation of form in divine creation, see Schlanger, *The Philosophy of Solomon Ibn Gabirol*, 239–42. See, for example, 3:13 [106.17–107.9].
165 *In Ioh.* n. 464 [LW 3, 397.1–3].
166 108.25–26: "Forma [...] compellitur in natura sua ad dandum se et ad conferendum formam suam."
167 *In Ioh.* n. 467 [LW 3, 399.6–10]. Cf. Palazzo's succinct description of Eckhart's specific usage of *Fons vitae* here, in "Eckhart's Islamic and Jewish Sources," 276 n. 60: "Furthermore, Avicebron['s] text is quoted to make it clear that operation predisposes matter not only for form, but also for whatever is connected or proper to form."

sive matter that it posits. Eckhart's strong metaphysical reading of Vg. John 8:34b overshadows the literal ethical meaning of the verse to such an extent that their link to one another becomes blurred.[168]

The subsequent reference to Avicebron's universal hylomorphism in Eckhart's *Commentary on the Gospel of John* takes place in his reading of Vg. John 14:8 ("Domine, ostende nobis Patrem, et sufficit nobis"). Already in *In Eccli.* n. 9, we saw how Eckhart adduces what he perceived to be Avicebron's transcendent concept of God within an exegetical framework, in which the God the Father of Vg. John 14:8, understood as the *principium sine principio*, occupies a significant role.[169] In *In Ioh.* n. 554, Eckhart analyses Vg. John 14:8 along similar lines, inasmuch as he attempts therein to throw light on different aspects of the transcendent oneness of God the Father. The one chief difference is that Eckhart does not directly refer to Avicebron's view of God, but rather to the superiority of form being a unifying principle over the passive and diversified matter. Eckhart inserts this Avicebronian viewpoint into his eighth reason, through which he accounts for the notion of God as the One. The general claim Eckhart advances is that "form and nature avoids and [...] detests matter."[170] To this effect, he establishes a worldview of God being the absolute One as opposed to matter being pure multitude. Eckhart supports this philosophical perspective by referring to no less than three excerpts from *Fons vitae:* "Therefore, he says, 'Show us' the One, 'and it is enough for us.' The One is opposed to matter, in that it is many and evil, or nothing, as said above. Hence the author of the *Fountain of Life* in 2.9 and 10 and 3.4 denies that any bodies composed of matter have activity, but [says] they are only passive due to matter."[171] A cursory examination of the three passages to which Eckhart made recourse hints at the fact that at their core lies the sharp ontological contrast between God's pure actualism, conferred on the universal form, and the passive substrate of matter. It would thus appear that Eckhart, when drawing on these examples from *Fons vitae*, does not merely have Avicebron's precept concerning the passivity of matter in mind, but also the conceptual framework in which Ibn Gabirol advanced it. Against this backdrop, one can better observe how Eckhart's overall programmatic presentation here, which attempts to prove that God is the transcendent unity, corresponds to Gabirolean philosophy.[172]

---

168 On Eckhart's "philosophic allegorism" in his LW, see Schwartz, "Meister Eckhart and Moses Maimonides," 398.

169 LW 3, 237.11–238.2.

170 *In Ioh.* n. 551 [LW 3, 481.5–6] (FLW 183).

171 *In Ioh.* n. 554 [LW 3, 483.12–484.2] (FLW 184). Cf. Flasch, *Meister Eckhart: Philosopher of Christianity*, 181–182.

172 The section from *Fons vitae* 2:9 that Eckhart probably has in mind is 40.6–27: the view that Ibn Gabirol outlines here is of God's creative productivity, mediated through the active rational substances, imposed on the passive and receptive substrate of matter. Zimmermann and Sturlese offer 40.6–10: LW 3, 484 n. 1. The image from *Fons vitae* 2:9 that Eckhart draws upon is 40.6–27. As Zimmermann and Sturlese suggest, the specific section from the subsequent chapter in *Fons vitae* 2:10 from which Eckhart draws support is 41.11–16. In this passage, Ibn Gabirol opposes passive matter that has been

By adducing this textual evidence from *Fons vitae*, Eckhart appears to want to actually weaken the sweeping thrust of the claims raised by Avicebronian hylomorphism. Instead of emphasising the exceptional ontological status of universal matter in *Fons vitae*—which Ibn Gabirol conceptualises as a basic noetic principle intrinsic to all of created reality and as even having priority over form[173]—Eckhart chooses to firmly relegate it to a secondary position. From this perspective, Eckhart can proceed to establish a baffling connection between Avicebron and Thomas Aquinas, as the latter also denies the possibility of matter being active ("Videtur quod nullum corpus sit activum"): "On this see Thomas in the [*Summa of Theology*], Ia, q. 115, a. 1."[174] As is known, Aquinas, in light of his incorporeal conception of angels, systematically grapples specifically with the radical assertions of Avicebronian hylomorphism in four subsequent chapters (5–8) of his *De substantiis separatis seu de angelorum natura.*[175]

Summary

This study maps Meister Eckhart's various explicit and implicit references to Avicebron's *Fons vitae*. What emerges from these invocations is an unusual image of Ibn Gabirol. Eckhart gives the eleventh-century Andalusian Jewish philosopher a significant position within his Christian body of thought, unmatched by most of his non-Christian sources of inspiration from the Middle Ages. Unlike other central non-Christian authorities cited by Eckhart, such as *Rabbi Moyses* (Maimonides) and the *heidenischer meister* (Avicenna), the ambiguity attached to the religious affiliation of the Avicebronian worldview appears to have actually contributed to its adoption. Eckhart, not entirely sure as to Avicebron's religious affiliation, could not only approach his innovative philosophy in a more objective way, but could also exploit it more flexibly for his own specific Christocentric interests.

Eckhart does not single out a specific theme from *Fons vitae*, but rather seems to endorse the outlook that Gabirolean thought provides. His fascination with *Fons vitae* centres on a wide array of topics covering the main ideas of Gabirolean philosophy. Aside from Ibn Gabirol's hylomorphism, Eckhart also draws on his notions

set in motion to God as the ultimate source of movement from which all of reality overflows. The editors of the critical edition specifically refer to 83.1–6. See LW 3, 484 n. 1 and cf. FV 1:14. The third section from *Fons vitae* 3:4 to which Eckhart may have wished to allude is 82.9–83.20. In these paragraphs, Ibn Gabirol depicts God as being the transcendent artificer of the universe, whose pure creative activity requires an intermediary substance in shaping passive matter. For further analysis, see Schlanger, *The Philosophy of Solomon Ibn Gabirol*, 132.

173 For matter's priority over form, see McGinn, "Ibn Gabirol: The Sage among the Schoolmen," 86–87; Schlanger, *The Philosophy of Solomon Ibn Gabirol*, 181–82.

174 *In Ioh.* n. 554 [LW 3, 484.2–3] (LWE 184).

175 For a more detailed analysis of this correlation, see Brunner, "Maître Eckhart et Avicébron," 148.

of divine creation, the intellectual ascent of the human intellect to the noetic realm, God's transcendent unity, negative theology, and the fourfold *ordo essendi*. Our survey has made it clear that in fact, it was not chiefly *Fons vitae's* well-known doctrine of universal hylomorphism that resonated with Eckhart's philosophy, but rather its outlined *Einheitsmetaphysik*, which conceptualises God as well as His creation through the determination of Oneness. The Gabirolean formulation that the Two follows from the One also does not lure him into thinking otherwise, for Eckhart knows that the category of the Two, denoting universal matter and form, is merely a construct to make Avicebronian hylomorphism more comprehensible. Strictly speaking, form and matter can only exist in their interdependent unity within the universal intellect.

At the same time, Eckhart seems to approach *Fons vitae* in a programmatic manner. This study has pointed to two of the inclinations underlying Eckhart's understanding of Avicebron: (1) Eckhart is reluctant to take Ibn Gabirol's radical hylomorphism at face value. Rather, he attempts to weaken the pivotal role of the Avicebronian *Materialprinzip* by decisively relegating it to the secondary status of a substrate for the creative productivity of form. (2) Although Eckhart was familiar with Ibn Gabirol's negative theology, he appears to want to conflate God with the universal form—as evidenced by his reading of *Fons vitae* 5:22 and his general presentation of Gabirolean hylomorphism—equating the universal form with God by virtue of its transcendence, unity, and creative power.

In Eckhart's reading of *Fons vitae,* the fifth chapter gains centre stage. The reason for this emphasis, as I have argued, lies in the fact that it harmonises well with Christian theological patterns of thought in general and with Eckhart's metaphysics of the *verbum* in particular. While it is certainly possible that Eckhart supposed Avicebron to be a Christian, I maintain that it was not this feature that attracted him to his universal philosophy, but rather the speculative thrust of his innovative metaphysics of the One. By incorporating such a unique philosophical voice as Ibn Gabirol into his thought, Eckhart could better account for the speculative elements intrinsic to the Christian faith and thus present them in a universal light. Given the scrutinising critique of *Fons vitae* levelled by the most celebrated Dominicans, Albertus Magnus and Thomas Aquinas, Eckhart's broad adoption of Avicebronian thought in his Christian philosophy stands out as being a rather subversive move. Inserting such a controversial philosophical source into his Christian philosophy must also factor in what Kurt Flasch assesses as Eckhart's "divergence from the traditional path of common scholasticism and his explicit polemical position against it."[176] Yet it would seem that precisely this subversiveness is that which renders Eckhart's Christian philosophy so appealing and worth exploring.

176 Flasch, *Meister Eckhart: Philosopher of Christianity*, 157.

Bibliography

Albertus Magnus. *Buch über die Ursachen und den Hervorgang von allem aus der ersten Ursache: Liber de causis et processu universitatis a prima causa.* Edited and translated by Henryk Anzulewicz, Maria Burger, Silvia Donati, Ruth Meyer, and Hannes Möhle. Hamburg: Felix Meiner Verlag, 2006.

Aristotle. *Posterior Analytics. Topica.* Translated by Hugh Tredennick and E.S. Forster. Cambridge, MA: Harvard University Press, 1960.

Augustine. *On the Trinity: Books 8–15.* Edited by Gareth B. Matthews. Translated by Stephen McKenna. Cambridge: Cambridge University Press, 2002.

Augustine. *St. Augustine's Confessions in Two Volumes.* Translated by William Watts. London: William Heinemann; Cambridge, MA: Harvard University Press, 1912.

Beccarisi, Alessandra. "Zwischen Averroes, Avicenna und Avicebron. Meister Eckhart und die Noetik im Islam und Judentum." In *Meister Eckhart* – interreligiös, edited by Christine Büchner, Markus Enders, and Dietmar Mieth, 223–40. Stuttgart: Kohlhammer, 2016.

Brunner, Fernand. "Maître Eckhart et Avicébron." In *Lectionum Varietates. Hommage à Paul Vignaux (1904–1987)*, edited by Jean Jolivet, Zenon Kaluza, and Alain de Libera, 133–54. Paris: Vrin, 1991.

Meister Eckhart. *Die deutschen Werke.* Edited by Josef Quint and Georg Steer. 5 vols. Stuttgart: Kohlhammer, 1936–2016.

Meister Eckhart. *Die lateinischen Werke.* Edited by Konrad Weiss, Loris Sturlese, Albert Zimmermann, Joseph Koch, Heribert Fischer, Bruno Decker, Ernst Benz, Karl Christ, Bernhard Geyer, and Erich Seeberg. 5 vols. Stuttgart: Kohlhammer, 1936–.

Meister Eckhart. *Meister Eckhart: Teacher and Preacher.* Edited and translated by Bernard McGinn. New York: Paulist Press, 1986.

Meister Eckhart. *Meister Eckhart: The Essential Sermons, Commentaries, Treatises, and Defense.* Translated by Edmund Colledge and Bernard McGinn. New York: Paulist Press, 1981.

Meister Eckhart. *Meister Eckhart: Werke II.* Edited by Niklaus Largier. Translated by Josef Quint. Frankfurt am Main: Deutscher Klassiker Verlag, 1993.

Flasch, Kurt. *Meister Eckhart. Die Geburt der "Deutschen Mystik" aus dem Geist der arabischen Philosophie.* 2nd ed. Munich: Beck, 2008.

Flasch, Kurt. *Meister Eckhart: Philosopher of Christianity.* Translated by Anne Schindel and Aaron Vanides. New Haven: Yale University Press, 2015.

Freudenthal, Gad. "Maimonides on the Scope of Metaphysics *alias* Ma'aseh Merkavah: The Evolution of His Views." In *Maimónides y su época*, edited by Carlos del Valle, Santiago Garcia-Jalón, and Juan Pedro Monferrer, 221–30. Madrid: Sociedad Estatal de Conmemoraciones Culturales, 2007.

Ibn Gabirol, Solomon. *Avencebrolis (Ibn Gabirol) Fons Vitae, ex Arabico in Latinum Translatus ab Johanne Hispano et Dominico Gundissalino.* Edited by Clemens Baeumker. In *Beiträge zur Geschichte der Philosophie des Mittelalters, Texte und Untersuchungen*, vol. 1.2, edited by Clemens Baeumker and Georg von Hertling. 2nd ed. Münster: Aschendorff, 1995.

Ibn Gabirol, Solomon (Avicebron). *The Font of Life (Fons Vitae).* Translated by John A. Laumakis. Milwaukee: Marquette University Press, 2014.

Guttmann, Jacob. *Die Philosophie des Salomon ibn Gabirol.* Göttingen: Vandenhoeck & Ruprecht's Verlag, 1889.

Guttmann, Jacob. *Die Scholastik des dreizehnten Jahrhunderts in ihren Beziehungen zum Judenthum und zur jüdischen Literatur.* Breslau: M. & H. Marcus, 1902.

Guttmann, Julius. "Zu Gabirols allegorischer Deutung der Erzählung vom Paradies." *Monatsschrift für Geschichte und Wissenschaft des Judentums* 80 (1936): 180–84.

Halfwassen, Jens. "Ein Meister der Philosophie. Neue Literatur zu Meister Eckhart." *Philosophische Rundschau* 58 (2011): 172–77.
Halfwassen, Jens. "Gibt es eine Philosophie der Subjektivität im Mittelalter? Zur Theorie des Intellekts bei Meister Eckhart und Dietrich von Freiberg." *Theologie und Philosophie* 72 (1997): 337–59.
Halfwassen, Jens. "Gott als Intellekt: Eckhart als Denker der Subjektivität." *Meister-Eckhart-Jahrbuch* 5 (2011): 13–25.
Harvey, Warren Zev. "Filosofía y poesía en Ibn Gabirol." *Anuario Filosófico* 33, no. 67 (2000): 491–504.
Israeli, Isaac. *The Book of Definitions*. In *Isaac Israeli. A Neoplatonic Philosopher of the Early Tenth Century*. Translation and commentary by Alexander Altmann and Salomon M. Stern. Chicago: University of Chicago Press, 2009.
Laumakis, John A. "Avicebron (Solomon ibn Gabirol) and Aquinas on Primary and Secondary Causality." PhD diss., Marquette University, 2001.
Kern, Udo. *"Gottes Sein ist mein Leben." Philosophische Brocken bei Meister Eckhart*. Berlin: De Gruyter, 2003.
Klein-Braslavy, Sara. "Bible Commentary." In *The Cambridge Companion to Maimonides*, edited by Kenneth Seeskin, 245–72. Cambridge: Cambridge University Press, 2005.
Klein-Braslavy, Sara. "The Philosophical Exegesis." In *Hebrew Bible/Old Testament: The History of Its Interpretation*, edited by Magne Sæbø, vol. 1, part 2: 302–20. Göttingen: Vandenhoeck & Ruprecht, 2000.
Maimonides. *The Guide of the Perplexed*. Translated by Shlomo Pines. 2 vols. Chicago: University of Chicago Press, 1963.
McGinn, Bernard. "Ibn Gabirol: The Sage among the Schoolmen." In *Neoplatonism and Jewish Thought*, edited by Lenn E. Goodman, 77–109. Albany, NY: SUNY Press, 1992.
McGinn, Bernard. *The Mystical Thought of Meister Eckhart: The Man from Whom God Hid Nothing*. New York: Crossroad, 2001.
Mičaninová, Mária. "Was Aquinas Right When He Claimed That Avicebron Departed from Aristotle?" *Archive of the History of Philosophy and Social Thought* 62 (2017): 281–305.
Mojsisch, Burkhard. *Meister Eckhart: Analogy, Univocity and Unity*. Translated by Orrin F. Summerell. Amsterdam: B.R. Grüner, 2011.
Palazzo, Alessandro. "Eckhart's Islamic and Jewish Sources: Avicenna, Avicebron, and Averroes." In *A Companion to Meister Eckhart*, edited by Jeremiah M. Hackett, 253–98. Leiden: Brill, 2013.
Pessin, Sarah. *Ibn Gabirol's Theology of Desire: Matter and Method in Jewish Medieval Neoplatonism*. Cambridge: Cambridge University Press, 2013.
Pessin, Sarah. "Solomon Ibn Gabirol [Avicebron]." In *The Stanford Encyclopedia of Philosophy*, Winter 2016 Edition, edited by Edward N. Zalta: https://plato.stanford.edu/archives/win2016/entries/ibn-gabirol/.
Polloni, Nicola. "Toledan Ontologies: Gundissalinus, Ibn Daud, and the Problems of Gabirolian Hylomorphism." In *Appropriation, Interpretation and Criticism: Philosophical and Theological Exchanges Between the Arabic, Hebrew and Latin Intellectual Traditions*, edited by Alexander Fidora and Nicola Polloni, 19–49. Barcelona: FIDEM, 2017.
Proclus. *The Elements of Theology*. Translated by Eric R. Dodds. 2nd ed. Oxford: Oxford University Press, 1971.
Pseudo-Aristotle. *The Theology of Aristotle*. Translated by Geoffrey Lewis. In *Plotini Opera. Tomus II: Enneades IV–V*. Edited by Hans-Rudolf Schwyzer and Paul Henry. Oxford: Oxford University Press, 1977.
Runia, David T. *Philo in Early Christian Literature: A Survey*. Assen: Van Gorcum; Minneapolis: Fortress Press, 1993.

Schlanger, Jacques. *The Philosophy of Solomon Ibn Gabirol.* [Hebrew.] Jerusalem: Magnes Press, 1980.
Schwartz, Yossef. "Die Rezeption philosophischer Schriften aus dem Judentum." In *Grundriss der Geschichte der Philosophie. Die Philosophie des Mittelalters IV: 13. Jahrhundert*, edited by Peter Schulthess and Christoph Flüeler, 1:206–16. Basel: Schwabe Verlag 2017.
Schwartz, Yossef. "Meister Eckhart and Moses Maimonides: From Judaeo-Arabic Rationalism to Christian Mysticism." In *A Companion to Meister Eckhart*, edited by Jeremiah M. Hackett, 389–414. Leiden: Brill, 2013.
Schwartz, Yossef. "Salomon Ibn Gabirol: Zwischen Schöpfungsmythologie und Geheimnis der Einheit." In *Platonismus im Orient und Okzident: Neuplatonische Denkstrukturen im Judentum, Christentum und Islam*, edited by Raif G. Khoury, Jens Halfwassen, and Frederek Musall, 141–60. Heidelberg: Universitätsverlag Winter, 2005.
Schwartz, Yossef. "Thirteenth Century Hebrew Psychological Discussion: The Role of Latin Sources in the Formation of Hebrew Aristotelianism." In *The Letter Before the Spirit: The Importance of Text Editions for the Study of the Reception of Aristotle*, edited by Aafke M.I. van Oppenraay and Resianne Fontaine, 173–94. Leiden: Brill, 2012.
Schwartz, Yossef. *"To Thee Is Silence Praise": Meister Eckhart's Reading in Maimonides' Guide of the Perplexed.* [Hebrew.] Tel Aviv: Am Oved Publishers, 2002.
Spade, Paul V. "Binarium Famosissimum." In *The Stanford Encyclopedia of Philosophy*, Fall 2008 Edition, edited by Edward N. Zalta: https://plato.stanford.edu/archives/fall2008/entries/binarium/.
Sterling, Gregory E. "Prepositional Metaphysics in Jewish Wisdom Speculation and Early Christological Hymns." *Studia Philonica Annual* 9 (1997): 219–38.
Vinco, Roberto. "Meister Eckhart's Non-standard Natural Theology." *Neue Zeitschrift für Systematische Theologie und Religionsphilosophie* 58, no. 4 (2016): 473–88.
Weisheipl, James A. "Albertus Magnus and Universal Hylomorphism: Avicebron. A Note on Thirteenth-Century Augustinianism." In *Albert the Great: Commemorative Essays*, edited by Francis J. Kovach and Robert W. Shahan, 239–60. Norman: University of Oklahoma Press, 1980.
Wittmann, Michael. *Die Stellung des hl. Thomas von Aquin zu Avencebrol (Ibn Gebirol) im Entwicklungsgang der arabischen Philosophie.* In *Beiträge zur Geschichte der Philosophie des Mittelalters, Texte und Untersuchungen*, vol. 3.3, edited by Clemens Baeumker and Georg von Hertling. Münster: Aschendorff, 1900.

Giada Coppola

Obadiah Sforno and the Individual Human Soul[1]

The philosophical activity of Obadiah Sforno, who is primarily known for his exegetical interpretations of the Holy Scriptures, has received little scholarly attention. Nonetheless, the few scholars interested in *Light of the Nations*—Sforno's (uniquely) philosophical work—were particularly critical of the author's position in this exposition, showing a sceptical attitude towards the uniqueness of the work.

Obadiah Sforno's philosophical treatise reproduces the classical schema of the medieval *quaestio disputata* ("disputed question"), in which there was no place for a "new" or "innovative" analysis characterising some of the most original interpreters of the Renaissance period. He published the Hebrew version of *Light of the Nations*, entitled *Or ʿAmmim*, in Bologna in 1537. In 1548, also in Bologna, he published the Latin version, with the title *Lumen Gentium*. The fact that Sforno himself translated his own work from Hebrew into Latin makes him particularly exceptional.[2]

The Hebrew and Latin versions do not appear to offer significant differences in their overall view. The order of the questions and the general approaches to each argument are identical (with very few exceptions), but nonetheless, we find several changes concerning the style and the breadth of topics, as will be clear from reading the passages presented in this paper.

Although the structure of *Light of the Nations* reproduces the medieval format of the disputed question, it seems that Sforno exercises a sceptical approach towards introducing the *quaestio* and establishing the solution. In *Lumen Gentium*, every question is preceded by *dubitatur utrum* ("it is questionable"), and in the last part —namely, the solution—of both versions, we read:

Ad dubium ergo respondetur	נשיב אם כן להתיר ספק החקירה הנזכרת
To dispel any doubt	

Doubt does not only play a dialectical function in this pamphlet. Sforno is following the Maimonidean tradition,[3] and he introduces "doubt and perplexity,"[4] which are

1 I would like to thank Professor Warren Zev Harvey for his advice, his valuable comments, and his suggestions on how to improve this article.

2 I will employ the appellation *Light of the Nations* to refer to both versions of the pamphlet in general terms; when appropriate, I will make a distinction between the Hebrew and Latin editions.

3 Cf. Warren Zev Harvey, "Maimonides' First Commandment, Physics, and Doubt," in *Hazon Nahum: Studies in Jewish Law, Thought, and History Presented to Dr. Norman Lamm*, ed. Yaakov Elman and Jeffrey S. Gurock (New York: Michael Scharf Publication Trust of the Yeshiva University Press; Hoboken, NJ: Ktav, 1997), 149–62.

https://doi.org/10.1515/9783110618839-008

inherent in the quest for knowledge. Nevertheless, Sforno introduces the term *safeq* to present the solutions to the questions, in which sense the doubt is a synonym of "objection" (with reference to the previous arguments), following Gersonides, who was the first to use it in this manner. In the Renaissance, Isaac Abravanel employed *safeq* to translate *obiectio* in his *Roš Amanah*, confirming its double function, as did David Messer Leon in his *Tehillah le-Dawid*. For this reason, the relationship to Jewish scepticism cannot be ruled out, as is also highlighted by Giuseppe Veltri in his book *Alienated Wisdom*.[5]

Two articles by Robert Bonfil on Sforno's *Light of the Nations* appeared in 1976 and 1995 respectively.[6] Both focused on the interpretation of the idea of the soul (the crucial points of which I will analyse in detail later). Bonfil highlights the lack of originality in the author's approach to philosophy; he considers Sforno not only a "terrible" philosopher—in comparison with his contemporaries, taking into account the brilliant work of Leone Ebreo—but also a woeful interpreter of the classical philosophical tradition.[7] As Warren Zev Harvey has emphasised, a comparison between Leone Ebreo and Sforno turns out to be inappropriate, as the first was a great philosopher, while Sforno was an erudite scholar. Nevertheless, Bonfil recognises a "modern approach" represented by the relevance of the topic of the soul in Sforno's work, as the question of the immortality of the soul was one of the most common issues discussed in philosophical treatises during the Renaissance period.[8]

---

4 Harvey, 161.

5 Giuseppe Veltri, *Alienated Wisdom: Enquiry into Jewish Philosophy and Scepticism* (Berlin and Boston: De Gruyter, 2018), 143–44: "The sixteenth and seventeenth centuries are foundational for several aspects connected both with the Jewish consciousness of and attitudes towards scepticism. At the end of the Humanist period and the Renaissance, we can observe the first changes in the perception and formulation of philosophical questions. After some seminal beginnings, mirrored by 'Ovadya Sforno, there could be no hesitation in recognising that the use of *dubitatur/dubium* and quaestiones *dubitatae* is more than a rhetorical tool. They became a dialectical strategy that played an important role in the Late Renaissance (*Accademia dei Dubbiosi*) and early modern period (*Accademia degli Incogniti*, Simone Luzzatto). The 'doubts' (*dubitationes*) are not primarily a literary instrument to introduce the dogmatic opinion of teachers, masters, or opponents, but a dialectical strategy to introduce very delicate questions that could have dangerous results for the majority. The topics handled involve the usual themes of sceptic origins and tradition: authority, morality, theodicy, collective society and individuality, immortality of the soul as a social and moral problem (doctrine of remunerations), etc."

6 Robert Bonfil, "The Doctrine of the Soul and Holiness in the Teachings of Obadia Sforno" [Hebrew], *Eshel Beer Sheva* 1 (1976): 200–257; Bonfil "Il Rinascimento: la produzione esegetica di 'O. Servadio Sforno," in *La lettura ebraica delle Scritture*, ed. Sergio J. Sierra (Bologna: EDB, 1995), 261–67.

7 Bonfil, 270: "Non dovrebbe far meraviglia che non lo si sia preso sul serio e che, come filosofo, a lungo andare egli sia stato condannato al dimenticatoio."

8 Bonfil, 271: "Eppure, anche nella forma nella quale si presentava, il *Lumen Gentium* offriva un curioso esempio di elementi decisamente innovatore innestati sulla struttura essenzialmente medievale. Se, infatti, al di là dell'elenco delle quaestiones disputate, riflettiamo sulla quantità di energia discorsiva e conseguentemente sulla quantità di spazio dedicata ad ognuna di quelle questioni, ci si accorge subito che l'interesse era decisamente spostato su quanto era di attualitá nel Rinascimento, e in particolare sulla questione dell'anima umana etc."

Apparently, Sforno's intention, as the author himself announces in the prologue of both treatises, was to refute some of the Aristotelian positions which were entirely at odds with the principles of the Jewish faith. Hence, the subtitle of *Lumen Gentium* is worthy of note: in it, Sforno expressly asserts that the book was written "against some opinions of Aristotle," and that it "demonstrates the doctrines of the creation of the universe from nothing, the providence God has for human beings, and the immortality of human soul, which they [i. e., Aristotelian philosophers] shamefully reject, deriding religion."[9]

Charles Manekin follows the pattern established by Professor Bonfil, albeit with a less critical approach: "Yet it is also an attack on those principles of Aristotelianism that conflict with religious dogma, placing Sforno in the intellectually awkward position of appealing to Aristotle in order to undermine his own philosophy!"[10] In fact, it may be better to state "appealing to the authority of Aristotle and Averroes in order to undermine their own philosophy." However, in my opinion, Sforno's particular approach represents a component of modernity, consisting of quoting a sentence not only for the argument, but also for the counterargument, which is the confutation of the topic presented in the first part of the *quaestio*. It is necessary to note that most of the quotations from Aristotle and Averroes in *Light of the Nations* are favourable and not critical, despite what was announced by the author himself. Sforno does not care about the principle of contradiction, according to which it is not possible that *p* and *-p* may be presented by the same authority; in this sense, he is not anchored in the classical medieval system.

Sforno was defined by the late Mauro Zonta as the last representative of what may be called "Hebrew Scholasticism,"[11] but also as an "anti-philosopher," as he characterised him in his contribution to the first international symposium on Obadiah Sforno.[12] Taking into account the latest discoveries of the DFG-funded Sforno

---

9 Obadiah Sforno, *Lumen Gentium* (Bologna, 1548); frontispiece: "Opusculum nuper editum contra nonnullas Peripateticorum opiniones demonstratiue docens presertim circa creationem et vniuersi nouitatem et diuinam de mortalibus curam et humanarum animarum immortalitatem, quas ipsi turpiter respuunt religionis penitus irridentes quare illud merito Lumen Gentium appello."

10 Charles Manekin, in *The Jewish Philosophy Reader*, ed. Daniel H. Frank, Oliver Leaman, and Charles H. Manekin (New York: Routledge, 2000), 284.

11 Mauro Zonta, "The Autumn of Medieval Jewish Philosophy: Latin Scholasticism in 15th-Century Hebrew Philosophical Literature," in *Herbst des Mittelalters? Fragen zur Bewertung des 14. und 15. Jahrhunderts*, ed. Jan Aertsen and Martin Pickavé (Berlin and New York: De Gruyter, 2004), 475: "In Italy, there is evidence of the persistence of an interest in Latin Scholasticism by some Jewish philosophers active in the 16th century, e. g. Obadiah Sforno"; Zonta, *Hebrew Scholasticism in the Fifteenth Century. A History and Source Book* (Dordrecht: Springer, 2006), 30.

12 This symposium, entitled "Lost and Found in Translation—Ovadiah Sforno and His World," took place in Hamburg on 18 and 19 January 2017. In his paper entitled "*Or 'Ammim/Lumen Gentium* and the *quaestio disputata*," Zonta stated: "Ovadiah Sforno is therefore not seen today as a simple 'philosopher' but rather as an 'anti-Aristotelian' thinker, and so, considering the key role played by Aristotle in the thinking of that time, almost as a sort of 'anti-philosopher.'" Zonta was probably follow-

Project at Hamburg University, as well as the studies of Giuseppe Veltri[13] and Warren Zev Harvey,[14] it is now possible to newly place Sforno in the history of Jewish thought: he may be defined not only as the last interpreter of Scholasticism, but also as the last Averroist. As Veltri phrased it, Sforno is "an Averroist *malgré lui.*" Harvey, during his fellowship at MCAS, proposed an enlightening explanation: that *Light of the Nations* is the result of Sforno having taught classes in Latin on Averroist philosophy. Sforno was presumably hired to teach this class by the professors at Bologna University, or perhaps by Christian theologians. In this sense, Averroes represents not only an "authority" in Sforno's treatise, but also the subject matter of his teaching. As Harvey suggests, Sforno could have been moved to teach the Christian scholars in Bologna about Averroes for the same reason he chose to teach Reuchlin Hebrew in Rome.

This general overview paves the way for new perspectives: Elijah del Medigo (1458–93)—who was most active in Padua and Venice—would no longer be held as the last "Averroist"[15] of the Jewish tradition, since Averroes would have continued to have been studied in Jewish communities into the sixteenth century. As Montada noted, this tendency persisted in the Latin world—with its ups and downs—until the Renaissance period. A similar phenomenon may be observed in Jewish thought: "Jewish Averroism" is a concept that usually refers to Jewish philosophers who philosophised within the context of Averroes's interpretation of Aristotelian philosophy. For this reason, Sforno should be considered as a Jewish Averroist; chronologically speaking, as the last one. The fact that Sforno does not conform to all of Averroes's philosophy is not an indication of an anti-Averroistic attitude.

Even though Bonfil and Manekin note the influence of Averroes throughout Sforno's work, they do not stress the importance of this aspect, not remotely considering the general overview of the Italian setting.[16] On the contrary, for them, the impor-

---

ing the interpretation of Cecil Roth in his *The Jews in the Renaissance* (Philadelphia: Jewish Publication Society, 1959), 80–81.

13 Veltri, *Alienated Wisdom*, 162–67.

14 The second international symposium on Obadiah Sforno and his world, entitled "The Philosophical Canon of Obadiah Sforno and His Contemporaries," took place in Hamburg on 5 and 6 November 2018.

15 Josep Puig Montada, "Eliahu del Medigo, the Last Averroist," in *Exchange and Transmission across Cultural Boundaries: Philosophy, Mysticism and Science in the Mediterranean World*, ed. Haggai Ben-Shammai, Shaul Shaked, and Sarah Stroumsa (Jerusalem: Israel Academy of Sciences and Humanities, 2013), 155–86.

16 Bonfil, "Il Rinascimento," 266: "Il metodo del ragionamento ello Sforno resta dunque ancorato alla pesante dotta struttura accademica che ci si accorge poi subito essere costruita tutta di "mattoni" di *auctoritates phylosophorum*, prima tra i quali naturalmente Aristotele e Avveroè. Insomma, dichiarando in cima all'opera latina, che l'intento era disputare con *nonnullas Peripateticorum opiniones*, lo Sforno non intendeva affatto respingere il discorso peripatetico, bensì precisamente inserirsi organicamente in esso, prendere parte in una discussione considerata attualmente in corso, sempre valida"; 269: "Questo sistema aristotelico-averroistico [...] costituiva la struttura di riferimento del pensiero medievale di tutti, ebrei e cristiani Esso era ancora al centro della discussione filosifica nelle

tance accorded to Averroes corroborates the idea that Sforno was anchored to an outdated medieval tradition. In the Italian universities and in academic environments, Averroes was still being studied and commented on by scholars and members of the intellectual community. Since the lack of new elements in Sforno's treatise, such as (Neo-)Platonism and Kabbalah, has been considered a negative factor up until now, in view of recent studies, is it possible to suppose that Sforno's intention was to become a "full" member of an élite and to be accepted by the Italian university milieu?

It is likely that *Light of the Nations* was written for a particular audience (Christian and academic) and this could also explain Sforno's *modus operandi*, according to which only Aristotle and Averroes are quoted, since they are still considered the major "authorities" for systematic philosophical thought. Sforno's interpretation and exposition of Aristotelian philosophy is highly complex and sophisticated; it is not only based on medieval Scholastic sources, such as Aquinas, Grosseteste, and Jandun, but also on contemporary ones, among them Paul of Venice, Gaetano da Thiene, and Augustine Nifo; namely, the representative figures of the "strong" Averroeian tradition. This is the tradition according to which Giglioni identifies the Averroistic tendency in European culture.[17] Sforno belongs to this trend, all the more so given that he decided to write his work in both languages for two different audiences.

This point also characterises Sforno's general attitude towards philosophy, which is embedded in his deliberate choice to refer only to Averroes throughout the entire *Light of the Nations*, with the exception of Maimonides, who is mentioned twice in *Or ʿAmmim*.[18] Moreover, another particular aspect—which has been neglected by scholars—is the use of Latin sources in the Hebrew version as well: Sforno does not employ the Hebrew translations of Averroes, but only the Latin editions,[19] and this is very surprising since all the texts were available in Hebrew translation.

---

universitá d'Italia durante il Quattrocento e gran parte del Cinquecento. Sennonché in questo periodo operavano sul sistema medesimo le tendenze caratteristiche dell'umanesimo e del rinascimento, rendendo possibile l'introduzioone in esso di componenti eclettiche, soprattutto platoniche, che finirono per trasformarlo più o meno radicalmente. Orbene, contrariamente a quanto si puó tuttora leggere qua e là neo manuali di storia degli ebrei nel Rinascimento, il *Lumen Gentium* restó fondamentalmente fedele al sistema aristotelico-averrostico."

17 Cf. Guido Giglioni, "Introduction," in *Renaissance Averroism and Its Aftermath: Arabic Philosophy in Early Modern Europe*, ed. Anna Akasoy and Guido Giglioni (Dordrecht: Springer, 2013), 2. "'Averroistic' refers to the generic cultural label denoting a pronounced rationalistic attitude, of a vaguely Aristotelian ilk, towards questions of philosophical psychology (in particular, the nature of the human mind and its survival after the death of the body), natural determinism and, above all, the relationships between philosophical freedom and dogmatic truths, often of a religious kind."

18 Obadiah Sforno, *Or ʿAmmim* (Bologna, 1537), 8, 10.

19 A clear example is given by the employment of the expression "contra Garicone" (נגד גאריקון) to indicate the Megarians in book Theta of *Metaphysics*, a misunderstanding which is only found in the Michael Scotus edition, which was published in Venice in 1473.

Sforno's erudition is nevertheless evident throughout, displayed in his use of Jewish, Arabic and Scholastic sources, which he cleverly disguises, as there is no mention of them in his treatise. His philosophical knowledge is quite clear when we compare the pamphlet with the exegetical and biblical commentary. An evident example is represented by his interpretation of Numbers 7:89,[20] in which he interprets מדבר (vocalised *middabber*) as God speaking to Himself, which is clearly a reference to the theory of the intellect in Maimonides's *Guide of the Perplexed* 1:68. However, Sforno does not mention this biblical verse in his *Light of the Nations*, even if he presents a similar argument in the sixth question, "On the Creation," according to Maimonidean principles. As Rabbi Kravetz supposes, the commentaries are chronologically later than *Or ʿAmmim.*

Sforno considers the relation between reason and revelation as an essential point in order to understand the ultimate goal of human beings, which is naturally the higher knowledge of God. His attitude towards philosophy follows the classical Jewish tradition, according to which it is possible to find a standpoint to reconcile the claims of reason and revelation. Sforno criticises the Aristotelian principles that are contrary to the Torah. Nonetheless, he accepts twenty-seven "philosophical propositions" as being in accordance with the Torah's teachings.[21]

However, as his starting point for *Light of the Nations*, Sforno assumes that the root of every philosophical and scientific thought is founded in the Torah: Moses, during the revelation on Sinai, received the Torah that is not only the Law, but also the original wisdom. *Light of the Nations* sticks to this principle: in order to resolve the philosophical issues, the "final" solution Sforno presents is always characterised by the exegetical interpretation of a biblical passage.

Sforno's Arguments regarding *De Anima*

The problem of the human soul and its perfection occupies the third part of *Light of the Nations*,[22] and, as mentioned above, it represents the most important section of

---

20 "And when Moses was gone into the tabernacle of the congregation to speak with Him, then he heard the voice of one speaking unto Him from off the mercy seat that was upon the ark of testimony, from between the two cherubims: and he spake unto Him" (King James Version).

21 Cf. *Or ʿAmmim*, 12–14, and *Lumen Gentium*, 9–11.

22 Summarising the topics of *Light of the Nations*, we can divide the work into three different sections. In the first part of the book—chapters 1–7—Sforno analyses the physical cosmos, denying any possibility of the pre-eternity of the sublunary and superlunary worlds with particular attention to some topics such as the movement of coming-to-be and passing away, the elements, prime matter, the heavens, and motion. In the second part—chapters 8–10—he analyses the metaphysical and ontological aspects, the nature of God, unicity, corporeality, knowledge, and God's will. In chapters 11–14, the author approaches some questions on the soul, the nature of the soul, the morality of man, and providence. Chapter 15 could be considered a sort of "outsider" on account of its principal argu-

Sforno's enquiry. Sforno demonstrates the immortality of the human soul and its incorruptibility by refuting four arguments, which are founded upon Averroes's Long Commentaries on Aristotle and the *Incoherence of the Incoherence.*[23]

Bonfil specifically criticises Sforno's approach to the question of the soul, which he considered to be an obsolete approach commonly found in the classical medieval Judeo-Christian tradition that was closed to the new influences typical of this era.[24] Sforno "saves" some principles of the Averroeian/Aristotelian system (the twenty-seven propositions presented at the beginning of *Light of the Nations*) which are in accordance with Jewish religious dogmas. Hence, Bonfil asserts that Sforno may be compared with "some thinkers [who are] supposed to be paradigmatic followers of Averroes without believing in the Averroeian doctrines," as Pomponazzi phrased it.[25]

At the beginning of chapter 12, the reader is faced with the first problem, which constitutes the heart of the question:

Utrum intellectiva hominis anima qua substantialiter dicitur rationalis sit quid immortale vel non.	אם הנפש האנושית השכלית אשר בה היה האדם מדבר בעצם היא בלתי נפסדת אם אין
Whether the intellectual human soul, by which man it is essentially called rational, is immortal or not.	Whether the intellectual human soul, by which man is essentially rational, is incorruptible or not.

ment, which includes and embraces all those three aspects—physics, metaphysics, and providence—with particular focus on the relationships between deity, essence, and humankind.

23 The primarily sources are the Long Commentaries on *On the Soul* 3.5, Book Lambda of *Metaphysics*, and *On the Heavens*, as well as the *Incoherence of the Incoherence.* As aforementioned, Sforno probably uses the Scot's version for the commentaries and for the original Averroes' works, he uses the Nifo's edition.

24 Bonfil, "Il Rinascimento," 269–70: "Lo sforzo precipuo del nostro viene quindi rivolto a correggere il sistema aristotelico-averroistico dal di dentro, a salvarlo dai propri errori e quindi anche dalle critiche che si sarebbero potute muovere ad esso. Non diversamente da consimili tentativi di pensatori cristiani, come per esempio Pietro Pomponazzi, lo Sforno intese salvare il principio averroistico dell'intelletto separato e eterno comune a tutto il genere umano senza peró rinunciare alla esigenza religiosa di credere nell'esestinza di un'anima individuale forma dell'uomo, sostanza separata e eterna, non pura potenzialità. Come Averroé, lo Sforno sosteneva che l'intelletto materiale non è *neque corpus neque virtus in corpore.* Secondo lui, però, l'anima umana è *sostanza individuale*, eterna, emanante da una non meglio definita essenza superiore."

25 Bonfil, 270: "Con questa posizione di apertura, lo Sforno si allineava decisamente con quei pensatori che a detta di Pomponazzi si credevano di essere averroisti senza peraltro esserlo davvero." Cf. Bruno Nardi, *Studi su Pietro Pomponazzi* (Florence: Le Monnier, 1965), 18–19: "Ideo teneo quod orrupt Averrois devitat maiores difficultates quam alia; nec me pudet amore orruptio me ipsum retractare. Unde qui dicunt me aliis orruptio ut contradicam, mentiuntur. Oportet enim in philosophia haereticum esse, qui veritatem invenire cupit."

In *Or ʿAmmim*, the statement of the problem focuses on the concept of the corruptibility of the soul (*nifsad*). In *Lumen Gentium*, we find the word "immortality" (*immortale*). The two ideas are naturally equivalent, because the notion of incorruptibility logically embraces the idea of immortality. Sforno uses the Hebrew term *niṣḥi* to translate the Latin *immortale*, even if it literally signifies "eternal." Nevertheless, the author himself explains that *nifsad* and *niṣḥi* and *immortale* and *incorruptibile* are synonyms. This kind of distinction is also presented by Thomas Aquinas, who, when alluding to Augustine in the *Summa Theologiae*, shows that some substances are incorruptible (*incorruptibilis*) and immortal (*immortalis*).[26] In this example, it is interesting to note that the Hebrew *ʿeṣem nivdal*—literally, "separate substance"—is translated as *abstractum* from matter, which normally indicates the pure form (נבדל מן החומר).

Quodlibet abstractum est immortale sive incorruptibile.[27]	שכל עצם נבדל הוא נצחי[28]
Every abstract substance is immortal or incorruptible.	Every separate substance is incorruptible.

Sforno, in defence of the immortality of the soul, points out four essential arguments according to which he explains the nature of the (human) soul, from the general definition of the faculty to the particular characterisation of the intellectual human soul:

- The (in)corruptibility of the soul (in *Or ʿAmmim*)[29]/(im)mortality of the soul (in *Lumen Gentium*)[30]

26 Cf. Thomas Aquinas, *Summa Theologiae* 1.97.1: "Respondeo dicendum quod aliquid potest dici incorruptibile tripliciter. Uno modo, ex parte materiae, eo scilicet quod vel non habet materiam, sicut Angelus; vel habet materiam quae non est in potentia nisi ad unam formam, sicut corpus caeleste. Et hoc dicitur secundum naturam incorruptibile. Alio modo dicitur aliquid incorruptibile ex parte formae, quia scilicet rei corruptibili per naturam, inhaeret aliqua dispositio per quam totaliter a corruptione prohibetur. Et hoc dicitur incorruptibile secundum gloriam, quia, ut dicit Augustinus in epistola ad Dioscorum, tam potenti natura Deus fecit animam, ut ex eius beatitudine redundet in corpus plenitudo sanitatis, idest incorruptionis vigor. Tertio modo dicitur aliquid incorruptibile ex parte causae efficientis. Et hoc modo homo in statu innocentiae fuisset incorruptibilis et immortalis. Quia, ut Augustinus dicit in libro de quaest. Vet. Et Nov. Test., Deus hominem fecit, qui quandiu non peccaret, immortalitate vigeret, ut ipse sibi auctor esset aut ad vitam aut ad mortem. Non enim corpus eius erat indissolubile per aliquem immortalitatis vigorem in eo existentem; sed inerat animae vis quaedam supernaturaliter divinitus data, per quam poterat corpus ab omni corruptione praeservare, quandiu ipsa Deo subiecta mansisset. Quod rationabiliter factum est. Quia enim anima rationalis excedit proportionem corporalis materiae, ut supra dictum est. Conveniens fuit ut in principio ei virtus daretur, per quam corpus conservare posset supra naturam corporalis materiae."
27 *Lumen Gentium*, 57v.
28 *Or ʿAmmim*, 89.
29 *Or ʿAmmin*, 84: שכל צורת נפסד היא אמנם נפסדת והנה הנפש האנושית השכלית היא צורת נפסד אם כן היא נפסדת.

- The nature of the intellectual human soul; i. e., whether each individual has their own soul or where there exists only one common soul for the entire human species.[31]
- The potentiality of the intellectual human soul;[32] namely, whether the intellect is a power of the soul.
- The perfection of the intellectual soul as the actualisation of the material intellect/the fallacy of the intellectual human soul.

The problem of the soul is presented together with another question intrinsically related to the last part of the argument Sforno presents in this chapter; namely, the perfection of the (human) soul. In this case as well, the Hebrew and Latin editions present some differences. The influence of Thomas Aquinas[33] is manifest in *Lumen Gentium*: Sforno's interpretation is easy to understand, as he translates the expression השלמות השני with *actum secundum*, a term that Aquinas uses in his commentaries.

Sforno's Solution: The Human Soul Is Emanated from a Spiritual Nature

Sforno proposes a solution, in line with the classical Jewish tradition: the intellectual human soul is immortal (incorruptible and eternal) and individual. It is interesting to focus on his analysis, which is naturally founded upon the Aristotelian and Averroeian system, but also characterised by some original elements, especially in the Latin version:

Intellectiva hominis anima qua scilicet homo dicitur substantialiter rationalis est quid immortale, ad individuorum nu-	הנפש האנושית השכלית אשר בה נאמר שהאדם הוא בעצם חי מדבר הוא עצם נצחי ויפול עליו ריבוי כמספר אישי המין האנושי ומקורו מטבע

30 *Lumen Gentium*, 55v: "Nulla mortalium forma est immortalis, intellectus hominis anima est huiusmodi, ergo non est immortalis."

31 Cf. Maimonides, *Guide of the Perplexed*, trans. Shlomo Pines (Chicago: Chicago University Press, 1963–78), 1:74 (seventh method).

32 Cf. Giovanni Pico della Mirandola, *Conclusiones sive Theses DCCCC: Romae anno 1486 publice Disputandae, sed non admissae*, ed. Bohdan Kieszkowski (Geneva: Droz, 1973), Conclusiones paradoxe numero LXXI. secundum opinionem propriam noua in philosophia dogmata inducentes: "Cuilibet seriei animarum unus correspondet purus intellectus."

33 Cf. Thomas Aquinas, *De Anima* II, lectio 4, 12, 14, 16; III, lectio 1, 2, 5–9, 11. Thomas Aquinas, *Aristotle's De Anima in the Version of William of Moerbeke and the Commentary of St. Thomas Aquinas*, trans. Kenelm Foster and Sylvester Humphries, repr. ed. (New Haven: Yale University Press, 1959).

34 *Lumen Gentium*, 58v.

35 *Or ʿAmmim*, 92.

merum dinumerabile, et hoc quia producitur a quadam spirituali natura a superis, sive abstractis intelligentiis emanante, quibus ipsa potentialiter inest, qui scilicet natura est humano generi communis prout videtur Aven R. de mente Philosophi inferred *Metaphysicorum* VII, 31 quod scilicet virtus intellectiva tanquam quid abstractus habet ab abstractio quodammodo produci. Et idem Platonis auctoritate infert *Metaphysicorum* XII, 18 ubi dicit: "Animam intellectivam fieri a Deis secundis" huiusmodi ergo nature proportio ad actuales hominum individuorum formas, talis est qualis est proportio materie prime ad actuales hominum individuorum subiecta.[34]

כוחיי משותף לכל המין האנושי שופע מנבדלים עליונים באמרו נעשה אדם בצלמנו ואמרו רבותינו ז"ל במלאכי השרת נמלך כלומר שנתן להם כוח לזאת ההשפעה ולזה אולי התעורר אריסטו' כפי מה שביאר אבן רשד בשמו בספר מה שאחר מאמר ז' פרק ל"א באמרו שהכוח השכלי יתהוה מהנבדלים וכן ביאר בספר הנזכר מאמר י"ב פרק י"ח בשם אפלטון באמרו שהנפש האנושית נאצלת מן האלהים השניים והנה יחס זה הטבע השופע הוא אל הצורות האנושיות האישיות כיחס החומר הראשון אל אישי נושאי המוחשות הנמצאים בפועל[35]

The intellectual human soul, by which man is essentially called rational, is immortal, and its plurality corresponds to the number of individuals because [the intellectual human soul] is produced by a spiritual nature emanating from the higher—namely, the separate intellects, in which it [the intellectual human soul] is found in potency, meaning that the nature [of the intellectual human

The intellectual human soul, by which it is said that man is essentially a rational living being, is an immortal substance and multiplicity is applicable to it, as is the number of the individuals of the human species. And [the intellectual soul] has its origin in the potential nature shared by the whole human species, which emanates from the higher separate [intellects], as it says [Genesis

36 Cf. Long Commentary on *Metaphysics* [1494 and 1497]; Averroes, *Ibn Rushd's Metaphysics: A Translation with Introduction of Ibn Rushd's Commentary on Aristotle's Metaphysics, Book Lām*, ed. Charles Genequand (Leiden: Brill 1984), 106: "This shows that these proportions are somehow inspired by a cause nobler, worthier and higher in rank than themselves, i.e. the world soul which Plato thought had been produced by the secondary gods and Aristotle by the sun and the ecliptic"; 108: "He also explained it here, saying that this soul is that postulated by Plato as being produced by the secondary gods and by Aristotle by the sun and the ecliptic. This question is extremely difficult and obscure and we shall explain it to the best of our abilities and according to the premises and principles which have been established, in our science by the doctrine of the man whose doctrine, in the words of Alexander, is the least subject to doubts, the most adequate to being, the most adapted and suited to it and the most free of contradictions."

37 Cf. Themistius, *On Aristotle On the Soul*, trans. Robert B. Todd (London: Bloomsbury, 2014), 20,19 paraphrasing Aristotle's *On the Soul* 407a2–11; p. 36; Plato, *Timaeus* 41a. Cf. Genequand, *Ibn Rushd's Metaphysics*, 106 n. 87.

soul] is common to the whole human species, Averroes seems to explain in the name of the Philosopher in *Metaphysics* 7.31, saying that since the intellectual power of the soul is a separate substance, therefore it is produced by a separate intellect. This is similar to what Plato says in *Metaphysics* 12.18,[36] in which he states: "The intellectual soul comes from the secondary gods."[37] And therefore, the relationship of this nature to the form of the individual human being existing in act is like the relationship of the prime matter to the substrate of the individual human being existing in act.	1:26]: "We made man in our image." And our Masters, of good memory, said [Genesis Rabba 8:8]: "He consulted the ministering angels," that is to say that He provided them with the power for this emanation. And this, perhaps, Aristotle recognised, according to what Averroes explains quoting him in *Metaphysics* 7.31, that the intellectual power came to be from the separate [intellects]. And he also explains in the aforementioned book 12, par. 18, quoting Plato, that the lofty human soul is emanated from the secondary gods and [that] the relation of this emanating nature to the human individual souls was like the relation of prime matter to the individuals of the substrates of the sensible which exist in act.

The most obvious difference is related to the biblical and Talmudic references, which do not appear in the Latin version. While in *Or 'Ammin*, the higher and separate intellects may be identified with the angels, according to Sforno's interpretation of Genesis Rabba, in *Lumen Gentium*, the separate intellects are emanated from "indeterminate" natures called the "higher spiritual natures." Moreover, in the Jewish philosophical tradition, the connection between the separate intellects—which are also intelligences—and the angels had already been noted by Maimonides[38] and Gersonides.[39]

A detailed explanation of the nature of the spiritual "substance" is given in the second counterargument, i.e., against the idea of the multiplicity of the soul. In that case, Sforno clearly states that the spiritual natures may be identified with Plato's secondary gods, mentioned by Averroes in his commentary on *Metaphysics Lambda*.[40] This tradition is a direct reference to Themistius's interpretation of the Aristo-

38 Cf. Maimonides, *Guide of the Perplexed* 2:4–6.

39 Cf. Gersonides, *The Wars of the Lord*, trans. Seymour Feldman (Philadelphia: Jewish Publication Society, 1999), 5:2: "Themistius, On Aristotle *On the Soul* which according to Plato is generated from the movers of the heavenly bodies, whereas according to Aristotle it is generated from the sun and the inclined sphere [i.e., the ecliptic]. Hence, the soul performs its operations accomplishing the [intended] goal, although not knowing the goal." P. 87,

40 Cf. Yoav Meyrav, "Spontaneous Generation and Its Metaphysics in Themistius' Paraphrase of Aristotle's *Metaphysics* 12," in *Aristotle Re-Interpreted. New Findings on Seven Hundred Years of the Ancient Commentators*, ed. Richard Sorabji (London and New York: Bloomsbury Academic, 2016), 195–210.

telian text. Sforno did not read Greek, and the Latin edition of Themistius's commentaries was printed after his death. We cannot rule out the possibility that he read the Hebrew translation of Themistius, but the most plausible source is Nifo, who uses Themistius and the other Greek commentators in his commentaries.[41]

Et tamen de ipsa spirituali natura quamvis antequam humano individuo copuletur habeat quodammodo esse scilicet potentiale, non tamen sequitur illa contradictionis implicatio, que sequitur ad sermonem dicentium ut detur materia prima existens a qualibet forma denudata, quia in quantum materia tantum est quid potentiale tantum, et pro quanto actu existeret esse quid actu ens quod implicat contradictionem sicut Aven R. docet in commento libro *De Caelo* III, 29 et quia ipsa intellectiva sive spiritualis natura a superis sive Deis secundis emanans, ut supra, non influitur ab eis actu nisi dato subiecto sibi proportionato et eiusdem capaci cuius sit actu forma, sicut accidit de lucido a quo non emanat actualis lux nisi dato diaphano eidem luci proportionate. Quare huiusmodi spiritualis natura habet suum potentiale esse in his secundis Deis antequam influant. Suum vero actuale esse acquirit postquam ab eis in subiecto sibi disposito influxa fuerit cum fiat actualis illius forma et per consequens plurificatur ad individuorum numerum ad quos emanare contingit.[42]

אמנם עם היות זה הטבע נמצא קודם שישפע על אישי המין לא תתחייב אותה הסתירה אשר היתה מתחייבת אם היה החומר הראשון נמצא קודם שתחול עליו צורה. וזה כי אמנם החומר הראשון בהיות מציאותו הכוחיי בלתי שופע מנמצא זולתו הנה אם היה נמצא קודם שתחול עליו צורה היה מתחייב מזה שהיה לו מציאות בפועל. ועם זה בהיות מציאותו בכח בלבד הוא מצד עצמו בלתי נמצא בפועל. ויצדקו עליו שני חלקי הסותר אשר לא יתכן כמו שביאר אבן רשד בספר השמים מאמר ג' פרק כ"ט אבל זה הטבע השכלי השופע מן השכלים ראוי שיהיה לו מציאות כוחיי במשפיע קודם שישפע בפועל. ולא ישפע בפועל זולתי בהמצא נושא מיוחד אליו ומוכן לקבלו. ויהיה אז מציאות וצורה בפועל לאותו נושא. כי כמו שיקרה לגשם המאיר שלא יגיח אורו זולתי בגשם ספיריי מוכן לקבלת אורו כמו כן לא ישפע זה הטבע השכלי זולתי על איש מהמין שיהיה מוכן לקבלו באופן שיהיה זה הטבע השכלי צורה בפועל לאיש הנזכר. ובכן יתרבה זה הטבע הנזכר כמספר האישים אשר ישפע עליהם[43]

And although this spiritual nature, before being in conjunction with the human individuals, is something in potency, it [this idea] does not involve a contradiction. And by this discussion, it follows

However, although this nature exists before it emanates on the individuals of the species, this does not have to mean that the contradictory exists which would be necessary if the first matter existed

---

41 Cf. Agostino Nifo, *Commentationes in librum Averrois de Substantia Orbis* (Venice, 1508), f. 2.
42 *Lumen Gentium*, 60r.
43 *Or ʿAmmim*, 95.

that there is prime matter devoid of form, because this matter is in potency only, and if existing in act is being in act, and this implies a contradiction, as Averroes explains in the commentary *On the Heavens* 3.29. And this is because the intellectual or spiritual nature is emanated from higher and secondary gods, as mentioned above. And it will only be emanated in act with the subject specified for it and disposed for it, and its form is in act because it will happen that a luminous body will emanate light in act only if there is a transparent body which is proportioned to its light. Likewise the intellectual nature, if it is in potency, will emanate before through the secondary gods; if it is in act it will receive things which have been emanated by the subject specified for it and it will produce a form in act and consequentially its plurality is the same as the number of individuals on which it will emanate.

prior to the initiation of a form on it. And this is so because the first matter, since it exists in potency only, does not emanate from another existent. If this existent existed prior to the initiation of form on it, it would have to be the case therefore that it had a reality in act. And since its essence is in potency only, it—with regard to its substance—does not exist in act. And two parts of this contradiction would be proven for it, which is not possible, as Averroes explains in *On the Heavens* 3.29. But this intellectual nature which emanates from the intellects is suited for a potential existence in the emanator before it emanates in act. And it will only emanate in act (together) with the forth-bringing of a substrate specified for it and disposed for its reception, and then the reality and the form will be in act for that substrate. Because, just as it is the case that luminous body will not emit his light except through a transparent body which is disposed for the reception of his light, likewise this intellectual nature will only emanate on an individual of the species which is disposed for its reception in such a way that this intellectual nature will be a form in act for the aforementioned individual. And therefore, the aforementioned nature will become a plurality which matches the number of the individuals on which it will emanate.

However, the author clarifies in this passage that the expression *spiritualis natura* means *intellectiva natura* (הטבע השכלי). The term belongs to classical philosophical terminology; in the Italian Renaissance Jewish context, a similar phrase is presented by Leone Ebreo in *Dialoghi d'amore*,[44] in which we also encounter the metaphor of the light and the transparent body.[45]

In *Lumen Gentium*, the correspondence between intellectual natures/secondary gods and angels is given a few lines later. Likewise, in the Platonic tradition, God orders the angels to create mortal beings:

"Et creavit Deus hominem ad imaginem suam etc." et fortasse hoc voluit Plato di-	"God makes man in our image, etc." Perhaps this is what Plato wants to attest

44 Leone Ebreo, *Dialoghi d'amore di Maestro Leone Medico Hebreo* (Roma: per Antonio Blado d'Assola, 1535), 169*r:* "Et tu non m'hai mostrato che l'intelletto humano qualche vota mene in tanta perfettione, che si puo sollevare a copularsi con l'intelletto divino, overo angelico separato da materia, e finirlo in atto vedendolo direttamente, e non per discorso potenztiale, ne mezzo corporeo." On the light cf. also 196*v:* "Ma questo è come vedere il lucido corpo del Sole in acqua, o in altro diafano; perchè la debile vista non può vedere de diretto in se stesso: che cosi il nostro intelletto humano nelle corporee vede l'incorporee." Cf. also Thomas Aquinas, *Summa contra Gentiles* 3.43 and *Scriptum super Sententiis*, lib. 2, d. 13, q. 1, a. 3.

45 This kind of metaphor was already familiar in the Scholastic milieu: cf. Robert Grosseteste, *De Luce* and *Hexameron*; Thomas Aquinas, *Summa contra Gentiles* 3.65.

46 *Lumen Gentium*, 62*v*, and cf. 62*r*–62*v:* "Et hec sapienter docet Sacrum Genesis documentum 1 cum dicit: 'Deum creasse hominem ad eiusdem imaginem quasi ad eius similitudinem' ubi per divinam imaginem intelligit intellectum quo intelligibilia imaginamur, quem dicit esse divine nature et per consequens est quid immortale, et cum hoc docet quod per eiusdem intellective nature cognitionem quam per suam propriam actionem percipimus, nonnullam de Deo nedum de reliquis abstractis possumus habere cognitionem, cum scilicet huiusmodi naturam Dei imaginem appellet, inferens quod sit quodammodo divine nature, ad quod sequitur ut per eiusdem intellective nature cognitionem ad aliqualem de Deo cognitionem pervenire possumus. Quare Aven R. in commento libro De Anima I, 2 ait: 'et debes scire que iuvamentum scientie de anima ad alias scientias invenitur tribus modis etc.' subdit ibidem dicens: 'divinus autem sive metaphysicum suscipit ab ea substantiam subiecti sui, hinc enim declaratur quoniam forme abstracte sunt intelligentes, et alia multa de cognitione dispositionum consequentium intelligentiam in eo quod est intelligentia et intellectus etc.' Et idem videtur inferre ibidem III, 5 ubi dicit: 'et nisi esset hoc genus entium quod scimus per scientiam anime non possemus intelligere multitudinem in rebus abstractis quemadmodum nisi sciremus naturam intellectus non possemus intelligere quod virtutes moventes debeant esse intellectus etc.' cumque prefatus Genesis textus ibidem scilicet circa humanam productionem tamen non autem circa reliquas dicat: 'et dixit Deus faciamus' quibus videbatur quodammodo angelos alloqui, docet quidem huiusmodi intellectivam naturam ab eisdem angelis Dei iussu emanare ut supra, et hoc quia tunc Deus dedit eis potentiam influendi huiusmodi naturam quam Dei imaginem appellat ab eodem Deo tunc creatam, quare sequitur dicens 'et creavit Deus hominem ad imaginem suam etc' et fortasse hoc voluit Plato dicens quod Deus iussit angelis ut crearent mortalia prout Aven R. refert in commento Metaphysicorum XII, 44. Cumque idem Genesis documentum ibidem dicat 'quasi ad similitudinem nostram' videtur inferre arbitrariam voluntatem, qua fit quodammodo Deo similis et angelis per quanto est agens cum cognitione, cuius oppositum accidit naturali agenti, similitudo enim per formam formam vero per actiones consideramus. Merito tamen ponitur ly quasi inferens quod non est omnimoda similitudo, ab angelis enim differt in suis actionibus quoad arbitrium, a Deo vero quo ad arbitrii nobilitatem. Quibus demonstrative docet eiusdem Dei imaginis immortalitatem, et hoc quia cum sit quid divine nature et per consequens presit corporeis etiam celestibus licet sint

cens quod Deus iussit angelis ut crearent mortalia prout Aven R. refert in commento Metaphysicorum XII, 44.[46]	when he says that God ordered the angels to create the mortal beings, as Averroes reports in the commentary on *Metaphysics* 12.44.

In the Hebrew edition, Sforno quotes verbatim the passage from the Long Commetary on *Metaphysics*[47] in which Averroes mentions Plato's opinion according to which the creator created the angels himself. As Harvey also notes, this a good example of Sforno's skill as a biblical commentator. He gives new meaning to Genesis 1:26 by reading it in light of the Platonic text. He also gives a new monotheistic interpretation to the Platonic text:

And if so, also with regard to the form which it contains, those activities will be nobler than all the rest of the natural existents, although some of them are permanent, like the heavens and all their hosts, as it says [Deuteronomy 10:14–15]: "They belong to *ha-Šem*, your God: the heavens, and the highest heavens, the earth, and everything which is in it. Only to your fathers did ha-Shem have affection." Indeed, it says: "Let us make man," as though He was consulting His "familia," as [in the case of] the "words of our fathers" [i.e., the Rabbanan] of good memory. It seems that it teaches [us] that God—may He be praised—decrees then that there will be the creation of this intellectual potency and its emanation on man	ואם כן גם מצד צורתו אשר בה תהיינה אותן הפעולות יותר נכבד מכל שאר הנמצאים הטבעיים עם היות קצתם נצחיים כמו השמים וכל צבאם כאמרו הן לה' אלהיך השמים ושמי השמים הארץ וכל אשר בה רק באבותיך חשק ה' אמנם באמרו נעשה אדם כנמלך בפמילייא שלו כדברי רבותינו ז"ל נראה שהורה שגזר אז האל יתברך שתהיה בריאת זה הכוח השכלי ושפעו באדם מאת משרתיו הנבדלים או מאחד מהם ונתן אז בנבדל כוח להשפיעו כמו שהתבאר למעלה וזה אולי רצה אפלטון במאמרו אשר הביא אבן רשד בביאור ספר מה שאחר מאמר י"ב פרק מ"ד באמרו שהאל ברא את המלאכים בידו וצוה שיבראו את הנפסדים ונשאר הוא במנוחה[48]

quid immortale a fortiori quidem sequitur ut ipsa Dei imago his prestantior, habeat esse quid immortale."

47 Cf. Long Commentary on *Metaphysics*, 1653: "This has been explained by Nicolaos the Peripatetic in his book on the *Metaphysics*. This is why we see that the man who has attained this science has achieved the most perfect state of existence, and that it is the best of his actions because it is the action in which the best being shares. What is said on the authority of Plato's myth, namely that the creator created the angels himself and then entrusted them with the creation of mortal animals and remained himself idle and inactive is a myth which it is wrong to take as truth. It may be that such an imputation on the creator is the cause of the obligation of the Sabbath in the religion of the Jews."

48 *Or ʿAmmim*, 101.

from His separate servants or from one of them, and He gave [it] then in a separate potency to emanate it, as has become clear above. And this is perhaps what Plato wanted [to say] in his statement which Averroes mentioned in the commentary on *Metaphysics* 12, par. 44: "God created the angels by His hand and commanded that they create the corruptible things. And He stayed at rest."

The same passage of the commentary on *Metaphysics* 12.44 is also presented in the eighth chapter of *Light of the Nations*, "On the Divine Unity":[49]

Et fortasse id voluit Plato in sermone ab Aven R. allegato in commento *Metaphysicorum* XII, 44 ubi dicit : "Hoc autem quod Plato dixit in suis verbis obscuris, quod Creator creavit angelos manu, deinde precepit eis creare alia mortalia et remansit ipse in quiete sine labore, non est intelligendus ad litteram etc." Et hoc fortasse putavit Plato ductum illa modernorum ratione ab Aven R. ibidem relata, qua scilicet dicitur quod: "Prima substantia habet esse una et simplex in fine, quoque ab uno et simplici non provenit nisi unum."[50]

ואולי זה רצה אפלטון במאמרו אשר הביא אבן רשד בביאור ספר מה שאחר מאמר י"ב פרק מ"ד באמרו אמנם זה שאמר אפלטון במאמריו הסתומים שהבורא ברא את המלאכים בידיו ואחר ציוה להם שיבראו ההווים ונפסדים ונשאר הוא במנוחה אין ראוי להבינו כפשוטו עכ"ל. ואולי הביא את אפלטון לחשוב זה סברת קצת אחרונים שאמרו שראוי שהעצם הראשון יהיה פשוט בתכלית הפשיטות ושמן האחד הפשוט לא יתהוה כי אם אחד אמנם כי זאת הסברא עם היותה בלתי מופתית אלא שהיא דעת אויצינא וזולתו מהאחרונים[51]

And this is probably what Plato wanted [to explain] in the speech which Averroes quoted in the commentary on *Metaphysics* 12.44, in which he says: "This is what Plato said in his obscure words: the creator created the angels by his hands, and afterwards he ordered them to create the mortal beings and he re-

And perhaps this is what Plato wanted [to explain] in his statement which Averroes mentions in the commentary on *Metaphysics* 12,44: "This is what Plato said in his obscure statements: the creator created the angels with His hands and afterwards He gave them the order that they should create the generable

49 *Or ʿAmmim*, 59–64; *Lumen Gentium*, 45*r* [49]–47*v* [53].
50 *Lumen Gentium*, 46*r* [51].
51 *Or ʿAmmim*, 62.

mained at rest without working. It is not appropriate to understand this literally." And this is probably what Plato thought, according to what Averroes explains considering the position of some later philosophers saying that the first substance will be one and absolutely simple, because from one comes forth the one and the simple.

and corruptible [things], and he remained at rest. It is not appropriate to understand this literally." And perhaps this brought Plato to consider the opinion of some later [philosophers] who said that it is appropriate that the first substance is absolutely simple, the simplicity, and that from a simple being nothing will be generated except [another] one. Yet this opinion is not demonstrative, but only the opinion of Avicenna, and, besides him, of other later [philosophers].

Sforno not only affirms that angels may be considered as the secondary gods (as in the Platonic tradition), but also that this vision is consistent with the account in Genesis. Hence, there is no contradiction between the Jewish tradition and the philosophical position, because the intellectual natures are emanated directly from the Creator.

Sforno uses the Platonic idea to interpret a biblical verse, since he quotes Plato (or better, the interpretation given by Averroes via Themistius) to clarify Genesis 1:26. In *Or ʿAmmim*, Sforno quotes not only Genesis 1:26, but also Psalm 148:2–6, distinguishing the activities of men from the Divine actions and additionally the difference between men and angels:

And when it says "like our likeness," it makes known that there will be an agent [acting] by knowledge and consciousness. And therefore, it will be similar to God—may He be praised—and His angels. And it is not right that this will only happen naturally to the rest of the acting existents, because the likening will be by form. And the form is known by the activities which follow from it. But it says "like our likeness," together with the [letter] "Kaf" of the resemblance ["like"], to teach [us] that the likening is not [a] perfect [identity]. And this is so

ובאמרו כדמותנו הודיע שיהיה פועל בידיעה והכרה ובזה ידמה לאל יתברך ולמלאכיו ולא כן יקרה לשאר הנמצאות הפועלות בטבע בלבד. כי אמנם ההדמות יהיה בצורה והצורה תוודע בפעולות הנמשכות ממנה אמנם אמר כדמותנו עם כ"ף הדמיון להורות שההדמות הוא בלתי שלם. וזה כי האדם הוא נבדל בפעולותיו בצד מה מהאל ית' עם היותו דומה לאל יתברך בענין הבחירה. וזה כי בחירת האל יתברך היא לעולם לטוב כי אמנם הטוב יהיה מצד התכלית. והנה אין תכלית יותר נכבד מהפקת רצונו יתברך אבל האדם עם היותו בחיריי לפעמים יהיה תכליתו בלתי טוב וזה בנטותו מהרצון האלהי הנ"ל. כאמרו הדרכי לא יתכנו בית ישראל הלא

52 *Or ʿAmmim*, 100.

דרכיכם לא יתכנו. ובצד מה הוא נבדל מהמלאכים בפעולותיו עם היותו דומה להם בענין היותו פועל בידיעה והכרה. כי אמנם פעולות המלאכים עם היותם בידיעה והכרה הם בלתי בחירה מאתם. וזה כי בהיותם שכל בפועל לעולם בלתי התנגדות כוח מתאוה תהיה פעולתם ישרה לעולם ולא יקרה בהם החטא הקורה בפעולות האדם לסיבת המתאוה והתרשלות השכל מלהתקומם מבלתי היותו שכל בפועל לעולם כמו שהעיד המשורר באמרו הללוהו כל מלאכיו כו'[52]

because man, by his activities, is separate in a certain way from God—may He be praised—although he is similar to God—may He be praised—in the case of free choice. And this is so because the free choice of God—may He be praised—is always for the good, because indeed, the good will be in accordance with the goal. And see: there is no nobler goal than the bringing out of His will—may He be praised. But man, although he also choses freely, sometimes has "a goal that is not good". And this is so because of his inclinations [to deviate] from the above-mentioned divine will, as it says [Ezekiel 18:29]: "Is my way not just, House of Israel? Is it not your ways that are unjust?" And in a certain way, he is separate from the angels because of his activities, although he is similar to them because he is an agent in knowledge and consciousness, because indeed, the activities of the angels, although they have knowledge and consciousness, are not freely chosen by them. And this is so because they are an eternally in act without the opposition of an appetitive faculty. Their activities will forever be right and the sin which occurs in the activities of man as caused by the appetitive [faculty] and the intellect neglecting to stand against [the state of] not forever being an intellect in act will not happen to them, as the poet testifies [Psalms 148:2, 6]: "Praise Him all His angels, etc."

On the one hand, the angels are the separate intellects[53] and the pure separate forms; on the other hand, men, who are similar to God and angels, sharing the intellectual

53 For the exegetical interpretation in Genesis, see Norbert M. Samuelson, *Judaism and the Doctrine*

soul with them, are superior to the latter, because human beings possess free will, which gives them a superior nature.[54]

In the *Kitvei*, Sforno shares the same ideas: God gave man an opening to acquire knowledge regarding entities that are separate from matter (i. e., angels) through the knowledge of our souls; for instance, the angels are substances separate from matter by definition:

In our image. [Who shall be] a substance that is everlasting [endowed with] reason; and thus God, the Blessed One, gave man an opening in His Torah to acquire knowledge regarding those separate from matter through the knowledge of our souls.	**בצלמנו.** שהוא עצם שכלי ונצחי ובזה פתח האל ית' פתח בתורתו לקנות ידיעה בעצמים הנבדלים בידיעת נפשנו[55]

Praise Him, all His angels, all substances which are separate from matter.	**הללוהו כל־מלאכיו**, כל שאר העצמים הנבדלים מחומר.
All His hosts, the stars. And both	**כל־צבאיו**, הכוכבים. ועל כולם
Sun and moon, whose actions are well-known, and after them	**שמש וירח**, שפעולתם יותר מפורסמת ואחריהם – **כל־כוכבי אור**
	שמי השמים, הגלגלים שהם שמים בגובה גם בערך אל

of Creation (Cambridge: Cambridge University Press, 1994), particularly 142: "The classical Jewish philosophers identify them with the separate intellects of the astronomy of their time. In the case of the sages and Rashi, these substances, who are second in power and perfection only to God, turn out to be reified virtues."

54 On the *imago Dei*, cf. the studies of Warren Zev Harvey, particularly his lecture at the second international symposium.

55 Genesis 1:26.

56 Cf. *Kitvei Rabbi Obadiah Sforno*, ed. Ze'ev Gottlieb (Jerusalem: Mossad ha-Rav Kook, 1987), Genesis 1:6–7:

יהי רקיע בתוך המים – יהי טבע בתוך המים היסודיים כמו גלגל בתוכם סביב, מבדיל בצורה קצתם מקצתם, באופן שאיזה חלק עליון מן המים לצד האויר ישוב לטבע אידיי. והנה בזה עלו בהכרח אל איזה גבול באויר היסודיי, וקבל האויר איזה עיסוי בהכרח להיות מקום לחלק שנהפך לאיד, והתפשט אל מקום רב הכמות ממקומו הראשון.

"Let the nature of the elemental waters become as though the form of a sphere is girdling it, separating one part from the other, in such manner that a portion of the upper waters adjacent to the air mass change once again to the nature of vapor, this air will now perforce somehow be made into a place for the portion of water which changed to vapor. It will expand considerably into a larger area than it was originally."

ויעש אלהים את הרקיע – ובהיות שכאשר סרו קצת המים היסודיים מתחת אותו החלק מהם שנהפך לטבע אידיי, כמו שהיה באמרו "יקוו המים מתחת השמים" (פסוק ט'), היה ראוי שירד אז החלק האידיי אל אותו המקום שסרו אותם המים משם, ועשה שאותו "הרקיע" המבדיל יהיה בו כח עוצר ומונע את החלק האידיי שלא ירד, והוא "המים אשר מעל

all your stars of light.
You heavens of the heavens, the spheres which are the highest heavens over the firmament are called heavens, **and you waters above the heavens**, the air [portion] of the waters which are the water above the firmament,
He also established them forever and ever.
Let them praise the name of the Lord, which is the perfect Name.
For He commanded and they were created, He did not speak before.
He also established them forever and ever, which are eternal and incorruptible forever.

הרקיע הנקרא שמים, והמים אשר מעל השמים,
האויר המימיי אשר הם המים אשר מעל לרקיע,
ויעמידם לעד לעולם[56]
יהללו את־שם הי, שיהיה אז השם שלם. ויהללו כמו כן
כי הוא צוה ונבראו, לא מדבר קודם ויעמידם לעד לעולם
שיהיו נצחיים בלתי נפסדים לעולם[57]

לרקיע", באופן שירד החלק האוירי המעוסה, ונשאר האידיי במקומו הראשון. ולזה בהגיע שם האיד הלח, יתעבה ויוליד הגשם והשלג והברד, ובהתעבותם יכבדו וירדו, כאמרו "לקול תתו המון מים בשמים" (ירמיהו י':י"ג). רצונו באמרו "השמים", הרקיע המעבה את החלק האוירי, כאמרו "ויקרא אלהים לרקיע שמים" (פסוק ח'). ובהגיע שם האיד הקיטורי הנלהב יוליד רעם וברק, כאמרו "ויעלה נשאים מקצה הארץ ברקים למטר עשה" (ירמיהו שם). ובהיות קצת היסוד המימיי הכבד למעלה מהאויר הקל אשר אצלנו נגד טבעם, יורה על פעולת פועל רצוניי מכוון אל תכלית בלי ספק, כאמרו "ומעשה ידיו מגיד הרקיע" (תהלים י"ט:ב').
ויהי כן – נשאר קיים כן נגד טבעו.

"Now, since the rest of the foundation waters—which remained below those waters that vaporize—gathered together, as it says, let the waters beneath the heaven be gathered, it should follow that the upper waters which became vapor would have filled the vacuum left by the departure of the waters. However, He made the firmament which separated (the higher and lower waters), in such manner that it was given the power of restraint, preventing the vapor portion—that is, the water which was above the firmament—from descending; in a manner that the fashioned atmosphere did fill the void while the vapor remained in its initial position. Now, when the moist vaporized water becomes dense, it brings forth rain, snow and hail as it becomes laden heavily (with water) and descends (to the earth), as it is said, 'at the sound of His placing an abundance of water in the heavens' (Jeremiah 10:13). What is meant by the term 'heavens': is the firmament which holds back the vapor portion, as it is said 'And God called the firmament "heavens" (Genesis 1:8). And when the smoke-laden mist comes there, it results in thunder and lightning, as it is said: 'He lifts up clouds from the ends of the earth, thunder from the rain He makes' (Jeremiah 10:13). Since the elemental waters, which are denser, are above the light air, this condition is contrary to nature, indicating that is an act performed by God's Will, directed without a doubt toward a purpose and end as it is said, 'the work of His hands shall the firmament declare' (Psalms 19:2)."

57 Psalm 148:1–2.

In the biblical verses, we find alternate explanations of the same philosophical concepts found in *Light of the Nations*. As in his philosophical work, Sforno considers the angels to be separate substances; they share their specific nature with the celestial bodies, similarly separate from matter. In his interpretation of the Psalm, Sforno focuses on the creation of the cosmos, i.e., the stars and the firmament, referring once again to Genesis 1:2 and 1:7, but also to *On the Heavens* 2:42. It is also interesting to note that Sforno uses a similar idea in *Or ʿAmmim.*[58]

And He provides the proof of this [Genesis 1:2]: "And the earth was Tohu and Bohu," that is to say: because the first form was necessarily in the prime matter specialised for it, and the composition of them was called "Tohu and Bohu." And from it, through change in different ways and in different parts of them, darkness, which is air deprived of light [ibid.], was generated "above the abyss," which is made up of the two heavy elements [i.e., earth and water], which are next to the centre. And he continues: "The spirit of God was hovering on the surface of the water," to convey elementary fire, that it was [generated] by the propulsion of a separate substance, moving and hovering over the darkness on the surface of the water, because the separate substances are called "the spirit of God," as told in the story of the work of creation [Psalms 104:4]: "He makes His angels spirits." And it testifies, if so, that He hovered over the air and move it on the water. And therefore, part of it became moist by its proximity to the water, and part of it (became) warm and dry, taking on a fiery form by its proximity to the sphere and its movement, as Aristotle explains the cause of elementary fire in *On the Heavens* 2.42.	ועל זה הביא מופת באמרה והארץ היתה תו ובהו כלומר כי בהכרח היתה הצורה הראשונה בחומר הראשון צורה מיוחדת לו נקרא המורכב מהם תהו ובהו וממנו נתהוה בהשתנותו בפנים מתחלפים ובחלקים מתחלפים ממנו חשך שהוא אויר נעדר האור על פני תחום שהם הב' יסודות הכבדים אשר אצל המרכז והוסיף ואמר ורוח אלהים מרחפת על פני המים להבין ולהורות הויית האש היסודיי שהיה בהנעת עצם נבדל מניע ומרחף את החושך על פני המים כי אמנם העצמים הנבדלים יקראו רוח אלהים באמרו בסיפור מעשה בראשית עושה מלאכיו רוחות. הודיע אם כן שרחף והניע האויר על המים ובכן היה קצתו לח מצד קרבתו אל המים וקצתו חם ויבש לובש צורה אשיית מצד קורבתו אל הגלגל ותנועתו כמו שביאר אריסטו' בסיבת האש היסודיי בספר השמים והעולם מאמר ב' פרק מ''ב.

58 *Or ʿAmmim*, 32.

Sforno tries to harmonise philosophy and the Torah in a double sense: from one side, in his *Kitvei*, he gives some philosophical doctrines belonging to the Aristotelian tradition in order to explain the biblical verses. On the other, he concludes each chapter of his philosophical enquiry with a biblical verse to show that only with the light of the Torah[59] is it possible to understand the ultimate end of the cosmos and of human beings:

But many beautiful words are found in the Holy Scripture which teach [us] great things about righteousness in ethics and politics to improve and orderly arrange both soul and body to turn them towards the [right] way. But it is good for those of clean hearts to present doubts, flaming arrows which disturb [us] and cause sorrow to the soul, at the beginning of the inquiry and to overwhelm them all with pure words and with clear arguments about all that is mysterious. These [words] are for the life of the [future] world. They bring forth justice in truth. This shall be told about ha-Shem to the [coming] generation.	הן רבים אמרי שפר נמצאים בספרי קדש מורים מדות והנהגות נוראו[ת] בצדק לתקן ולסדר מנפש ועד בשר להטותם הדרך אך טוב לברי לבב[60] לערוך ספקות מאזרי זיקות מציקות ומדיבות נפש בתחלת העיון ולהכריע את כלם באמרות טהורות עם ראיות ברורות על כל נעלם אלה לחיי עולם לאמת יוציאו משפט יספר לה' לדור

Bibliography

Primary Sources

Aquinas, Thomas. *Aristotle's De Anima in the Version of William of Moerbeke and the Commentary of St. Thomas Aquinas*. Translated by Kenelm Foster and Sylvester Humphries. Reprint ed. New Haven: Yale University Press, 1959.

Averroes. *Aristotelis Metaphysicorum libri XIIII. Cum Averrois Cordubensis in eosdem commentariis et epitome Theophrasti metaphysicorum liber*. Venice: s.n. 1552.

Averroes. *Aristotelis Physica, De celo et mundo, De generatione et corruptione, De anima, De sensu et sensato, De memoria et reminiscentia, De somno et vigilia, De longitudine et brevitate vitae, De meteoris, cum Averrois Cordubensis expositionibus; Liber Averrois de substantia orbis*. Translated by Michael Scotus. Venice: s.n. 1473.

59 *Or ʿAmmim*, 9.

60 As Harvey suggests, this is an implicit reference to Psalm 73:1 ("Surely God is good to Israel, even to such as are pure in heart") and Isaiah 50:11.

Averroes. *Destructio destructionum philosophiae Algazelis*. Agostino Nifo. Venice, s.n. 1497.
Averroes. *Ibn Rushd's Metaphysics: A Translation with Introduction of Ibn Rushd's Commentary on Aristotle's Metaphysics, Book Lām*. Edited by Charles Genequand. Leiden: Brill 1984.
Ebreo, Leone. *Dialoghi d'amore di Maestro Leone Medico Hebreo*. Rome: per Antonio Blado d'Assola, 1535.
Gersonides (Levi ben Gershom). *The Wars of the Lord*. Translated by Seymour Feldman. 3 vols. Philadelphia: Jewish Publication Society, 1984–99.
Maimonides, Moses. *The Guide of the Perplexed*. Translated by Shlomo Pines. 2 vols. Chicago and London: University of Chicago Press, 1963–78.
Nifo, Agostino. *Commentationes in librum Averrois de Substantia Orbis*. Venice, s.n.1508.
Pico della Mirandola, Giovanni. *Conclusiones sive Theses DCCCC: Romae anno 1486 publice Disputandae, sed non admissae*. Edited by Bohdan Kieszkowski. Geneva: Droz, 1973.
Sforno, Obadiah. *Kitvei Rabi Obadiah Sforno*. Edited by Ze'ev Gottlieb. Jerusalem: Mossad ha-Rav Kook, 1987.
Sforno, Obadiah. *Commentary* on the *Torah*. Translated by Raphael Pellcovitz. Brooklyn, New York: Mesorah Publications, 1997.
Sforno, Obadiah. *Lumen Gentium*. Bologna, s.n. 1548.
Sforno, Obadiah. *Or ʿAmmim*. Bologna, s.n. 1537.
Themistius. *On Aristotle On the Soul*. Translated by Robert B. Todd. Paperback ed. London: Bloomsbury, 2014.

Secondary Sources

Akasoy, Anna, and Guido Giglioni, eds. *Renaissance Averroism and Its Aftermath: Arabic Philosophy in Early Modern Europe*. Dordrecht: Springer, 2013.
Bonfil, Robert: "Il Rinascimento: la produzione esegetica di 'O. Servadio Sforno." In *La lettura ebraica delle Scritture*, edited by Sergio J. Sierra, 261–77. Bologna: EDB, 1995.
Bonfil, Robert. "The Doctrine of the Soul and Holiness in the Teachings of Obadia Sforno." [Hebrew.] *Eshel Beer Sheva* 1 (1976): 200–257.
Frank, Daniel H., Oliver Leaman, and Charles H. Manekin, eds. *The Jewish Philosophy Reader*. New York: Routledge, 2000.
Harvey, Warren Zev. "Maimonides' First Commandment, Physics, and Doubt." In *Hazon Nahum: Studies in Jewish Law, Thought, and History Presented to Dr. Norman Lamm*, edited by Yaakov Elman and Jeffrey S. Gurock, 149–62. New York: Michael Scharf Publication Trust of the Yeshiva University Press; Hoboken, NJ: Ktav, 1997.
Meyrav, Yoav. "Spontaneous Generation and Its Metaphysics in Themistius' Paraphrase of Aristotle's *Metaphysics* 12." In *Aristotle Re-Interpreted. New Findings on Seven Hundred Years of the Ancient Commentators*, edited by Richard Sorabji, 195–210. London and New York: Bloomsbury Academic, 2016.
Montada, Josep Puig. "Eliahu del Medigo, the Last Averroist." In *Exchange and Transmission across Cultural Boundaries: Philosophy, Mysticism and Science in the Mediterranean World*, edited by Haggai Ben-Shammai, Shaul Shaked, and Sarah Stroumsa, 155–86. Jerusalem: Israel Academy of Sciences and Humanities, 2013.
Nardi, Bruno. *Studi su Pietro Pomponazzi*. Florence: Le Monnier, 1965.
Roth, Cecil. *The Jews in the Renaissance*. Philadelphia: Jewish Publication Society, 1959.
Samuelson, Norbert M. *Judaism and the Doctrine of Creation*. Cambridge: Cambridge University Press, 1994.
Veltri, Giuseppe. *Alienated Wisdom: Enquiry into Jewish Philosophy and Scepticism*. Berlin and Boston: De Gruyter, 2018.

Zonta, Mauro. *Hebrew Scholasticism in the Fifteenth Century. A History and Source Book.* Dordrecht: Springer, 2006.

Zonta, Mauro. "The Autumn of Medieval Jewish Philosophy: Latin Scholasticism in 15th-Century Hebrew Philosophical Literature." In *"Herbst des Mittelalters"? Fragen zur Bewertung des 14. und 15. Jahrhunderts*, edited by Jan Aertsen and Martin Pickavé, 474–92. Berlin and New York: De Gruyter, 2004.

José María Sánchez de León Serrano

Arnold Geulincx: Scepticism and Mental Holism

1 Introduction

The Flemish thinker Arnold Geulincx (1624–69) is known among historians of philosophy as one of the founders of modern occasionalism, the view that God alone, not creatures, exerts causal efficacy.[1] Being counted among the modern occasionalists, Geulincx has also been associated with a wide variety of philosophical tendencies. Whereas for some, he is merely a minor disciple of Descartes, for others, he is the missing link between Descartes and Spinoza.[2] A certain line of interpretation (notably represented by Cassirer) sees Geulincx as a precursor of Kant.[3] According to de Vleeschauwer, Geulincx is the first thinker who explicitly equated knowing and doing, thus paving the way for the constructivism of Giambattista Vico and Benedet-

1 Compare this brief characterisation of occasionalism with that of Clatterbaugh in his *The Causation Debate in Modern Philosophy 1637–1739* (New York: Routledge, 1999), 97. The other founders of modern occasionalism were Louis de La Forge (1632–66) and Géraud de Cordemoy (1626–84), although Geulincx developed his own occasionalistic conception independently of them. See Jean-Christophe Bardout, "Occasionalism: La Forge, Cordemoy, Geulincx," in *A Companion to Early Modern Philosophy*, ed. Steven Nadler (Boston: Blackwell Publishers, 2002), 140–51. Modern occasionalism would reach its apex with Nicolas Malebranche (1638–1715); regarding the affinities and divergences between Geulincx's and Malebranche's occasionalism, see Steven Nadler, "Knowledge, Volitional Agency and Causation in Malebranche and Geulincx," *British Journal for the History of Philosophy* 7, no. 2 (1999): 263–74. I have explicitly said "modern" occasionalism, for a variation of this view (or group of related views, as occasionalism is not a homogeneous trend or school) already existed in medieval Muslim philosophy. See Ludwig Stein, "Antike und mittelalterliche Vorläufer des Occasionalismus," *Archiv für Geschichte der Philosophie* 2, no. 2 (1888): 193–245; also Dominik Perler and Ulrich Rudolph, *Occasionalismus: Theorien der Kausalität im arabisch-islamischen und im europäischen Denken* (Göttingen: Vandenhoeck & Ruprecht, 2000). For a thorough reconstruction of the emergence of occasionalism after Descartes, see Rainer Specht, *Commercium mentis et corporis. Über Kausalvorstellungen im Cartesianismus* (Stuttgart-Bad Canstatt: Friedrich Frommann Verlag [Günther Holzboog], 1966).

2 This is Bernard Rousset's main thesis in his posthumous work *Geulincx entre Descartes et Spinoza* (Paris: Vrin, 1999). In the early eighteenth century, Geulincx was accused of Spinozism by Christian Thomasius and Ruarda Andala. His philosophy has been suspected of Spinozism ever since. See H. J. de Vleeschauwer, "Three Centuries of Geulincx Research: A Bibliographical Survey," *Mededelings van die Universiteit van Suid-Afrika* 1 (1957): 1–72, and especially Han van Ruler, "Geulincx and Spinoza: Books, Backgrounds and Biographies," *Studia Spinozana* 15 (2006): 89–106.

3 However, Cassirer also stresses the large gap between Kant and Geulincx. See Ernst Cassirer, *Das Erkenntnisproblem in der Philosophie und Wissenschaft der neueren Zeit*, 3rd ed. (Berlin: Verlag Bruno Cassirer, 1922), 1:552.

https://doi.org/10.1515/9783110618839-009

to Croce.[4] Despite these multiple affiliations and fatherhoods—or perhaps because of them—Geulincx's philosophy has rarely been studied for its own sake and remains widely unknown.[5] With the exception of de Lattre's seminal work,[6] we are still awaiting a thorough study of Geulincx's multifaceted thought.

Without pretending to fill this lacuna, the present study will focus on Geulincx's remarkable approach to human cognition and its extent, which constitutes the basis of his entire philosophy, including the occasionalism for which he is known. Geulincx's position in this regard is at first glance extremely ambivalent, if not incoherent. On the one hand, he seems to deny the possibility that the human mind may attain genuine knowledge and to embrace a hopeless scepticism. On the other, he presents his own philosophy as a true metaphysics, which he defines as knowledge of things as they are in themselves, independent of our consideration. Moreover, in his metaphysics, he endorses a quasi-pantheistic conception—quite reminiscent of Spinoza's—that *prima facie* is the antipode to scepticism. I will try to make sense of these disparate claims and argue that they actually form a coherent whole. In particular, I will show how Geulincx's sceptical distrust of our cognitive powers leads him to embrace a form of mental holism, according to which the human mind is just a parcel or sub-region within an infinite intellect. Furthermore, I will examine how our containment within God's mind enables us to grasp necessary truths but prevents us from understanding the created world, whose existence is a contingent fact that is unaccountable by reason. The example of Geulincx will thus reveal that a fecund scepticism can, by discarding certain objects as unknowable, open up areas of exploration that were hitherto considered impenetrable. I will structure my exposition as follows. In the next section, I will examine Geulincx's general position on human cognition and its apparent inconsistencies. In the third section, I will reconstruct Geulincx's reasoning process in the first part of his *True Metaphysics* (*Metaphysica vera*), which seamlessly moves from an initial scepticism to ascer-

---

4 See H.J de Vleeschauwer, "Les antécédants du transcendentalisme. Guelincx et Kant," *Kant-Studien*, 45, 1–4 (1953): 246–47.

5 Geulincx's only significant follower was Richard Burthogge (1637/38–1705), who studied at the University of Leiden in 1661 and almost certainly attended Geulincx's lectures. Geulincx's influence is clearly recognisable in his *Essay upon Reason and the Nature of Spirits* (London: Dunton, 1694), which he dedicated to John Locke. Regarding Burthogge's thought in connection with Geulincx's, see Cassirer, *Erkenntnisproblem*, 1:543–53; Michael R. Ayers, "Richard Burthogge and the Origins of Modern Conceptualism," in *Analytic Philosophy and History of Philosophy*, ed. Tom Sorell and G.A.J. Roger (Oxford: Clarendon, 2005), 179–200. Furthermore, scholars have unsuccessfully attempted to establish a link between Geulincx and Leibniz, mainly due to Leibniz's simile of the two clocks, which also appears in Geulincx, despite the fact that Eduard Zeller conclusively demonstrated in 1884 that there is no such influence. See Eduard Zeller, "Über die erste Ausgabe von Geulincx' Ethik und Leibniz' Verhältnis zu Geulincx' Occasionalismus," *Sitzungsberichte der Königlich Preußischen Akademie der Wissenschaften zu Berlin* 29, no. 2 (1884): 673–95.

6 Alain de Lattre, *L'occasionalisme d'Arnold Geulincx* (Paris: Éditions de Minuit, 1967).

taining our containment in God's mind.[7] In the fourth section, I will examine Geulincx's demonstration of the existence of a corporeal world by means of the notion of time. In the fifth and final section, I will focus on some aspects of Geulincx's critique of Scholastic philosophy which indirectly illuminate his own philosophical approach.

2 Scepticism and the Priority of the Knower over the Known

Geulincx's epistemological position is summarised in an explanatory note from his short treatise *Metaphysics according to the Peripatetic Mind* (*Metaphysica ad mentem peripateticam*).[8] Neither the nature of this treatise (which I will briefly discuss in the last section) nor the context in which Geulincx inserts this explanatory note is important for us now. The text in question reads as follows:

> We should not consider things insofar as they are sensible (i. e., insofar as they enter into the senses under a certain aspect), nor insofar as they are intelligible (i. e., insofar as we think of them in a certain way). Yet we cannot consider them as they are in themselves [*ut sunt in se*], from which we see our great imperfection. Therefore, it only remains for us to do (which we can and must do) that in the judgment of the mind, whenever we apprehend something under a certain mode of our thought (which we always do, and we cannot do otherwise while we are human beings), we always keep in mind that the thing is not in itself as it is apprehended by us. Even if we always attribute the appearances of the senses and the intellect to the things themselves, there is nevertheless something divine in us that always tells us that it is not thus, and only in it does our wisdom [*sapientia*] consist, insofar as we are human beings.[9]

This passage distinguishes three ways of considering things: as sensible, as intelligible, and as they are in themselves. The first two ways show how things are *in relation to us*, both to our senses and to our intellect. Due to their relative character, these two ways of considering things do not count as real knowledge. Only the third way can

---

7 All references to Geulincx's works are from the standard edition: *Arnoldi Geulincx Antverpiensis Opera Philosophica*, ed. Jan Pieter Nicolaas Land, 3 vols (Den Haag: M. Nijhoff, 1891–93); henceforth OP. The English translations of Geulincx in this paper are mine.

8 OP 2:300–301; ad pag. 200.

9 OP 2:300–301; ad pag. 200: "Nos non debemus res considerare prout sunt sensibiles (id est, sub certa specie incurrunt in sensum); neque ut sunt intelligibiles (id est, sub certo modo a nobis cogitantur). Sed ut sunt in se, non possumus eas considerare; unde videmus magnam nostram imperfectionem. Hoc unum igitur restat nobis faciendum (quod et possumus et debemus facere), ut judicio mentis, quotiescunque rem aliquot sub modo aliquot cogitationis nostrae apprehendimus (quod equidem semper facimus, nec possumus aliter dum homines sumus), semper hoc teneamus, rem non esse ita in se, ut apprehenditur a nobis. Etiamsi nos semper phasmata sensus et intellectus ipsis rebus tribuamus; tamen est aliquid divinum in nobis, quod semper dicit nobis, non esse sic; et in hoc unico consistit nostra, quatenus homines sumus, sapientia."

rank as knowledge proper, insofar as it apprehends things as they are in themselves; i.e., independent of our consideration. Yet this description already shows that knowledge is out of our reach, for we cannot, as it were, go beyond ourselves and consider things from a standpoint that differs from ours. We seem, therefore, to be trapped in a deadlock and to be unable to overcome our distorted view of things. In view of this predicament, Geulincx argues, there is only one thing that we can (and must) do; namely, to always bear in mind that things are not the way that we perceive and conceptualise them. This awareness (or scruple) is the only knowledge available to us and is something divine in us (*aliquid divinum in nobis*), for in a way, it raises us above our cognitive imperfection, yet without thereby remedying our condition.

Judging by this passage, Geulincx's position can be described as a *resigned scepticism*. Yet his statements in other passages, as well as his attempt to develop a metaphysical system, seem to contradict the epistemological stance described above. In another explanatory note, he affirms that a true metaphysics considers things as they are in themselves, independent of us.[10] Moreover, he titles the exposition of his own philosophy *True Metaphysics*, thus suggesting that he is in possession of the knowledge that—according to the passage quoted above—human beings are constitutively unable to attain. Does this mean that Geulincx's views concerning human knowledge are simply inconsistent? In what follows, I will show that this is not the case and that Geulincx's metaphysical endeavour is actually a sagacious solution to the predicament described above. Between the mere consideration of things as they appear to us and the (unattainable) apprehension of things as they are in themselves, there is, for Geulincx, a third possibility. This third way refrains from considering *things* and focuses on the consideration of *ourselves*, i.e., of our minds regardless of their particular contents. Given the general uncertainty of our knowledge, our only certainty is the mere fact that we think and perceive, irrespective of the objects that are thought and perceived. Geulincx calls this fact the "prerogative of the knower over the known."[11] Certainly, knowledge requires both a knower and a known thing. Yet whereas the latter can somehow be *absent*—for instance, when we err—the former must necessarily exist. Therefore, the only possible way of expanding our knowledge, once we realise that things are *not* the way we apprehend them, is by taking our own thought as the point of departure for our philosophical inquiry. For Geulincx, sceptical doubt is thus the negative flipside of the knower's priority over the known things. Knowledge does not have its source in the known objects, but in the knower's *self-certainty*. Thus, our entire knowledge of the world must exclusively derive from a careful inspection of ourselves, without (unjustifiably) assign-

10 OP 2:287.
11 OP 2:267.

ing any reality to our perceptions beyond our consciousness.[12] This will be Geulincx's strategy in his *True Metaphysics*, as I will show in the next section.

Before examining Geulincx's self-explorative approach more closely, we must clear up a possible misunderstanding. By proposing an exploration of our own thought, Geulincx might seem to be engaging in the second kind of consideration of the three listed above; i.e., the consideration of intelligible things. Yet Geulincx does not simply dismiss sensible knowledge in favour of intellectual knowledge, understanding the latter as a knowledge by means of concepts. For him, both of them are equally unreliable, insofar as they are concerned with particular contents—be they sensible or intellectual—thus giving precedence to the known over the knower. Therefore, the self-inspection that Geulincx proposes does not have the meaning of an inquiry into our basic set of categories or our conceptual apparatus. For the same reason, for him, the phrase "self-inspection" does not designate an investigation of the human soul as a specific object among others or as a specific realm of reality differentiated from others, for such an investigation would presuppose a prior classification of beings, a preexistent map of reality, as it were, which we uncritically take for granted before exploring our own thought. This kind of implicit assumption is precisely what Geulincx's "prerogative of the knower over the known" calls into question.[13] Hence, the only valid starting point of knowledge is the consideration of our thought *simpliciter*, without further qualifications. One might rightly observe that if we put aside our entire set of basic concepts, the philosophical investigation is left with nothing. If our only certainty is the mere fact that we think, regardless of *what* we think and *how* we think, we can justifiably claim that our initial certainty is almost nil. Yet this objection, far from undermining Geulincx's endeavour, actually strengthens it. At the beginning of the philosophical inquiry, Geulincx demands: "Before entering metaphysics, you must consider yourself empty of all knowledge."[14] The knowledge that Geulincx is seeking cannot depend on a prior knowledge, for that would contradict the prerogative of the knower over the known. Therefore, and in

---

12 In Geulincx's *Ethics*, self-inspection is—together with self-contempt—one of the constituents of humility, which he views as the most important of the four cardinal virtues (that is: diligence, obedience, justice, and humility). Self-inspection thus constitutes the foundation of both true cognition and ethics. See OP 3:30–37. Regarding the ethical significance of self-inspection in Geulincx, see Andrea Sangiacomo, "Defect of Knowledge and Practice of Virtue in Geulincx's Occasionalism," *Studia Leibnitiana* 46, no. 1 (2014): 46–63.

13 One might wonder to which extent Geulincx can actually carry out this questioning of given categories, for he presupposes the real distinction between the corporeal and the mental, in which respect he still moves within a Cartesian framework. See the conclusion to this paper. For a thorough comparison between Descartes and Geulincx regarding their criticism of received categories and notions, see Mark Aalderink's *Philosophy, Scientific Knowledge, and Concept Formation in Geulincx and Descartes* (Utrecht: Zeno, 2010).

14 OP 2:140: "[...] necessum est, ut ante ingressum in Metaphysicam consideres the ipsum ut vacuum ab omni scientia [...]"

order to attain indisputable certainty, the philosophical inquiry must start from a *zero-point of knowledge*.

3 Geulincx's Autology: Self-Inspection and Mental Holism

Geulincx's main philosophical work, the treatise *True Metaphysics*,[15] begins with an *Autology* (or self-knowledge), continues with a *Somatology* (or knowledge of the corporeal world), and concludes with a *Theology* (or knowledge of God).[16] In this chapter, I want to focus on the first sections of his *Autology* and show how Geulincx's initial scepticism seamlessly expands into the affirmation of an all-encompassing mind, of which the human mind is a limited parcel or sub-region. As already noted, the priority of the knower over the known dictates that a true metaphysics cannot be derived from a previous knowledge. Thus, in a similar way to Descartes in his *Meditations* (and obviously following his example), Geulincx establishes that we must doubt everything before entering the path of true knowledge. Yet he insists that this procedure has nothing to do with the usual attitude of the sceptics, who do not aspire to knowledge and are content with remaining in permanent doubt. In contrast to the sceptics, Geulincx intends to attain genuine knowledge while at the same time remaining within the limits of the sceptical scruple described above. Therefore, the general doubt that he proposes at the beginning of his *True Metaphysics* is only the threshold to a fundamental, radical knowledge.[17]

Geulincx loosely adopts Descartes's meditative approach in the first reflections of his *Autology*, only to derive quite different conclusions from Descartes, as we will soon see. Thus, he starts by observing that even if I consider myself empty of knowledge and uncertain of everything, it is indubitable that I exist as a thinking being. Therefore, *cogito, ergo sum*. Even the stubbornest sceptic would have nothing to counter this certainty, which Geulincx calls the "prima scientia" (i.e., first discernment or first knowledge) of his *Autology*.[18] All further knowledge that we might attain must entirely derive from this first certainty, without surreptitiously introducing foreign elements. It is noteworthy that Geulincx avoids—as much as he can—using the word "res" with respect to the *ego*, for the fact that I think does not entail that I am a self-sufficient thing or a substance. To draw such a hasty conclusion would be a relapse into the "reifying" thinking that Geulincx is trying to overcome and to privilege known things over the knower. As we will see, it is precisely this strict adherence to

15 OP 2:139–98; annotations in the same volume, 266–310.

16 The treatise also contains an introduction in which Geulincx elucidates, among other things, the relationship between metaphysics and the rest of the sciences.

17 See Rousset, *Geulincx entre Descartes et Spinoza*, 50.

18 OP 2:147.

his method of self-inspection and the refusal to reify thoughts that will enable Geulincx to overcome the limited scope of the *ego*. In any case, having ascertained that I exist as a thinking being, I observe an infinite variety of thoughts or *modi cogitandi* in myself. I am aware that I see, touch, hear, etc. Besides these kinds of sensations, I also observe feelings in myself such as love, fear, and hatred, as well as mental operations such as affirming, associating, denying, inferring, and the like. All these things are modes of my thought of which I am aware and certain. We can thus establish a second discernment, immediately derived from the first, to wit: I have an infinite variety of modes of thought.[19] Geulincx observes that the ascertainment of this truth does not commit us to affirming the existence of things corresponding to these diverse *modi cogitandi*. If I perceive a light, for instance, that does not mean that there is indeed a light beyond my perception, or, in the event that there is, that the thing in question is the way I perceive it. The same applies to all the things I hear, smell, touch, etc. Nonetheless, it is undeniable that I have such perceptions, which are nothing but modes of my thought.

Having established that I think and that I contain an infinite variety of *modi cogitandi*, Geulincx goes a step further by observing that despite the plurality of thoughts I find in myself, I am something (*quaedam*) that is simple, undivided, and devoid of parts.[20] I may have a body that is divisible into parts—which is still uncertain—but as a thinking being, I have no parts. Certainly, I think and perceive a multitude of diverse things, yet the one that thinks and perceives this diversity is *one and the same*, and hence simple. When I see, touch, and even when I philosophise, I am, in all these different actions, the same "I." I will return to the simplicity of the *ego* in the next section, when I examine Geulincx's demonstration of the existence of a corporeal world. The ascertainment of my simplicity leads to the next, decisive discernment; namely, that there are many thoughts in me that do not depend on me (*multae sunt in me, quae a me non dependent cogitationes*).[21] I observe a plurality of thoughts and sensations in myself that do not arise whenever I want. The simple act of observing the world around me fills me with impressions that take place in me independently of my will. For Geulincx, the *given* or *found* character of these thoughts is incompatible with the possibility of my mind being the author of them. "Givenness"[22] and efficacy exclude each other, for efficacy brings into existence something that was not there before, i.e., that was not given. Now, Geulincx argues, if I am not the cause of these thoughts, something other than me must cause them, for it is impossible that they come from nothing or merely happen spon-

19 OP 2:148: "Varios habeo cogitandi modos in infinitum."
20 OP 2:149: "Ego sum res una atque simplex."
21 OP 2:149.
22 I employ the notion of the "given" here in a similar way to how Salomon Maimon would characterise it a century later in his *Versuch über die Transzendentalphilosophie* (Berlin: C.F. Voss, 1790), 203: "Eine Vorstellung, deren Enstehungsart in uns, uns unbekannt ist" ("a representation whose means of origination in us is unknown to us").

taneously. It might seem at first that they come from my body, but I still do not know whether a body can produce such sensations in me and, if so, how.

We thus arrive at the central insight of Geulincx's self-explorative approach, the one that will enable him to go beyond the limited scope of the *ego*. We have just seen that the infinite variety of thoughts in me that I have not produced must come from something other than me. Now, the otherness that causes all these thoughts in me must be aware that it causes them and of how it causes them; knowledge must necessarily accompany the production of all these *modi cogitandi*. Geulincx observes that this requirement—i.e., no causation without knowledge—is actually a self-evident principle, which we only fail to grasp due to deeply entrenched prejudices. For instance, we are used to thinking that when we approach a fire and feel warmth, the fire produces that warmth. In like manner, we assume that the sun produces light and that a waterfall produces a particular noise. Put differently: we unjustifiably assume that things that are devoid of thought can be efficacious and bring about certain effects. Yet this prejudice dissolves as soon as we pay heed to the following self-evident truth: "That which you do not know how to do, you do not do" ("quod nescis quomodo fiat, id non facis"; henceforth, the *quods nescis* principle).[23] To make this principle clearer, recall that consciousness is the only measuring stick or valid criterion in Geulincx's inquiry and that we are not allowed to illegitimately introduce extra-mental elements. Since we do not know of any efficacy except the one we immediately experience in ourselves—i.e., one accompanied by knowledge—the notion of efficacy only has meaning within the mental sphere.[24] From this, it follows that the

---

23 OP 2:150. Concerning the historical roots of the principle *quod nescis*, which go back to Galen's *De Foetus Formatione*, see Emanuela Scribano, "'Quod nescis quomodo fiat, id non facis.' Occasionalism against Descartes?," *Rinascimento* 51 (2011), 63–86; see also Andrea Sangiacomo, "Geulincx and the Quod Nescis Principle," in *The Oxford Handbook of Descartes and Cartesianism*, ed. Steven Nadler, Tad M. Schmaltz, and Delphine Antoine-Mahut (Oxford: Oxford University Press, 2019), 450–64. In his illuminating study, Sangiacomo shows that the *quod nescis* principle is alien to Cartesian philosophy and rather has its roots in late Scholastic thought. According to Sangiacomo, this principle can equally be used to support the autonomy and efficacy of natural agents vis-à-vis God (namely, by attributing knowledge to them) and to restrict causal efficacy to God alone (by depriving natural beings of knowledge). Geulincx is clearly interested in the second use, and he thus attempts to integrate the principle into Descartes's dualistic framework, which reduces the natural world to mere extension (deprived of thought). This operation enables him to philosophically underpin the Calvinistic view according to which God alone governs all events. Thus, Sangiacomo argues, Geulincx's endorsement of the *quod nescis* principle serves a specific agenda dictated by his Calvinistic creed. Concerning Geulincx's attempt to elaborate an ethical system (based on Cartesian premises) consistent with Calvinism, see van Ruler, "Geulincx and Spinoza: Books, Backgrounds and Biographies."

24 Regarding Geulincx's "subjective" or "mental" approach to the idea of causation—which distinguishes him from the rest of the occasionalists—see Ursula Renz & Han van Ruler, "Okkasionalimus," in *Enzyklopädie Philosophie*, ed. Hans Jörg Sandkühler (Hamburg: Felix Meiner, 2010), 2:1843–46. Hume's rejection of volitional efficacy features striking parallels with Geulincx's use of the *quod nescis* principle. Regarding the possible connection between the two thinkers, see Jason Jordan, "Voli-

otherness that produces all these *modi cogitandi* in me cannot be something that is brutish and devoid of thought. The fire cannot produce the feeling of warmth in me, just as the sun is *per se* unable to illuminate. The cause of all these effects must be a being endowed with consciousness and will, which knows exactly how to produce them. We must thus conclude (fifth discernment): "There is someone who knows and wills who is different from me" ("est sciens aliquis et volens diversus a me").[25] Aside from my own existence as a thinking being, I must admit the existence of another thinking being that induces those perceptions in me for which I am not responsible.

This fifth discernment is full of consequences and contains Geulincx's entire philosophy in embryonic form. Let us briefly consider some of its far-reaching implications. We have seen that the "givenness" of certain thoughts excludes the possibility of me being their cause. If I were their cause, I would know how to produce them, and therefore I would not merely *find* them in myself, independent of my will. Thus, the given character of certain thoughts in me indicates the limitation of my intellect, which would be perfect if it were the cause of all its thoughts. A perfect mind does not *find* anything—i. e., it is not passive or receptive—and therefore it is the only author of its thoughts and perceptions. I am therefore limited and imperfect, insofar as I am passive and receptive. Yet I am not passive with regard to something that is not, in turn, a mind. Something devoid of thought—such as a body—cannot act upon me, for, as repeatedly observed, there is no causation without knowledge. Behind the impressions I have not produced, there is not an unknown *thing*, but the unknown action of another mind. Therefore, if I am limited and passive, it is so necessarily with regard to an unlimited mind. Another way of saying this is that I am a limited intellect within an unlimited, all-embracing intellect, and my being part of it accounts for both my thinking power and my cognitive imperfection.[26] Guelincx can thus affirm: "We are from God and belong to God; we are not God Himself, because of our limitations and imperfections, with respect to which we do not belong to God,

tional Efficacy and the Paralytic's Arm: Hume and the Discursus of Occasionalism," *Intellectual History Review* 25, no. 4 (2015): 401–12.

25 OP 2:150.

26 We find a somewhat similar explanation of our cognitive imperfection in Spinoza. See his *Treatise on the Emendation of the Intellect*, in *Opera*, edited by Carl Gebhardt (Heidelberg: Carl Winter, 1925), 2:28; *The Collected Works of Spinoza*, edited and translated by Edwin Curley (Princeton: Princeton University Press, 1985–2016), 1:33: "But if it is—as it seems at first—of the nature of a thinking being to form true, or adequate, thoughts, it is certain that inadequate ideas arise in us only from the fact that we are part of a thinking being, of which some thoughts wholly constitute our mind, while others do so only in part" ("Quod si de natura entis cogitantis sit, uti prima fronte videtur, cogitationes veras sive adaequatas formare, certum est, ideas inadaequatas ex eo tantum in nobis oriri, quod pars sumus alicuius entis cogitantis, cuius quaedam cogitationes ex toto, quaedam ex parte tantum nostrum mentem constituunt").

for they do not belong to our being, but to non-being."[27] Observe that just as our limitations and imperfections do not belong to God, nor do they belong to us, for limitation is simply a lack of being, and hence our being is not different from God's. Geulincx illustrates this point immediately afterwards with the following simile (which drew him accusations of Spinozism): just as a vast prairie is divided into parcels, our minds can be considered parcels or sub-regions of God's unlimited mind.[28] If you remove the parcels—i. e., the limitations—there remains only a continuous and unlimited space. Similarly, if you remove our cognitive limitations, there is only an infinite mind.[29] We can thus see how Geulincx's self-explorative approach, together with the *quod nescis* principle, enables him to reestablish the continuity between the infinite and the finite and to embrace a form of mental holism. It is noteworthy that Geulincx arrives at this bold conclusion without ever abandoning the element of thought; i. e., by focusing on our consciousness alone and without presupposing anything previously known.[30] I observed in the previous section that for Geulincx, scepticism is the negative flipside of the knower's priority over the known. Now we can affirm that mental holism is the positive flipside of Geulincx's scepticism.

Yet although Geulincx advocates mental holism—i. e., that all minds are but one mind—he does not thereby advocate idealism; i. e., that the whole of reality is of a mental nature. As I will show in the next section, he contends that God causes an infinite variety of thoughts in me through the mediation of the body. Hence, there is after all something external to the mind, which is not just another mind, but something cor-

---

27 OP 2:269: "Nos sumus ex Deo et pertinemus ad Deum; non sumus tamen ipse Deus, propter limitationes et imperfectiones nostras, respectu quarum ad Deum non pertinemus, quia non pertinent ad nostrum esse, sed ad non-esse."

28 OP 2:269; also 293. Rousset argues (*Geulincx entre Descartes et Spinoza*, 86) that our mind's containment within God's mind is not to be taken literally, but rather in the sense that our minds are imperfect instantiations of God's perfect mind. Rousset's main reason for rejecting mental holism in Geulincx is the fact that Geulincx still conceives the relationship between God and His creatures in terms of "creation," a notion that in Rousset's opinion is clearly incompatible with holism or monism. Against Rousset, it must be noted that Geulincx explicitly equates "creation" with God's "self-limitation." See the following passage from his *Annotata latiora in Principia Philosophiae Renati Descartes* (OP 3:381): "Quod Deus creare possit, ab aeternitate sua habet; cum enim inde primus sit, atque ideo etiam illimitatus, potest etiam ille aliis quibusdam, quod in se eminentissime et illimitate habet, limitate communicare, dando illis portionem istius naturae, quam ille sibi totam et illimitatam vindicaverat. Limites enim ponendo certis suis perfectionibus, eas quodammodo alienat et extrase ponit, easdem tamen sibi retinens quatenus illimitatae sunt." Regarding Geulincx's account of creation as self-limitation, see Brian Cooney, "Arnold Geulincx: A Cartesian Idealist," *Journal of the History of Philosophy* 16, no. 2 (1978): 167–80.

29 OP 2:239: "Remove enim a te limitationem illam, qua intellectus tuus ita circumscriptus est, ut in futurum prospiciat nihil, de praeteritis pauca recolat, de praesentibus non multa sciat, multa de rerum natura ambigat, multa ignoret; his omnibus limitationibus [...] ablates, quid nisi Deum ipsum, infinitam mentem a qua praecisus fueras, apprehendis?"

30 He himself stresses this (OP 2:239; italics mine): "Et certe cum bene hanc rem perpendimus, *nos ipsos examinando*, cum id quod ad praecisionem, abstractionem, limitationemque pertinet, a nobis removerimus, *clarissime Deum ipsum in nobis agnoscimus et nos in illo.*"

poreal and devoid of consciousness. Yet this does not change any of the implications just drawn concerning the fifth discernment. Since causation requires knowledge, I cannot properly say that fire causes warmth in me, but rather that God uses fire *as an instrument* to produce warmth in me. Here, we can recognise Geulincx's occasionalism; i.e., the view that only God has causal efficacy. By establishing that there is no efficacy without consciousness, Geulincx is committed to affirming the existence of an omniscient being that concentrates all efficacy in itself. Certainly, the mere principle of *quod nescis* does not deprive me of all efficacy: although I am passive regarding certain thoughts, I am the actual source of others, such as my intentions, volitions, consents, denials, and the like. Yet, in any case, I cannot bring about anything beyond my own mental sphere without the assistance of another (knowing) being. For instance, I cannot move my own legs solely in virtue of myself, for I cannot exert an action on a thing that I have not produced, such as my body, even if I feel it to be "mine." My body is something *given to me*, and therefore something that is foreign to my knowledge and as such a sign of my mental limitations. In every action of mine in which something foreign to my mind is involved, I cannot be the only cause, and the concurrence of another knowing being therefore is indispensable. Consider now the entire created world as an infinite aggregate of finite beings—be they minds or bodies—interacting with one another. Their interaction is inconceivable without God effecting it, for bodies are devoid of efficacy and particular minds do not cause anything beyond the sphere of their consciousness. Hence, we must conclude that an unlimited omniscient being connects all particular things with one another and constitutes their unifying principle. The words of the apostle thus receive their proper philosophical meaning (Acts 17:28): "For in him we live, and move, and have our being" ("in ipso enim vivimus, et movemur, et sumus").[31]

4 Multiplicity, Time, and Corporeality

After demonstrating that an infinite mind causes those thoughts in me for which I am not responsible, Geulincx goes on to demonstrate that it produces these thoughts by means of a body, not immediately. To show this is the purpose of the sixth and seventh discernments of the *Autology*, in which Geulincx's reasoning becomes somewhat obscure. His argument, in broad strokes, runs as follows. Recall that the human mind is something simple and undivided: I am one and the same throughout my various mental states. God is also a mind, and hence is equally simple and undivided. Now, since variety cannot emanate from something simple (at least *directly*), the variety of thoughts I find in myself cannot emanate either from me or from God. Such variety is only explainable by means of something intrinsically multiple or susceptible to multiplicity. Such is the case with matter (or extension),

31 OP 2:239; also 293.

which—in contrast to the mental—admits division and multiplicity. Yet the mere notion of extension does not entail division, but merely susceptibility to *being divided.* Considered in itself, extension is something homogeneous and uniform. In order for multiplicity to emerge from mere extension, there must also be motion, which, together with matter, generates the diversity of figures we encounter in the corporeal world.[32] Since mere corporeality is unable to generate motion, God must be the one who—by a contingent decree of His will—introduces motion, and hence variety, into the corporeal world.[33] The diversity of my thoughts is thus due to the diversity introduced by God in inert extension. It might seem here that Geulincx betrays his own self-explorative approach by surreptitiously introducing an unjustified assumption; namely, the existence of a corporeal world besides the mental. If, up to this point, Geulincx has remained within the limits of a mere self-inspection, the sudden talk of corporeality in the sixth and seventh discernments seems to violate this restriction. At first glance, it is also unclear why Geulincx needs to introduce corporeality (and motion) in order to account for the diversity of our thoughts. We have seen that the "givenness" of our thoughts indicates the existence of another mind, which —in contrast to ours—is unlimited and able to generate impressions in us without our will. If our containment in God's mind already accounts for our limitation and passivity, and hence for the existence of "given" thoughts in us, why does Geulincx additionally need to invoke something extra-mental? Why can he not attribute the variety of our thoughts directly to God's efficacy? Clarifying this question is the goal of the following considerations.

Despite appearances, the introduction of the corporeal in this context is far from gratuitous (although its actual justification appears much later in the text, in the *Somatology*).[34] The key to the present difficulty lies in the notion of time. Time and succession presuppose motion, for if everything were at rest, it would make no sense to talk about time. Motion, in turn, presupposes corporeality, for only bodies are susceptible to motion. Now, I am conscious of time and there is a certain succession of mental states in me. Hence, since time presupposes motion, there must be bodies in motion, and therefore a physical world! Note that Geulincx is not saying that physical motion (from which time derives) causes a variety of mental states in me, but rather that God uses physical motion *as a means* to produce this variety of mental states. As already noted, entities devoid of thought such as bodies are unable to produce anything by themselves, either among themselves or in me as a thinking being. God is therefore ultimately responsible for the succession of mental states in me. Yet since time is inconceivable without physical motion, the temporality of my mental states necessarily requires the mediation of the corporeal. The gist of Geulincx's argu-

---

32 For Geulincx, division and motion are the same thing. See OP 2:279, ad pag. 176.

33 The idea that physical motion can only be caused by a mind would reappear, outside of occasionalist circles, in George Berkeley's essay *De motu*, written and published in 1721. See *The Works of George Berkeley*, ed. Alexander Campbell (Oxford: Clarendon Press, 1901), 1:487–527.

34 OP 2:176–77.

ment is the following: if the mind were the only existing reality, there would be neither plurality nor mutability, only a rigid and undifferentiated simplicity.[35] Difference and mutability are foreign to pure thought, which—as the third discernment has already shown—is one and simple. Therefore, besides pure thought, we must assume the existence of a being capable of being modified, differentiated and divided, in order to account not only for the temporality of mental states, but also for the existence of difference and multiplicity as such. We only find such susceptibility in the corporeal. Certainly, God's creation of a corporeal world and the suffusion of motion in it are incomprehensible events for us. The emergence of multiplicity and mutability from the divine mind, although an undeniable fact, is a mystery beyond our cognitive grasp.[36] I will return to this point in the next section.

The demonstration that God acts on us by means of the corporeal allows Geulincx to explain other things, such as the union of my mind with a particular body —i.e., "my" body—as well as the dissolubility of such a union (which opens the door to a proof of the mind's immortality). Moreover, the introduction of the corporeal into the philosophical inquiry sets the basis for a physical doctrine of a Cartesian nature, which Geulincx presents in his *Somatology* and develops in his *True Physics* (*Physica vera*).[37] However, dealing with these issues would go beyond the boundaries of this paper. In any case, Geulincx's demonstration of the existence of a corporeal world by means of the notion of time appears to be if not forceful, then at least plausible. As a good Cartesian, Geulincx takes extension and thought as the most fundamental notions and the supreme genera of things.[38] Just as the mental is a primary reality, unexplainable by something other than the mind itself, the physical has essential features that cannot be deduced from the mental alone.[39] In this respect,

35 Here, Geulincx touches upon a difficulty that would become crucially important in German Idealism; namely, the origin of difference and non-identity. Whereas thinkers like Fichte would take non-identity as external to thought, Hegel would attempt to conceive non-identity as intrinsic to thought (i.e., identity) itself.

36 We should not forget, however, that Geulincx attempts to provide a rational account of the idea of "creation." See note 28 above.

37 OP 2:368–457.

38 See Descartes's *Principles of Philosophy*, § 48, in *Œuvres de Descartes*, ed. Charles Adam and Paul Tannery (Paris: Vrin, 1974–1989) 8:23; *Philosophical Writings of Descartes*, ed. and trans. John Cottingham, Robert Stoothoff, Dugald Murdoch (vols. 1 and 2), and Anthony Kenny (vol. 3) (Cambridge and New York: Cambridge University Press, 1984–1991), 1:208: "But I recognize only two ultimate classes of things: first, intellectual or thinking things, i.e. those which pertain to the mind or thinking substance; and secondly, material things, i.e. those which pertain to extended substance or body" ("Non autem plura quam duo summa genera rerum agnosco: unum est rerum intellectualium, sive cogitativarum, hoc est, ad mentem sive ad substantiam cogitantem pertinentium; aliud rerum materialium, sive quae pertinent ad substantiam extensam, hoc est, ad corpus"). On this point (as in many others), Spinoza also appears to be very close to his contemporary and compatriot Geulincx, as they both endorse a holism with respect to both the mental and the corporeal.

39 It should be clarified that the irreducibility of the corporeal to the mental only holds for us (*quoad nos*), finite minds, insofar as the infinite's mind creation of extension is beyond our comprehension.

for Geulincx, thought and extension are—as de Vleeschauwer fittingly puts it—"transcategorial realities" independent of our knowledge.[40] Note that that which initially looked like an unattainable goal—i. e., knowledge of things as they are in themselves—now appears to be the necessary outcome of a self-inspection that is distrustful of our cognitive powers. The initial scepticism has thus developed into a limited, yet indisputable knowledge. From the objective vantage point thus attained, it is now possible to look at the common misconceptions of philosophers and unmask the mechanism behind them. As Spinoza repeatedly stresses, "veritas index sui et falsi": the true accounts for itself and for its opposite, the false. In accordance with this principle, Geulincx attaches to his *True Metaphysics* an annex entitled *Metaphysics according to the Peripatetic Mind*,[41] where he critically examines the basic notions of Scholasticism, which for him is the philosophical expression of common sense. In order thus to complete our brief exposition of Geulincx's epistemological views, it is pertinent to take a brief look at this work.

5 Anti-Dogmatism and the Critique of Scholasticism

In his *Metaphysics according to the Peripatetic Mind*, Geulincx declares Peripatetic philosophy false from the very outset.[42] In accordance with the classification presented in the second section, this kind of philosophy is a consideration of things *relative to us*, hence devoid of objective validity. For Geulincx, the falsity of Peripatetic philosophy comes from its naive pretension of directly exploring things without previously exploring the knower itself. Peripatetic philosophy barely problematises our cognitive access to reality, and thus it takes mere appearances—both sensitive and intellectual—for the things themselves. Insofar as it assigns to the things themselves that which is only relative to us, this kind of philosophy is dogmatic. Geulincx's main purpose in this text is thus to surmount dogmatism *from within* by showing the de-

---

Quoad se, however, the corporeal entirely derives from the mental, for the corporeal is passive, and the passive is necessarily posterior (not in temporal terms, but ontologically and conceptually) to the active. See OP 2:160–61.

40 H.J. de Vleeschauwer, "Les antécédants du transcendentalisme." Thus, Geulincx claims (OP 2:240): "Videmus itaque duas res singulares, mentem inquam atque corpus, creditas hactenus ut universales: illam Deum [...], hoc vero creaturam." As already indicated in the previous note, we should avoid construing the distinction between thought and extension as a rigid dualism, for Geulincx equally maintains that extension is "eminently" contained in God (see OP 2:300). In this regard, Cooney convincingly argues (see "Arnold Geulincx: A Cartesian Idealist") that extension is only "external" to finite minds, not to God, who comprehends everything.

41 OP 2:199–265.

42 Concerning Geulincx's criticism of Aristotelianism in conjunction with that of Bacon and Descartes, see Aalderink, *Philosophy, Scientific Knowledge, and Concept Formation in Geulincx and Descartes*, 17–64.

ficiencies of its conceptual apparatus. I want to focus on Geulincx's critical examination of two conceptual pairs that are characteristic of Scholasticism: universality and singularity on the one hand and essence and accident on the other. The upshot of Geulincx's critical assessment will roughly be the following: that which is the most impenetrable and obscure for Scholastic philosophy is actually the most knowable for the human mind. Conversely, that which Scholasticism takes as the most knowable is actually the most obscure. Thus, for Geulincx, God's intellect before creating the world is the most accessible for the human mind, whereas the created world is the most impenetrable.

In the section dedicated to universality and singularity,[43] Geulincx focuses his attention on the main intellectual operation of Scholastic philosophy: abstraction.[44] It is through abstraction that Peripatetic thought generates its basic concepts, such as being, substance, the one, etc. Abstraction consists in separating by means of the intellect a particular content from the integral whole to which it belongs and considering it in an isolated manner. When the separated content is a feature common to various individuals, the product of abstraction is a universal concept. In the case of the most general concepts, such as being or substance, the common feature is not a particular content (for such concepts are devoid of content), but an objectified mental operation.[45] There are cases in which abstraction does not separate something common to many, but instead simply one thing from another, such as when we abstract a point from a line or a line from a surface. In any case, abstraction is nothing but a mental operation that does not change anything in the nature of things. Yet Geulincx also observes—quoting Aristotle—that "those who abstract do not lie" ("abstrahentium non est mendacium").[46] That is, we must concede that universals exist in the things themselves, for abstraction selects "something" existing in reality (otherwise it would not be *something*). However, we should not thereby assume that the thing thus abstracted exists *in exactly this way*; i.e., differentiated from the rest. Thus, the actual risk of abstraction consists in attributing a separate and independent existence to something that can only exist as part of something else.

To illustrate this point, take, for instance, the notion of body. Particular bodies are finite and have a specific shape. If we select these features and universalise them, we might erroneously conclude that the body or the corporeal *as such* (*simpliciter*) is essentially finite and has a shape. Yet corporeality in itself is not the result of artificially separating (or abstracting) certain features common to particular bodies; rather, par-

43 OP 2:235–40.

44 Regarding Geulincx's theory of abstraction in the context of seventeenth-century debates on abstraction, see again Aalderink, *Philosophy, Scientific Knowledge, and Concept Formation in Geulincx and Descartes*, 323–49.

45 Gabriel Nuchelmans traces Geulincx's reflections on this issue back to the scholastic distinction between *actus exercitus* and *actus significatus*. See Gabriel Nuchelmanns, *Judgment and Proposition from Descartes to Kant* (Amsterdam: North-Holland Publishing Company, 1983), 100–104.

46 OP 2:236.

ticular bodies are abstractions of the only existing body—i. e., extension—which as such is unlimited and without shape. As we can see, the validity of abstraction depends on how we apply the distinction between the absolute and the relative. Since abstraction artificially separates things that originally belong together, the abstracted content is necessarily relative to the integral whole from which it has been separated. Abstraction can thus never provide access to an absolute and self-contained reality. Put differently: we do not know what corporeality as such is by isolating features from particular bodies, but by removing all particularities and delimitations that we have superimposed onto the corporeal. The corporeal nature manifests itself when we eliminate its finite appearances. The same applies to the mind: particular minds are limited and imperfect, but if we remove all particularity, we will grasp the mind as such, which is unlimited and perfect (recall the analogy with the prairie and its parcels in the second section). In this respect, Geulincx observes that properly speaking, we should not talk about bodies and minds, but about corporeal things and mental things.[47] In sum: abstraction separates the finite from the infinite, whereas in fact the finite—i. e., the delimited, the abstract—exists only *within* the infinite and must be conceived through it. Note that the removal of all particularity that Geulincx commends as the right way of grasping reality—as opposed to abstraction—coincides with the sceptical scruple discussed in the second section. Only the distrust towards our own subjective forms and *modi dicendi* ("ways of speaking") enables us to examine thought *as such*—i. e., in an absolute manner—thus transcending all relativity.

Let us now pass to the notions of essence and accident.[48] Essential is a predicate that necessarily belongs to a thing, so that it is impossible to deny it of it. Thus, the essence of the body is to be extended, and the essence of the mind is thought. There are also certain "secondary" properties that we can deduce through reasoning from a thing's essence, such as infinity from extension, its infinite divisibility, etc. There are, furthermore, predicates of a thing that do not necessarily belong to it, so that we can conceive the thing without them. Motion, for instance, presupposes extension, but does not immediately derive from it, as the fact that we can conceive of extension without motion makes clear. The connection between these two notions is contingent—we could also say "accidental"—for we cannot move from the one to the other through an interrupted reasoning. For Geulincx, there cannot be any science of the contingent due to this gap in the reasoning or *scala rationis*. The existence of contingent things depends solely on God's will, which Geulincx—following Descartes—clearly distinguishes from God's intellect. To the class of contingent things belong the world, the sun, the earth and its parts, human beings, etc. All these things fall outside of the realm of the rational, for no demonstration can explain *why* they exist. Yet the impossibility of "deducing" the contingent is not incompatible with tracing the root of contingent things. As already shown, Geulincx can demonstrate

47 OP 2:286–87.
48 OP 2:261–65.

the existence of a corporeal world by virtue of the notion of time, which presupposes motion, which in turn presupposes extension. Although the upwards reasoning from time to extension is valid, the downwards reasoning from extension to time is not. From the existence of contingent things, we can infer the existence of a creator, but the existence of contingent things as such is rationally unaccountable, for it exclusively depends on God's decree.

If we now connect these considerations with Geulincx's reflections on abstraction, we can see that in a way, the contingent and the finite (or abstract) coincide. As already observed in the previous section, motion generates the variety and multiplicity of things in the corporeal world. Moreover, particular things—be they minds or bodies—are abstractions from unlimited realities without which they cannot be properly conceived, yet from which they cannot be deduced. The whole realm of the finite and contingent is thus rationally unaccountable for us. This realm constitutes the point of departure of Peripatetic philosophy, from which it attempts to ascend to the apprehension of the absolute and unconditioned. Yet we have seen that the particular cannot provide access to the absolute. For this reason, Geulincx affirms that Scholastic philosophy has an obscure notion of infinite realities. By contrast, for Geulincx, infinite realities—such as thought and extension—are the only ones that are transparent to the human mind, as long as it refrains from grasping the particular and focuses on the inspection of its own self. Whereas Scholastic philosophy remains imprisoned in a world fashioned by our subjective view of things, the intellectual itinerary proposed by Geulincx elevates the mind to the "transcategorial," yet at the cost of alienating it from the existing world.

6 Conclusion

Geulincx's rich thought comprises further aspects that require a separate examination, such as a full-blown ethical system,[49] a relatively innovative logic,[50] and an out-

---

49 See *Ethics* (*Ethica*) in OP 3:1–271. The first part of Geulincx's *Ethics* appeared in 1665 under the title *De Virtute et Primis Ejus Proprietatibus.* Ten years later (six years after Geulincx's death), the complete *Ethics* was published. There is an English translation (which includes Samuel Beckett's notes): *Ethics, with Samuel Beckett's Notes*, trans. Martin Wilson, ed. Han van Ruler, Anthony Uhlmann, and Martin Wilson (Leiden and Boston: Brill, 2006). Geulincx's ethics features striking parallels with Stoicism that have been explored by Ruben Buys, "Between Actor and Spectator: Arnout Geulincx and the Stoics," *British Journal for the History of Philosophy* 18, no. 5 (2010): 741–61. Regarding Spinoza's and Geulincx's respective approaches to human passions, see Mark Aalderink, "Spinoza and Geulincx on the Human Condition, Passions, and Love," *Studia Spinozana* 15 (2006): 67–87.

50 His main works in this area are *Logica fundamentis suis restituta* (OP 1:165–454) and *Methodus inveniendi argumenta* (OP 2:1–111), the first of which appeared the same year as *La logique ou l'art de pensar* by Arnauld and Nicole (1662). For instructive accounts of Geulincx's logic, see Nuchelmans, *Judgment and Proposition from Descartes to Kant*, 99–120; also Karl Dürr, "Arnold Geulincx und die klassische Logik des 17. Jahrhunderts," *Studium Generale 18* (1965–68): 520–41.

dated yet quite elaborate physics.[51] In this paper, I have confined myself to what I take to be the core of his philosophy; namely, the self-explorative approach developed in his *Autology*. This approach is a fine example of how scepticism, far from hindering philosophical inquiry, can in fact significantly expand its scope. Hegel famously stated that scepticism is "the free side of every philosophy," insofar as it raises the mind above the limitation and one-sidedness of common sense.[52] In Geulincx's case, scepticism presumably liberates the mind from its subjective forms and its *modi dicendi*, thus revealing that the mind is one and infinite and that our finite mind is just a parcel of it. Furthermore, the temporality of our mental states indicates the existence of bodily motion—without which there would be no temporality—and hence the existence of a physical realm beyond the mind itself. Geulincx's self-inspection thus results in the attainment of two "transcategorial" realities, the mental and the physical. One could justifiably ask here whether these realities are indeed as "transcategorial"—i.e., independent of our knowledge—as Geulincx pretends. Just as Geulincx sharply criticises the conceptual apparatus of Scholasticism for being merely "subjective," he could equally have questioned the objectivity of the (Cartesian) distinction between the mental and the corporeal and subjected it to a critical examination. Yet here we encounter—as Cassirer rightly put it—"the inner limit" of Geulincx's anti-dogmatism.[53] Despite his efforts to disembarrass himself from uncritical assumptions, Geulincx still thinks in Cartesian terms, even if he departs from Descartes in crucial respects. Thus, he cannot be critical of Cartesian philosophy to the same extent that he is of Scholasticism, precisely because Descartes provides him with the conceptual resources to rebut Scholasticism. This does not mean, however, that Geulincx accepts the Cartesian understanding of the mental and the physical without further ado. As has been shown, he attempts to purify these notions of all traces of "reifying" thinking, a purification that results, among other things, in the *de-substantialisation* of the human mind and its localisation within the divine intellect. Spinoza arrived at the same conclusion, based on the same Cartesian premises. A separate study is required to show the differences and similarities of their respective approaches.

51 See his *Physica Vera* (OP 2:368–446) and his *Physica Falsa s. ad mentem Peripateticorum* (OP 2:313–67). For a good summary of Geulincx's physics, see Rousset, *Geulincx entre Descartes et Spinoza*, 101–28.

52 Georg Wilhelm Friedrich Hegel, *Verhältnis des Skeptizismus zur Philosophie*, in *Jenaer Schriften 1801–1807*, ed. Eva Moldenhauer and Karl Markus Michel (Frankfurt am Main: Suhrkamp Verlag, 1986), 229.

53 Cassirer, *Erkenntnisproblem*, 1:540–41.

Bibliography

Aalderink, Mark. *Philosophy, Scientific Knowledge, and Concept Formation in Geulincx and Descartes.* Utrecht: Zeno, 2010.

Aalderink, Mark. "Spinoza and Geulincx on the Human Condition, Passions, and Love." *Studia Spinozana* 15 (2006): 67–88.

Ayers, Michael R. "Richard Burthogge and the Origins of Modern Conceptualism." In *Analytic Philosophy and History of Philosophy*, edited by Tom Sorell and G.A.J. Roger, 179–200. Oxford: Clarendon, 2005.

Bardout, Jean-Christophe. "Occasionalism: La Forge, Cordemoy, Geulincx." In *A Companion to Early Modern Philosophy*, edited by Steven Nadler, 104–51. Malden, MA: Blackwell, 2002.

Berkeley, George. *The Works of George Berkeley.* Edited by Alexander Campbell Fraser. 4 vols. Oxford: Clarendon Press, 1901.

Burthogge, Richard. *An Essay upon Reason and the Nature of Spirits.* London: Dunton, 1694.

Buys, Ruben. "Between Actor and Spectator: Arnout Geulincx and the Stoics." *British Journal for the History of Philosophy* 18, no. 5 (2010): 741–61.

Cassirer, Ernst. *Das Erkenntnisproblem in der Philosophie und Wissenschaft der neueren Zeit*, 3rd ed. 3 vols. Berlin: Bruno Cassirer, 1922.

Clatterbaugh, Kenneth. *The Causation Debate in Modern Philosophy 1637–1739.* New York: Routledge, 1999.

Cooney, Brian. "Arnold Geulincx: A Cartesian Idealist." *Journal of the History of Philosophy* 16, no. 2 (1978): 167–80.

Descartes, René. *Œuvres de Descartes.* Edited by Charles Adam and Paul Tannery. 11 vols. Paris: Vrin, 1974–1989.

Descartes, René. *The Philosophical Writings of Descartes.* Translated by John Cottingham, Robert Stoothoff, Dugald Murdoch (vols. 1 and 2); Anthony Kenny (vol. 3). 3 vols. Cambridge: Cambridge University Press, 1984.

Dürr, Karl. "Arnold Geulincx und die klassische Logik des 17. Jahrhunderts. " *Studium Generale* 18 (1965–68): 520–41.

Geulincx, Arnold. *Arnoldi Geulincx Antverpiensis Opera Philosophica.* Edited by Jan Pieter Nicolaas Land. 3 vols. Den Haag: M. Nijhoff, 1891–93.

Geulincx, Arnold. *Ethics, with Samuel Beckett's Notes.* Translated by Martin Wilson. Edited by Han van Ruler, Anthony Uhlmann, and Martin Wilson. Leiden and Boston: Brill, 2006.

Hegel, Georg Wilhelm Friedrich. *Jenaer Schriften 1801–1807.* Edited by Eva Moldenhauer and Karl Markus Michel. Frankfurt am Main: Suhrkamp Verlag, 1986.

Jordan, Jason. "Volitional Efficacy and the Paralytic's Arm: Hume and the Discursus of Occasionalism." *Intellectual History Review* 25, no. 4 (2015): 401–12.

de Lattre, Alain. *L'occasionalisme d'Arnold Geulincx: étude sur la constitution de la doctrine.* Paris: Éditions de Minuit, 1967.

Maimon, Salomon. *Versuch über die Transzendentalphilosophie.* Berlin: C.F. Voss, 1790.

Nadler, Steven. "Knowledge, Volitional Agency and Causation in Malebranche and Geulincx." *British Journal for the History of Philosophy* 7, no. 2 (1999): 263–74.

Nuchelmans, Gabriel. *Judgment and Proposition: From Descartes to Kant.* Amsterdam: North-Holland Publishing Company, 1983.

Perler, Dominik, and Ulrich Rudolph. *Occasionalismus: Theorien der Kausalität im arabisch-islamischen und im europäischen Denken.* Göttingen: Vandenhoeck & Ruprecht, 2000.

Renz, Ursula, and Han van Ruler. "Okkasionalimus." In *Enzyklopädie Philosophie*, edited by Hans Jörg Sandkühler, 2:1843–46. Hamburg: Felix Meiner, 2010.

Rousset, Bernard. *Geulincx entre Descartes et Spinoza*. Paris: Vrin, 1999.
Ruler, Han van, "Geulincx and Spinoza: Books, Backgrounds and Biographies." *Studia Spinozana* 15 (2006): 89–106.
Sangiacomo, Andrea. "Defect of Knowledge and Practice of Virtue in Geulincx's Occasionalism." *Studia Leibnitiana* 46, no. 1 (2014): 46–63.
Sangiacomo, Andrea. "Geulincx and the Quod Nescis Principle." In *The Oxford Handbook of Descartes and Cartesianism*, edited by Steven Nadler, Tad M. Schmaltz, and Delphine Antoine-Mahut, 450–64. Oxford: Oxford University Press, 2019.
Scribano, Emanuela. "'Quod nescis quomodo fiat, id non facis.' Occasionalism against Descartes?" *Rinascimento* 51 (2011): 63–86.
Specht, Rainer. *Commercium mentis et corporis. Über Kausalvorstellungen im Cartesianismus.* Stuttgart-Bad Cannstatt: Friedrich Frommann Verlag (Günther Holzboog), 1966.
Spinoza, Benedictus de. *Opera*. Edited by Carl Gebhardt. 4 vols. Heidelberg: Carl Winter, 1925.
Spinoza, Benedictus de. *The Collected Works of Spinoza*. Edited and Translated by Edwin Curley. 2 vols. Princeton: Princeton University Press, 1985–2016.
Stein, Ludwig. "Antike und mittelalterliche Vorläufer des Occasionalismus." *Archiv für Geschichte der Philosophie* 2, no. 2 (1889): 193–245.
de Vleeschauwer, H.J. "Les antécédants du transcendentalisme. Guelincx et Kant." *Kant-Studien* 45, 1–4 (1953): 245–73.
de Vleeschauwer, H.J. "Three Centuries of Geulincx Research: A Bibliographical Survey." *Mededelings van die Universiteit van Suid-Afrika* 1 (1957): 1–72.
Zeller, Eduard. "Über die erste Ausgabe von Geulincx' Ethik und Leibniz' Verhältnis zu Geulincx' Occasionalismus." *Sitzungsberichte der Königlich Preußischen Akademie der Wissenschaften zu Berlin* 29, no. 2 (1884): 673–95.

Avraham Rot

Spinoza's Affective Scepticism[1]

Spinoza is known to have made it the business of his philosophy to present us with the most immediate alternative to Descartes with regard to virtually every fundamental philosophical subject: while Descartes is a dualist, Spinoza is at once a monist (with respect to substance) and a pluralist (with respect to attributes); while for Descartes, philosophy begins with the I, for Spinoza, it begins with God or the self-caused; while Descartes sharply distinguishes between the human intellect and the will, considering the former finite and the latter infinite, Spinoza insists that they, like all the other faculties, are one and the same thing; like others, notably Malebranche, Spinoza rejects Descartes's mind–body interactionism, ridicules the notion that the pineal gland is the seat of the soul, and argues instead for a parallelist conception of the psychosomatic relation; while for Descartes, wonder (*admiration*) is "the first of all the passions," for Spinoza, wonder (*admiratio*) is not even a passion, but merely "an imagination of a thing in which the mind remains fixed because this singular imagination has no connection with others";[2] and the list can be further extended.

On the face of it, the subject of scepticism appears to be yet another case in point: while Cartesianism proceeds from doubt, Spinozism proceeds from certainty. Indeed, the received scholarly view is that in direct opposition to Descartes's method of doubt, which is commonly considered a radical form of scepticism that marks the

1 I am very grateful to the two anonymous reviewers for their generous comments and suggestions on a draft of this essay. I also thank Máté Veres, Rachel Aumiller, Behnam Zolghadr, Nancy Abigail Nuñez Hernández, Timothy Franz, José María Sánchez de León Serrano, Yoav Meyrav, Jonathan Garb, Michah Gottlieb, Haim Kreisel, Julie Klein, and other friends and colleagues at the Maimonides Centre for Advanced Studies for helpful discussions about the themes discussed here.

2 For the quoted texts, see René Descartes, *The Passions of the Soul*, § 53, and Benedict de Spinoza, *Ethics*, part 3, definitions of the affects, definition 4. Citations from Descartes's *Meditations on First Philosophy*, *Principles of Philosophy*, and *The Passions of the Soul* are taken from *The Philosophical Writings of Descartes*, trans. John Cottingham, Robert Stoothoff, and Dugald Murdoch, 2 vols. (Cambridge: Cambridge University Press, 1984–85), hereafter abbreviated as CSM. Citations from Spinoza's *Treatise on the Emendation of the Intellect* (*Tractatus de Intellectus Emendatione*, abbreviated as TdIE), *Political Treatise* (*Tractatus Politicus*, abbreviated as TP), *Ethics* (*Ethica*, abbreviated as E), and *Theological-Political Treatise* (*Tractatus theologico-politicus*, abbreviated as TTP) are taken from *The Collected Works of Spinoza*, trans. Edwin Curley, 2 vols. (Princeton: Princeton University Press, 1985–2016). The abbreviations of subsections of Spinoza's *Ethics* are as follows: 1–5 (part number); d (definition, if positioned directly after digit); a (axiom); p (proposition); s (scholium); d (demonstration, if not positioned directly after digit); exp (explanation); pref (preface); app (appendix); l (lemma). Original language references are to *Spinoza Opera*, ed. Carl Gebhardt, 4 vols. (Heidelberg: Carl Winters, 1925).

https://doi.org/10.1515/9783110618839-010

beginning of modern philosophy, Spinoza's philosophy is thoroughly antisceptical.[3] The purpose of this paper is to question this view and to demonstrate that the opposite is in fact the case; namely, while Spinoza indeed presents us with the immediate alternative to Descartes with regard to this subject as well, this alternative is not antiscepticism, but a more radical, truthful, and realistic form of scepticism. (I am using these qualifications interchangeably.) More particularly, Cartesian scepticism is limited in that it is dichotomous, transitory, dualistic, and narrowly cognitive; Spinozist scepticism is radical, since it is continuous, ongoing, and all-encompassing, that is, at once—or, to use Heideggerian terminology, equiprimordially (*gleichursprünglich*)—cognitive, embodied, imaginative, mnemonic, motivated, and affective. It is true that under the species of eternity, there appears to be no room for doubt in Spinoza's system. God or nature is indeed not subject to doubt. But this does not mean that there is no doubt *in* God or nature. On the contrary, according to Spinoza, doubt is ubiquitous, for God or nature is necessarily affected in a great many ways, these affections in turn are also necessarily affected in a great many ways, and these latter, secondary affections are necessary features of finite existence that designate the realm of both doubt and affectivity.

I shall proceed as follows. In the next section, I will consider the reasons that Spinoza has been regarded as an antisceptical thinker by examining some of his most famous explicit remarks on the subject. I will demonstrate that while it is true that Spinoza appears to be speaking against scepticism, he does not reject the reality or even the desirability of doubt as such, but rather merely takes issue specifically with the sense in which Descartes says that he doubts. Spinoza's contention is that Cartesian doubt is either insincere or misconceived, and inasmuch as Descartes's very conception of doubt is untruthful or fallacious, he employs this term with equal inadequacy both when he considers the validity of clear and distinct ideas and when he considers the reality of the external world. Spinoza's positive statements about the nature of doubt are scarce and often perplexing. The subsequent sections are devoted to the elucidation of the meaning of doubt in Spinoza's philosophy. In sections 2 and 3, I will demonstrate that rather than a purely intellectual act involving the operation of the faculty of reason alone, as it is for Descartes, Spinozist doubt involves the operation of all the faculties in concert and in an equiprimordial manner. In section 4, I will show that Spinoza's discussion of doubt in the second part of the *Ethics*, which appears to amount to a few bizarre remarks on the

---

3 For accounts pursuing this line of interpretation, see Willis Doney, "Spinoza on Philosophical Skepticism," *The Monist* 55, no. 4 (1971): 617–35, and Michael Della Rocca, *Spinoza* (London: Routledge, 2008), esp. 127–34, 263–64, 281. I would like to thank one of the two reviewers of this paper, who pointed out that one should take into consideration the distinction between epistemological and religious scepticism. It is of course only with respect to the former kind of scepticism that Spinoza has been considered an antisceptical thinker, whereas with respect to the latter kind, which can be understood in terms of secularisation or critical hermeneutics, it is widely agreed that Spinoza is not just a sceptic, but a major forerunner of modern scepticism.

topic, is completed only in the third part of this book. Here doubt suddenly makes sense as Spinoza explicates its affective dimension in addition to its cognitive, perceptual, mnemonic, imaginative, and motivational dimensions. While this is easily overlooked, Spinoza's most specific and readily intuitive description of doubt is given in the definitions of the affects of hope, fear, confidence, and despair, which I will examine in sections 5 and 6. By way of conclusion (section 7), I will return to Descartes and consider the various senses in which it can be said that Spinozist scepticism is more radical than Cartesian scepticism.

1

Spinoza presents his critique of Cartesian scepticism in what is most probably his earliest surviving work, the *Treatise on the Emendation of the Intellect.* His point of departure in this context is the argument that it is ridiculous to assume that in order to know something, one should know that one knows it. The problem with this kind of reasoning, he explains, is that it leads to an infinite regress. "In order for me to know, it is not necessary to know that I know, much less necessary to know that I know that I know [...]. For to know that I know, I must first know. [...] For certainty and an objective essence are the same thing." Based on this argument, Spinoza formulates his famous conception of truth as *index sui et falsi* in the subsequent passages:

> Since truth, therefore, requires no sign, but it suffices, in order to remove doubt, to have the objective essences of things, or, what is the same, ideas, it follows that the true Method is not to seek a sign of truth after the acquisition of ideas, but the true Method [*vera Methodus*] is the way that truth itself, or the objective essences of things, or the ideas (all those signify the same) should be sought in the proper order. [...] The Method must speak about Reasoning, or about the intellection [...]. It is understanding what a true idea is by distinguishing it from the rest of the perception; by investigating its nature, so that from that we may come to know our power of understanding and so restrain the mind that it understands, according to that standard [*ut ad illam normam*], everything that is to be understood; and finally by teaching and constructing certain rules as aids, so that the mind does not weary itself in useless things.[4]

It is clear from these passages why Spinoza has commonly been regarded as an antisceptical thinker. The true method proceeds from certainty, from the condition of having a true idea that can serve as a standard with which to assess all other ideas. As Spinoza writes later in the treatise, "doubt always arises from the fact that things are investigated without order."[5] Accordingly, what is crucial is that the foundational certainty be established as firmly as possible. However, this goal is only attainable in virtue of the power of an idea whose certainty has already been

4 TdIE §§ 36–37.
5 TdIE § 80.

established. In the *Ethics*, Spinoza designates this idea as *causa sui*, that whose "essence involves existence"; here, he designates it as God.[6]

This indeed appears to be an unequivocal formulation of a strong antisceptical position. Although what Spinoza explicitly rejects here is only the notion that philosophical inquiry should begin with doubt, it is obvious that he does not consider doubt the proper end of it, nor does he appear to consider it a necessary intermediary stage. He thus appears to suggest that if we proceed from the proper foundation and in the right order, no doubt should arise at any stage. Even more fundamentally, if "certainty and an objective essence are the same thing," the very possibility to doubt appears to be negated. Given that objective essence belongs to the realm of representation, what Spinoza is in effect saying here appears equivalent to Berkeley's dictum that *esse est percipi*. Under the constraints of such an epistemology, it is as difficult to make sense of doubt as it is of falsehood.

In the subsequent passages of the treatise, however, it becomes evident that Spinoza does not reject the methodical use or reality of doubt as such, but specifically the Cartesian conception of doubt. Without mentioning Descartes by name, though while clearly thinking of him as his immediate interlocutor, Spinoza speaks here not only of "true Method," but also of "true doubt" (*vera dubitatio*), distinguishing both of them in the same breath from the method and doubt that one finds in Descartes:

> I am speaking of true doubt in the mind [*vera dubitatione in mente*], and not of what we commonly see happen, when someone says in words that he doubts, although his mind does not doubt. For it is not the business of the Method to emend that. That belongs rather to the investigation of stubbornness, and its emendation.[7]

That the one who stubbornly "says in words that he doubts, although his mind does not doubt" is no other than Descartes becomes abundantly evident a few lines later, where Spinoza rejects what Descartes famously presents as the reason for the most radical doubt; that is, the notion of a deceiving God:

> Only so long as we have no clear and distinct idea of God, can we call true ideas in doubt by supposing that perhaps some deceiving God exists, who misleads us even in the things most certain. [...] But if we have the kind of knowledge of God that we have of the triangle, then all doubt is removed. And just as we can arrive at such a knowledge of the triangle, even though we may not know certainly whether some supreme deceiver misleads us, so we can arrive at such a knowledge of God, even though we may not know whether there is some supreme deceiver.[8]

6 E1d1.
7 TdIE § 77.
8 TdIE § 79.

In the *Ethics*, this argument is summarised as follows: "He who has a true idea at the same time knows that he has a true idea, and cannot doubt the truth of that thing."[9] Spinoza further explains in the scholium to this proposition:

> For no one who has a true idea is unaware that a true idea involves the highest certainty. For to have a true idea means nothing other than knowing a thing perfectly, or in the best way. And of course no one can doubt this unless he thinks that an idea is something mute, like a picture on a tablet, and not a mode of thinking, i.e., the very [act of] understanding.

We can read these passages as follows. It is impossible to call clear and distinct ideas into doubt; hence, Descartes cannot be sincere when he tells us that he does so. It is in this sense that one can make the judgment that when Descartes says that he questions the truthfulness of clear and distinct ideas, he "says in words that he doubts, although his mind does not doubt." However, this interpretation appears to introduce a redundancy. Why should Spinoza infer from the fact that it is altogether impossible to doubt clear and distinct ideas, no matter in what way, that it is impossible to doubt such ideas sincerely? The answer is that by attacking the very meaning of Cartesian doubt, Spinoza's critique equally targets both Descartes's use of the concept of doubt with regard to clear and distinct ideas and his use of this concept with regard to unclear and indistinct ideas, including those that furnish the so-called external world. Regardless of whether or not they are clear and distinct, in saying that he doubts them, Descartes treats ideas not as they truly are—that is, as "mode[s] of thinking" and the "very [act of] understanding"—but as "something mute, like a picture on a tablet." As we shall see, it is not the case that Spinoza thinks that it is also impossible to doubt unclear and indistinct ideas; that is, that doubt is altogether meaningless or pointless regardless of whether it is exercised with respect to unclear and indistinct ideas or clear and distinct ideas. The true method is not one that eliminates doubt but one that involves the recognition of what "true doubt in the mind" is. What is impossible is not to doubt such unclear and indistinct ideas as the idea of the I or any other idea of a finite being or of an affection of a finite being; what is impossible is to be certain of such ideas. Since one can only pretend or imagine oneself to be certain about such ideas, calling them into doubt as if they were not necessarily a matter of doubt equally involves the workings of pretence or the imagination.

2

We have thus far focused on what doubt is not, according to Spinoza. It is now time to consider his positive account of doubt. What does Spinoza mean when he speaks of "true doubt in the mind"? The first thing that Spinoza notes in this regard in the

9 E2p43.

Treatise on the Emendation of the Intellect is that doubt is a state that necessarily involves a conflict of ideas and *ipso facto* more than one idea. "There is no doubt in the soul [...] through the thing itself concerning which one doubts. [...] But doubt will arise through another idea which is not so clear and distinct."[10] Thus, the reason that doubt cannot be artificially produced or generated, employed at will with respect to both doubtful and indubitable ideas as if they were pictures on a tablet, is that doubt can only take place insofar as it has already taken place; that is, when one already has another idea "which is not so clear and distinct." In short, in order to be in doubt, one should always already be in doubt. As we have seen, the same holds for certainty: in order to know that one has a true idea, one should always already have a true idea. It is accordingly impossible to doubt clear and distinct ideas, just as it is impossible to be certain of unclear and indistinct ideas. And it is likewise as meaningless to seek after a sign of truth for what one already knows as it is redundant to doubt what one already doubts.

But what is radical about doubting that which is already dubitable and remaining certain of that which is certain?[11] My argument is that what makes Spinoza's scepticism radical is precisely its realism; that is, its concrete limitedness instead of fanciful universal applicability. That the most profound philosophical discoveries may appear to amount to trivialities is to be expected. After all, why do we need a Parmenides to tell us that "there is Being, but nothing is not"?[12] This oscillation between the trivial and the profound, the everyday and the ontological, is key to understanding Spinoza's philosophy. Our investigation can therefore benefit from a consideration of a philosophy in which this oscillation takes centre stage; namely, that of Martin Heidegger. The affinities between Spinoza's and Heidegger's philosophies are remarkable, yet for various reasons—a prominent one of which is Heidegger's conspicuous lack of engagement with Spinoza's works—underexplored.[13] This

---

10 TdIE § 78.

11 Or, as one of the anonymous reviewers puts it: "This being the case, the presumable 'radical skepticism' that the author attributes to Spinoza amounts to the following triviality: Spinoza does not doubt that which is indubitable and is skeptical concerning that which is doubtful."

12 Parmenides, *Parmenides*, trans. Leonardo Tarán (Princeton: Princeton University Press, 1965), 54.

13 Kasper Lysemose notes: "Heidegger writes about Spinoza only twice. The first time is in his 1926/27 Marburg lectures on *History of Philosophy from Thomas Aquinas to Kant*. These lectures appear to have been purely a matter of duty. We encounter here long quotes not penetrated philosophically but accompanied with an unusual amount of biographical information. The second time is in the 1936 lectures on *Schelling's Treatise on the Essence of Human Freedom*. Spinoza is presented here as part in [*sic*] the emergence of the modern concept of system and Heidegger's evaluation of his contribution is as short as it is unfavorable." See Lysemose, "The Force of Existence: Looking for Spinoza in Heidegger," *Sophia* (2019). Heidegger nevertheless made at least one more remark about Spinoza; namely, in his 1929/30 Freiburg lecture course *Die Grundbegriffe der Metaphysik: Welt—Endlichkeit—Einsamkeit*, ed. Friedrich-Wilhelm von Herrmann, vols. 29 and 30 of *Gesamtausgabe*, ed. Vittorio Klostermann (Frankfurt: Klostermann, 2010), 437, translated by William McNeill and Nicholas Walker as *The Fundamental Concepts of Metaphysics: World, Finitude, Solitude* (Bloomington: Indiana University

is not the place to give a systematic treatment of this issue. Nonetheless, certain points of correspondence are noteworthy as they can orient us in our search for the meaning of doubt in Spinoza's system.

What Spinoza says can easily be rendered in Heideggerian terms. When he argues that ideas should not be considered to be like pictures on a tablet, he rejects the consideration of things as what in Heideggerian parlance is called *Vorhandenheit*; that is, as beings that are merely "present at hand."[14] Ideas do not just present themselves as indifferent objects in relation to which our faculties can operate and on which we can cast doubt at will. Rather, ideas always already imply the involvement of the mind that has them. Heidegger, as is well known, is particularly fond of using the expression, "always already," *immer schon*. The question of being always already involves the nature of the kind of being for whom being is an issue.[15] Similarly, being is always already mine (*Jemeinigkeit*); it always has a here and a there (*da*); it is always in-the-world; it always has a for-the-sake-of-which (*Worumwillen*); and it is always already somehow attuned (*gestimmt*).[16] Just as there is always already an intention behind a question, the essential characteristics of existence are disclosed as things that are always already there, as necessary presuppositions. But what kind of argument is it to say that something is always already such and such? How does it save us from infinite regress or vicious circularity? To address this question, let us consider more closely the manner in which Spinoza follows a similar line of reasoning.

As we have seen, Spinoza understands doubt as something that takes place only inasmuch as it has already taken place. Just as in order to have a clear and distinct idea, one must always already have a clear and distinct idea, in order to doubt, one must always already have an unclear and indistinct idea; that is, an idea of which one can never be certain. Now, in order to have such an unclear and indistinct idea, or what Spinoza calls a "confused idea," one has to have not just the capacity to think, but also the capacity to imagine, remember, and feel, which implies, in turn, that one must also be a corporeal, finite, and temporal being. To capture all these preconditions at once, let us consider another Heideggerian notion, one which Heidegger most probably borrowed from Emil Lask; namely, that of equiprimordiality (*Gleichursprünglichkeit*).[17] Heidegger's notion of equiprimordiality and

---

Press, 1995), 302. Given Heidegger's almost complete lack of explicit engagement with Spinoza's work, it appears safer to speak here of affinities rather than of direct influence.

14 Martin Heidegger, *Sein und Zeit*, 19th ed. (Tübingen: Niemeyer, 2006), division 1, ch. 2, translated by John Macquarrie and Edward Robinson as *Being and Time*, repr. ed. (Oxford: Blackwell, 2001), hereafter abbreviated as BT. For the first occurrence in this book of the expression *Vorhandenheit*, translated as "presence-at-hand," see BT 7 [original]/26 [translation].

15 This is what Heidegger refers to as the "es geht um" of being (BT 191/236).

16 BT 52/78, 53/78, 86/116, 134/173.

17 See BT 131–33/170–73, 137/176, 142/182, 220/263. On Emil Lask's influence on Heidegger with regard to *Gleichursprünglichkeit* and related concepts, see Theodore Kisiel, "Heideggers Dankesschuld an

the logic implied by the expression "always already" directly follow from each other. One important occasion on which Heidegger presents the notion of equiprimordiality is his rejection of the traditional hierarchical division of the faculties into the general domains of thinking, willing, and feeling, where thinking is typically given precedence over the others and feeling is relegated to the lowest status. In opposition to this traditional understanding, Heidegger argues that the main characteristics of finite existence—including, in his own terminology, understanding (*Verstehen*), affectedness (*Befindlichkeit*), and thrownness (*Geworfenheit*)—are all equiprimordial; that is, equally originary. Accordingly, in our lived experience, we never encounter things in a single dimension as if they were "present at hand" (*vorhanden*); that is, as something that is entirely objective and about which we can think or not think or think in different ways regardless of the other dimensions of our experience.

As we have seen, Spinoza likewise rejects the notion that things appear to our minds as if they were present at hand when he says that "no one can doubt [a true idea] unless he thinks that an idea is something mute, like a picture on a tablet," etc. For Spinoza, too, this rejection entails the rejection of the traditional hierarchical division of the faculties. And, as is more specifically pertinent to our investigation, it also entails the rejection of the possibility of spontaneously willing to doubt something that the intellect perceives in a clear and distinct manner. As Spinoza famously contends, "the will and the intellect are one and the same."[18] Thus, in identifying the will with the intellect, Spinoza defines the will as "a faculty of affirming and denying [*affirmat vel negat*], and not desire."[19] He makes this assertion in the context of rejecting the Cartesian conception of the will, particularly the distinction between the will (as infinite) and the intellect (as finite).[20] Spinoza's contention is not that there is no will, as his common classification as a necessitarian is often taken to imply, but that "there is no absolute faculty of willing and not willing."[21] In his account, and this is the crucial point, no faculty enjoys this status; that is, of being absolutely distinguishable from the other faculties. In his words: "There is in the mind no absolute faculty of understanding, desiring, loving, and the like. [...] These and similar faculties are either complete fiction or nothing but metaphysical beings, *or* universals, which we are used to forming from particulars."[22] As in Heidegger's case, the alternative to the traditional view with which Spinoza presents

---

Emil Lask: Sein Weg vom Neufichteanismus zu einer Hermeneutik der Faktizität," *Studia Phaenomenologica* 1, nos. 3/4 (2001): 221–47.

18 E2p49c.

19 E2p48s.

20 See corollary and scholium of E2p49. On the unity of the intellect and the will in Spinoza, see John Cottingham, "The Intellect, the Will, and the Passions: Spinoza's Critique of Descartes," *Journal of the History of Philosophy* 26, no. 2 (1988): 239–57, and Stephan Schmid, "Spinoza on the Unity of Will and Intellect," in *Partitioning the Soul: Debates from Plato to Leibniz*, ed. Klaus Corcilius and Dominik Perler (Berlin: De Gruyter, 2014), 245–70.

21 E2p48s.

22 E3p48s.

us is not a monolithic conception of the mind, but a complex description of it. Although for Spinoza, the finite mind is nothing but an idea, this idea is not an indifferent or undifferentiated form of representation, a bundle of confused ideas, but a form of expression that manifests itself on different levels and in multiple dimensions.

Using contemporary terminology, one may say that in arguing that the will is not desire, but rather "a faculty of affirming and denying," Spinoza does not wish to downplay the significance of that dimension of mentation that has often been discussed in late modern philosophical psychology in terms of intentionality. Rather, while Spinoza rejects the concept of desire, he introduces the corresponding term "striving" (*conatus*), which can be regarded as the most immediate counterpart of the notion of intentionality in his philosophy.[23] Now, striving itself is also regarded by Spinoza as a kind of affirmation. As he writes: "The first and principle [tendency] of the striving of the mind [...] is to affirm the existence of the body"[24] and "the essence of the mind consists in this [...] that it affirms the essence of the body."[25]

Spinoza's notion of affirmation designates the common ground, or equiprimordial source, of the various faculties. Famously highlighted by Gilles Deleuze in his interpretation of Spinoza's works, this affirmation is expressive (i.e., dynamic and multilayered) rather than, like a picture on a tablet, representational (i.e., static and flat).[26] We have already seen that thinking and willing are nothing but forms of affirmation, and this appears to apply to perception as well.[27] What is crucial for the purposes of our investigation is that affectedness—that is, the capacity of finite modes to be affected, or what can be described as their power to be affected—is also defined in terms of affirmation: "An affect [*affectus*] which is called a passion of the mind is a confused idea, by which the mind affirms [*affirmat*] of its body, or of some part of it, a greater or lesser force of existing than before, which, when it is given, determines the mind to think of this rather than that."[28] In particular, the final clause of this definition confirms that we can understand Spinoza's argument as similar to Heidegger's contention that affectedness (*Befindlichlkeit*) is equiprimor-

23 However, unlike Franz Brentano, who is thoroughly committed to Cartesian dualism and representationalism and who in fact does not go much beyond Descartes, notwithstanding the significance that is attributed to his reintroduction of the concept of intentionality, Spinoza does not use his concept of striving to differentiate mental phenomena from physical phenomena, but rather to describe their fundamental unity. See Franz Brentano, *Psychology from an Empirical Standpoint*, ed. Linda L. McAlister, trans. Antos C. Rancurello, D.B. Terrell, and Linda L. McAlister, 2nd ed. (London: Routledge, 1995), 68. On Brentano's Cartesianism, see Tim Crane, "Intentionality as the Mark of the Mental," *Royal Institute of Philosophy Supplements* 43 (1998): 229–51.

24 E3p10d.

25 E3, General Definition of the Affects, exp.

26 See Gilles Deleuze, *Expressionism in Philosophy: Spinoza*, trans. Martin Joughin (New York: Zone Books, 1990).

27 TdIE § 72; cf. E2d3exp.

28 E3, General Definition of the Affects.

dial with understanding (*Verstehen*), or, in Heidegger's words, that "affectedness always has its understanding" while "understanding always has its mood."[29]

To sum up this point, in arguing that "there is in the mind no absolute faculty of understanding, desiring, loving, and the like," Spinoza does not reject the significance of conceiving of the mind in terms of distinguishable faculties, but rather argues for the equiprimordial status of these faculties. Thinking, willing, and feeling are equiprimordial because they are all fundamentally nothing but forms of affirmation. Doubt accordingly cannot arise from some originary discord of the faculties. Whence then does it arise?

3

Crucial for the understanding of Spinoza's concept of doubt are his interrelated concepts of the imagination, memory, and temporality, all of which are explicated in terms of the embodied nature of finite modes, of which a systematic exposition is given in the so-called physical digression, that is, the set of lemmata about the body presented early in the second part of the *Ethics*.[30] While up to this point, Spinoza speaks only of the affections of substance, his initial remarks about the imagination provide the first occasion in the *Ethics* on which we encounter the expression "the affections of the human body."[31] Spinoza's initial remarks about memory provide the second such occasion. The crucial proposition for our purposes is E2p17:

> If the human Body is affected with a mode that involves the nature of an external body, the human Mind will regard the same external body as actually existing, or as present to it, until the Body is affected by an affect that excludes the existence or presence of that body.

It is only "to retain the customary words" that Spinoza goes on to explain in the scholium to this proposition that "the affections of the human body whose ideas present external bodies as present to us, we shall call images of things [...]. And when the mind regards bodies in this way, we shall say that it imagines." Similarly, Spinoza explains here that "memory is [...] nothing other than certain connection of ideas involving the nature of things which are outside the human Body." Thus rooted in temporality and indissolubly related to memory, imagining is not creating an idea *ex nihilo*, but perceiving something as if it were in the present even though it is in fact in the past, or, as is perhaps less intuitive but also true for Spinoza, in the future. We should further note that Spinoza suggests that time itself is "a product of the

---

29 BT 142/182: "Befindlichkeit hat je ihr Verständnis [...]. Verstehen ist immer gestimmtes."
30 The phrase "physical digression" is taken from David R. Lachterman, "The Physics of Spinoza's *Ethics*," *The Southwestern Journal of Philosophy* 8, no. 3 (1977): 71–111.
31 E2p17s.

imagination."[32] It is this conceptual framework that accounts for the possibility of doubt in Spinoza's system. In this context, Spinoza clarifies again why he considers narrowly cognitive doubt to be meaningless: "It is the nature of reason to regard things as necessary, not as contingent."[33] Evidently, it is impossible to regard things as necessary and doubt them at the same time.[34] "From this it follows," Spinoza says in the corollary to the proposition just quoted, "that it depends only on the imagination that we regard things as contingent, both in respect to the past and in respect to the future."

This may lead one to think that Spinoza conceives of the imagination as the source of all doubt; however, this is not quite the case. For one thing, contingency is a precondition for doubt, but not yet doubt itself. For another, in accordance with his conception of the faculties as equiprimordial, Spinoza regards the capacity to imagine, too, as one which shares a common origin with all other human capacities. It is on these grounds that he affirms that "the imaginations of the mind, considered in themselves, contain no error."[35] Just as doubt cannot arise from the intellect, the will, or perception alone, it also cannot arise from the imagination alone. But it also cannot be regarded as a product of some interplay between the faculties, such as the kind that Descartes assumes that exists between the finite intellect and the infinite will. So, whence does it arise?

4

Doubt does not arise from any faculty individually, nor from some interplay between them, but precisely from their unity, from the fact that the understanding is always already affected or that thinking is always already wishful, which amounts to the same thing. I shall now demonstrate that this holistic approach is only fully articulated in the third part of the *Ethics*, where Spinoza discusses the affects. In the second part of this book, which is devoted to the mind, this approach is merely stated, or, at best, indicated by that which is lacking from the description of doubt that Spinoza presents at that stage. Let us first consider that incomplete description and then see how it is completed in the third part.

In the second part of the *Ethics*, we do not yet find a fully-fledged notion of doubt, but only a closely related one: the "vacillation of the imagination." Spinoza provides a complex and rather strange example to explain this vacillation. A "child" is accustomed to seeing "Peter," "Paul," and "Simon" one after another

32 E2p44s; translation is taken from *Spinoza: Complete Works*, ed. Michael L. Morgan, trans. Samuel Shirley (Indianapolis: Hackett, 2002), 269.

33 E2p44.

34 But note well that, as we shall see in more detail below, Spinoza does not contrast doubt with necessity or certainty, but only with the absence of doubt. See E2p49s.

35 E2p17s.

in the morning, at noon, and in the evening respectively. Suddenly, one evening, he sees "James instead of Simon." As a result, Spinoza explains, the association between the coming of the evening and the appearance of Simon is now extended to the appearance of James, such that the child's imagination vacillates; namely, between the possibilities that James and Simon will appear in the evening. He used to feel certain about something—that every day, Simon would be the last to come; now, as the day proceeds according to the old order, his imagination vacillates. He thinks: maybe this time, James will come.

This example is strange for several reasons, one of which is particularly pertinent for our purposes. Why tell a whole story about how Peter appears in the morning and Paul at noon and only then tell us about Simon's regular appearance in the evening? Why not make the same point by focusing solely on the interruption of the child's routine encounters with Simon in the evening? The answer is suggested by Spinoza's reasoning that "the human Mind will regard the same external body as actually existing, or as present to it, until the Body is affected by an affect that excludes the existence or presence of that body." This notion of the persistence of the object, which is founded on physical-biological principles pertaining to the interaction of particles constituting composite bodies, is strange. For instance, it implies that if the child had never seen Peter and Paul, but only Simon, he would have imagined Simon as permanently present regardless of the time of day, and also that similarly, upon seeing James instead of Simon, he would have continued to imagine James in the same way, such that there would have been no vacillation and no room for doubt. Spinoza himself confirms this when he gives the example of a child imagining a winged horse. Regarding this example, Spinoza reasons that insofar as "the child does not perceive anything else that excludes the existence of the horse, he will necessarily regard the horse as present." Spinoza explicitly states that this child will be unable "to doubt [the horse's] existence, though he will not be certain of it."[36]

The description of doubt becomes more complete in the third part of the *Ethics*, after it has been established that "the mind [...] strives to imagine those things that increase the body's power of acting."[37] The implication here is that just as thinking is indissolubly related to willing, imagining is not merely a passive experience, but a matter of striving. The persistence of the imagined object can accordingly be regarded as a product not of a merely indifferent fixation of the imagination, but of a motivated one. The absence of an "affect that excludes the existence" of the object is now joined to the presence of an affect that affirms the existence of the individual perceiving or imagining it.[38] It is only once this is established that the place of doubt in Spinoza's system becomes apparent.

---

36 E2p49s.

37 E3p12.

38 I assume here that the individual's power of acting and force of existing are the same thing, as Spinoza never clearly distinguishes between the two.

Here, Spinoza describes another "vacillation," one that ensues not from perceptual conflict, but from affective conflict: "If we imagine that a thing which usually affects us with an affect of Sadness is like another which usually affects us with an equally great affect of Joy, we shall hate it and at the same time love it."[39] In this context, Spinoza makes a telling remark concerning how this "vacillation of the mind" is related to the formerly described "vacillation of the imagination": "This constitution of the Mind which arises from two contrary affects is called vacillation of mind, which is therefore related to the affect as doubt is to the imagination." While this merely appears to suggest that the conflict of affects (love and hate) and the conflict of ideas (Simon and James) are analogous, the next sentence confirms that what Spinoza has in mind here is not sheer analogy but originary unity: "Nor do vacillation of mind and doubt differ from one another except in degree."[40] I cannot explain what difference in degree Spinoza has in mind here, but I believe that this remark confirms that he regards cognitive and affective doubt as ultimately, because qualitatively, indistinguishable. This means not only that affective conflict is rooted in cognitive conflict, but also that the cognitive or perceptual conflict described in part 2 in terms of the vacillation of the imagination also necessarily has an affective dimension. The reason that this is not specified at that earlier stage is that at that point, the nature of the affects is yet to be explicated.

To sum up this point, the capacity to doubt has multiple dimensions: perceptual, mnemonic, imaginative, motivational, and affective. Only from such a holistic perspective does doubt have significance according to Spinoza. Given that this multiplicity has not yet been explicated in the second part of the *Ethics*, the workings of the imagination appear unmotivated at this stage, as no more than a mental fixation on a single idea. The description of doubt is complemented in part 3 as follows. While the imagination is a precondition for doubt, it is not a free-floating faculty. One strives to imagine certain things. The child continues to imagine a winged horse because imagining a winged horse has become part of his nature. In affirming the existence of the horse, he affirms his own nature. He would be disappointed to discover that there is no winged horse out there. Similarly, the child who is used to seeing Simon in the evening must have a particular feeling towards him, and another one towards James when he sees the latter. Otherwise, he would have not noticed the difference between the two in the first place. For "there are as many species of joy, sadness, and desire, and consequently of each affect composed of these (like vacillation of mind) or derived from them (like Love, Hate, Hope, Fear, etc.), as there are species of objects by which we are affected."[41]

39 E3p17.
40 E3p17s.
41 E3p56.

5

Accordingly, we find a particularly clear articulation of Spinoza's understanding of doubt in the third part of the *Ethics*, especially in the proposition that immediately follows the one we have just considered on affective conflict. It is here rather than in the *Theological-Political Treatise* or in the second part of the *Ethics* that the place of doubt in Spinoza's system is fully specified. Doubt is nonetheless not announced as the main subject matter of this proposition. Rather, Spinoza is primarily concerned here with the manner in which temporality colours the affective dimension of finite existence. This concern yields the definitions of hope, fear, confidence, and despair, all of which are explicated in terms of the exercise or "removal" of doubt. Let us consider these definitions as they are formulated in the list of definitions of the affects provided towards the end of the third part of the *Ethics*:

> XII. Hope is an inconstant Joy, born of the idea of a future or past thing whose outcome we to some extent doubt.
>
> XIII. Fear is an inconstant Sadness, born of the idea of a future or past thing whose outcome we to some extent doubt.
>
> XIV. Confidence is a Joy born of the idea of a future or past thing, concerning which the cause of doubting has been removed.
>
> XV. Despair is a Sadness born of the idea of a future or past thing concerning which the cause of doubting has been removed.

The psychological observations presented here, qua psychological observations (rather than metaphysical or epistemological ones), are far from original. Similar observations can readily be found in the writings of Spinoza's ancient, scholastic, and modern predecessors, including Aristotle, Seneca, Aquinas, Hobbes, and Descartes.[42] Nonetheless, Spinoza's formulation of them is unique. For one thing, it is particularly clear and elegant, as expressed in its perfect formal and temporal symmetry and in its conceptual parsimony. Especially notable with regard to the latter point is Spinoza's avoidance of the encumbering concept of desire, which is used, for instance, by both Aquinas and Descartes (although in conflicting ways).[43] For another, and this is arguably the reason why Spinoza's formulation is so clear and el-

42 See Aristotle, *The "Art" of Rhetoric*, trans. John Henry Freese (Cambridge, MA: Harvard University Press, 1926), 2.5; Seneca, *Letters from a Stoic: Epistulae Morales ad Lucilium*, trans. Robin Campbell (New York: Penguin, 1969), letter 5, p. 38; Thomas Aquinas, *The "Summa Theologica" of St. Thomas Aquinas* 1–2.40–44, trans. Fathers of the English Dominican Province, vol. 6 (London: Burns Oates and Washbourne, 1927), 455–98; Thomas Hobbes, *Leviathan*, 3rd ed. (London: George Routledge and Sons, 1887), ch. 6; CSM 1:350–51, 389, 391.

43 For Aquinas, despair (*desperatio*) necessarily involves desire (*Summa Theologica* 1–2.40.4, 1–2.41.2), while Descartes's position is that despair (*désespoir*) "entirely extinguishes desire" (CSM 1:389).

egant, his treatment of these affects, as well as that of the affects in general, is presented in a unified framework which brings together their ontological, epistemological, and ethical significance. In this, Spinoza differs from most other philosophers, including all the major figures I have just mentioned, who as a rule are primarily if not exclusively concerned with the passions as a moral or practical problem. In other words, what makes the observations presented here unique is that they are not merely psychological or ethical, but also metaphysical, which, given that Spinoza identifies existence with intelligibility, means that they have both ontological and epistemological significance. What is particularly significant for our purposes is their epistemological implications that pertain to the nature of doubt. Let us then consider these observations as they are presented by Spinoza and then explore their implications.

Hope and fear are the two complementary and inextricable aspects of the same affective state; namely, that of having doubt (*dubito*) with regard to the outcome of the idea in relation to which they are experienced. Hope is the pleasurable aspect of this experience; fear the unpleasurable. In other words, fear itself is nothing but the hope that that in the face of which the fear is experienced will not (and, which is perhaps less intuitive, also did not) take place, and vice versa. Simply remove the cause of doubt and one of the following transformations will take place. In the positive case, the hope aspect of the emotional state will take over, such that the overall state will transform into that of confidence, a pleasurable state that is altogether devoid of fear. In the negative case, the fear aspect of the state will take over, to the effect that all hope will be eliminated and the fear itself will transform into despair, a wholly unpleasurable state. Thus, Spinoza further writes in his explanation of the definitions just quoted:

> Confidence [...] is born of Hope and Despair of Fear, when the cause of doubt concerning the thing's outcome is removed. This happens because man imagines that the past or future thing is there, and regards it as present, or because he imagines other things, excluding the existence of the things that put him in doubt. For though we can never be certain of the outcome of singular things [...] it can still happen that we do not doubt their outcome. As we have shown [...] it is one thing not to doubt a thing, and another to be certain of it.

It is only at this stage that the nature of doubt is fully specified. This is done in a manner that complements the partial (and therefore abstract) description of doubt that we find in the second part of the *Ethics*. Here, we learn that doubt is not only necessarily embodied, imaginative, mnemonic, temporal, motivated (i.e., related to striving), and affective, but that it also necessarily finds expression in specific kinds of affects: fear and hope, which are really a single affect. However stubborn one may be in one's insistence that the possibility that we are all brains in vats constitutes a serious philosophical problem, this does not necessarily express an actual fear that this is the case.[44] We are now in a position to see that the problem with Cartesian scepticism from a Spinozist viewpoint is that it lacks a proper affective dimension. In his *Meditations*, Descartes presents us with no more than a thinking exercise; real thinking does not work that way.

However, given what Spinoza says about hope and fear—namely, that these affects "cannot be good of themselves";[45] that they "show a defect of knowledge and lack of power in the mind";[46] that "the more we strive to live according to the guidance of reason, the more we strive to depend less on hope, to free ourselves from fear";[47] that hope and fear are linked to superstition and the disciplining of the multitude through negative reinforcement[48]—is it not precisely Spinoza's negative evaluation of doubt that we find in the passages just cited? Spinoza's remark about hope and fear at the opening of the *Theological-Political Treatise* appears to make this point explicit:

> If men could manage all their affairs by a definite plan, or if fortune were always favorable to them, no one would be in the grip of superstition. But often they are in such a tight spot that they cannot decide on any plan. Then they usually vacillate wretchedly between hope and fear, desiring immoderately the uncertain goods of fortune, and ready to believe anything whatever. While the mind is in doubt, it's easily driven this way or that—and all the more easily when, shaken by hope and fear, it comes to a standstill. At other times, it's over-confident, boastful and presumptuous.[49]

Even if it is granted that Spinoza has solid grounds for dismissing Descartes's cognitive doubt as narrow and insincere, what warrants the claim that he positively presents us with a more radical conception of doubt? Furthermore, Spinoza says relatively little about doubt; as a rule, he deals with it in the *Ethics* not in the definitions and propositions, but in the demonstrations and scholia, most of which I have already partly cited or referenced. This appears to imply that if he considers doubt to be significant at all, it is only secondarily so.

In response to these claims, one should first note that Spinoza's negative appraisal of the state of doubt and of the corresponding passion of hope-fear cannot be understood as a trivial and wholesale rejection of their desirability. For to "depend less on hope" and "free ourselves from fear" cannot mean something similar to the "removal" of doubt which leads us from hope-fear to confidence/despair. Spinoza makes this point abundantly clear as he takes care to note that whatever he says

---

44 Similarly, contemporary philosophers often use the words "worry" and even "anxiety" to describe a concern with philosophical problems without ever establishing that anyone, including themselves, actually worries or feels anxiety about the problem at hand.

45 E4p47.

46 E4p47d.

47 E4p47s.

48 E4p63s.

49 TTP III/5.

about fear and hope also applies to confidence, despair, and related passions, since all the latter passions depend on hope and fear.[50] As we have seen in the passage just quoted from the *Theological-Political Treatise*, Spinoza speaks against hope-fear and confidence in the same breath. It is for this reason that it is inadequate to say that the child imagining a winged horse and nothing else has confidence in the existence of the horse in the present moment, for confidence must be acquired: one must first doubt in order to be able to remove that doubt. As we have seen, Spinoza makes this point explicit: the child imagining the winged horse who "does not perceive anything else that excludes the existence of the horse [...] will necessarily regard the horse as present. Nor will he be able to doubt its existence, though he will not be certain of it."[51] As we have also seen, these examples remain abstract as they belong to a section of the book in which the affective dimension of doubt is not explicated. In reality, no one is affected by a single idea, neither children nor adults. Rather, we are all "liable to a great many affects."[52] Lacking doubt in the sense of simply being unaware of any other possibility is certainly not what Spinoza has in mind when he speaks of freedom from fear. Such a state is not just undesirable, but also abstract and unreal.

Being "liable to a great many affects," we are always already subject to all sorts of doubts. It follows from this that we necessarily experience a mixture of hope and fear (or what the ancient Greeks called *elpis*; that is, expectation).[53] Now, while Spinoza makes it clear that his negative evaluation of hope and fear applies to confidence and despair, it is not clear whether it applies to them equally. All other things being equal, is it better to have confidence or be in despair than to be governed by hopes and fears? The answer to this question hinges on whether the "removal" of doubt that leads from hope-fear to confidence/despair is a good thing. This removal, Spinoza tells us, does not lead to certainty, for "we can never be certain of the outcome of singular things" and "it is one thing not to doubt a thing, and another to be certain of it." While it is impossible to doubt clear and distinct ideas, it is also impossible to be certain of confused ideas; that is, of the ideas of singular things (which Spinoza interchangeably qualifies as finite, particular, and contingent).

But if we can never be certain of singular things, then should we not avoid thinking about them altogether? Now, many commentators on Spinoza have mistaken this

---

50 E4p47s.
51 E2p49s.
52 E4p44s.
53 On the ambivalent significance of *elpis* in classical Greek literature, see *Hope, Joy, and Affection in the Classical World*, ed. Ruth R. Caston and Robert A. Kaster (New York: Oxford University Press, 2016), especially the essays by Douglas Cairns ("Metaphors for Hope in Archaic and Classical Greek Poetry," 13–44), Damien Nelis ("Emotion in Vergil's *Georgics:* Framing and the Politics of Hope," 45–74), and Laurel Fulkerson ("'Torn between Hope and Despair': Narrative Foreshadowing and Suspense in the Greek Novel," 75–94). See also Claudia Bloeser and Titus Stahl, "Hope," in *The Stanford Encyclopedia of Philosophy* (Spring 2017 Edition), ed. Edward N. Zalta, https://plato.stanford.edu/archives/spr2017/entries/hope/.

to be the case; namely, that the imperative of Spinoza's philosophy is to learn how to ignore finitude. In particular, this is the implication of the so-called acosmist reading of Spinoza. According to this reading, famously presented by Hegel in his *Lectures on the History of Philosophy*, but first articulated by Maimon in 1792, Spinoza is not an atheist (as suggested by Jacobi, for instance), but an acosmist. On the one hand, he affirms the reality of God as an infinite, unified, and static whole, but on the other, he rejects the reality of the cosmos as the manifestation of finitude, multiplicity, and change.[54]

While this reading may appear plausible based on the first part of the *Ethics*, where Spinoza states that "in nature there is nothing contingent,"[55] it is incongruent with the rest of the text. It is either the case that the *Ethics* is inconsistent or that this book demands a more nuanced reading. Considering "all particular things [...] contingent and corruptible,"[56] Spinoza nonetheless provides us with numerous definitions of such things, including definitions of finite minds and bodies and of their various affections. And the function of definition in Spinoza's system is not to capture or designate possibly inexistent entities, but rather to "express," "affirm," or "explain" the "nature," "reality," or "essence" of the thing defined.[57]

Furthermore, not only does Spinoza devote a great deal of attention to the nature of finite beings, but at no stage does he suggest that all we should be doing is to contemplate God in isolation or alongside other clear and distinct (i. e., necessary) ideas by cultivating an indifferent or apathetic attitude towards all contingencies or by denying their existence altogether. This would amount to confusing Spinozism with Buddhism, Stoicism, or Christian asceticism. Rather, Spinoza argues that "the more we understand singular things [*res singulares*]"—that is, contingent things such as ourselves and our passions—"the more we understand God."[58] However, contingencies can never be understood if the faculty of reason is taken as the sole point of departure. We must begin not only with an "I think," but also with an "I remember," "I imagine," "I strive," and "I feel." Though "a thing is called contingent only because of a defect of our knowledge"[59]—that is, because of our affects, which, unlike God's affections, render our thinking confused—the point Spinoza is making is not that we should overcome the affects entirely, "for each one governs everything

---

54 In Hegel's words: "The world has no true reality, and all this that we know as the world has been cast into the abyss of the one identity. There is therefore no such thing as finite reality, it has no truth whatever; according to Spinoza what is, is God, and God alone." Georg Wilhelm Friedrich Hegel, *Lectures on the History of Philosophy*, trans. Elizabeth Sanderson Haldane and Frances H. Simson, repr. ed. (London: Routledge and Kegan Paul, 1974), 3:281. Regarding Maimon's first articulation of this reading, see Yitzhak Melamed, "Why Spinoza Is Not an Eleatic Monist (Or Why Diversity Exists)," in *Spinoza on Monism*, ed. Philip Goff (Houndmills: Palgrave Macmillan, 2011), 218 n. 22.

55 E1p29.

56 E2p31c.

57 E1p8s2i–ii, E1p16d, E3p4d, E3da6, TdIE § 95, Ep. 34.

58 E5p24.

59 E1p33s1.

from his affects; those who are torn by contrary affects do not know what they want, and those who are not moved by any affect are very easily driven here and there."[60] In practice, moreover, a state of not having any affect is not only undesirable, according to Spinoza, but also impossible, for "man" as he writes, "is necessarily always subject to passions."[61] And even if a state of utter indifference were possible or desirable, any attempt at controlling or reducing the affects by means of anything that is external to them, such as the power of the will or that of reason, is still necessarily futile in Spinoza's view, given that "an affect cannot be restrained or taken away except by an affect opposite to, and stronger than, the affect to be restrained."[62]

Since we are "necessarily always subject to passions," we can never form clear and distinct ideas without experiencing accompanying feelings and confusion. In this reading, the *summum bonum* of the third kind of knowledge (also referred to as "intuitive knowledge," *scientia intuitiva*) is not achievable—at least not as long as one is alive. As long as one is alive, one is necessarily affected by a great many singular things, and this capacity to be affected expresses the power of the individual, not its weakness: "The human Mind is capable of perceiving a great many things, and is the more capable, the more its body can be disposed in a great many ways."[63] And: "Whatever so disposes the human Body that it can be affected in a great many ways, or renders it capable of affecting external Bodies in a great many ways, is useful to man."[64] Just as God's affections are expressive of God's absolute infinite nature, the affections of human individuals, including their active and passive affects, their joyful and sad passions, are expressive of their finite nature. Thus, even the basic passive and negative affect of sadness, defined as "man's passage from a greater to a lesser perfection," is not a privation. "For a privation," Spinoza reasons, "is nothing, whereas the affect of sadness is an act." [65] In short, being affected is not only inevitable; one *should* be affected by as many singular things as one is capable of being affected by without changing one's own nature, which boils

---

60 E3pref.

61 E4p4c. Spinoza repeats this argument in the first chapter of the *Political Treatise* (1.5): "For this much is quite certain, and proved to be true in our *Ethics*, that men are necessarily subject to passions."

62 E4p7; emphasis added. Nietzsche similarly argues: "The will to overcome an emotion is ultimately only the will of another emotion or of several others" ("Der Wille, einen Affekt zu überwinden, ist zuletzt doch nur der Wille eines anderen oder mehrerer anderer Affekte"). See Friedrich Nietzsche, *Jenseits von Gut und Böse: Vorspiel einer Philosophie der Zukunft* (Leipzig: Naumann, 1886), 94, translated by Walter Kaufmann as *Beyond Good and Evil: Prelude to a Philosophy of the Future*, new ed. (New York: Vintage, 1989), 98. Likewise, Heidegger writes in *Being and Time* (136/175): "When we master a mood, we do so by way of a counter-mood" ("Herr werden wir der Stimmung nie stimmungslos, sondern je aus einer Grundstimmung"). On Spinoza's affirmation of the passions, see Avraham Rot, "The Ontological Status of the Affects in Spinoza's Metaphysics: 'Being in,' 'Affection of,' and the Affirmation of Finitude," *The Review of Metaphysics* 71, no. 4 (2018): 667–701.

63 E2p14.

64 E4p38.

65 E3da3.

down to preserving a certain proportion of motion and rest in the body and, correspondingly, a certain order of actions and the passions in the mind.[66]

6

We can now reconsider our question: given that we must and ought to be affected by singular things, of which we can never be certain, and given that even when we have clear and distinct ideas, we must and ought to also have other interrelated ones with regard to which our thinking is imaginative, passionate, and confused in correspondence with our inherently finite and temporal and essentially embodied nature—that is, in short, given that absolute certainty is unattainable—is it better to be in a state of doubt and correspondingly experience hope and fear, or to be in the states of confidence or despair, in which the doubt has been "removed"?

To answer this question, we should give a closer examination of what constitutes the removal of doubt that leads from hope-fear to confidence/despair. Spinoza presents this removal as an operation of the imagination: "This happens because man imagines that the past or future thing is there, and regards it as present, or because he imagines other things, excluding the existence of the things that put him in doubt." We have established that this imagining is not whimsical, but rather motivated and intentional, for "the mind [...] strives to imagine those things that increase the body's power of acting."[67] The purpose of this imagining is for one to persevere in one's being. Accordingly, inasmuch as it fulfils this purpose, it must be a good thing.

But why should anyone imagine something as present even if it is in fact in the past or in the future or imagine something to the effect of excluding one's doubt concerning some other thing (which Spinoza suggests amounts to the same thing)? The answer is for the same reason that one might be motivated to undergo the transition from hope-fear to confidence/despair. This reason is not difficult to grasp. We can never be absolutely certain of our capacity to attain something that we hope for, and so, in having the hope, we are also afraid that we will never attain it. But sometimes, in order to persevere in our being, we have to put an end to the vacillation of hope and fear and transpose ourselves instead, through the imagination, into a stable state of either confidence or despair. In other words, since it can be impractical to experience the vacillation between hope and fear, we may imagine that things are certain even if they cannot be so. Thus, we may comport ourselves in confidence even if we are not sure about what we are doing and likewise despair of certain things simply because we cannot take on the responsibility that hoping for them entails. There is nothing wrong about this. Insofar as they help us persevere in our

66 E2 l7s; E3p2s.

67 E3p12.

being, confidence and despair and the kind of imagining they involve are in fact good things.

However, we also see that confidence and despair are derivative states. What comes first is hope-fear. Given that we are necessarily affected in a great many ways, we always already doubt. In other words, doubting is something that we do by default. Sometimes, we stop doubting. But the only reason for us to do so is to keep our self-integrity; that is, to preserve the unique proportion of motion and rest in our bodies and the unique order and connection of actions and passions in our minds, those relations that define each of us as a unique individual. We thus imagine certain things *in order* to become confident or to sink into despair, and we do so as a means of coping with certain situations or with our lives in general. We all use this coping mechanism in certain ways and to certain extents. We cannot doubt the outcome of all the singular things by which we are affected, even though we can never be entirely certain of the outcome of a single one of them.

This kind of not-doubting, as I have already noted, is crucially different from the kind of not-doubting exemplified by the child exclusively imagining a winged horse. For that child is unable to doubt and therefore cannot properly be said to have confidence in the existence of the horse. In fact, such a child can never exist and such a lack of doubt is impossible, as one is always already affected in many ways, some of which must appear to conflict with others. By contrast, the lack of doubt that we find in confidence and despair is very real. While no one is exclusively affected by a single particular thing, confidence and despair are prevalent feelings. The crucial point is that these feelings involve not the elimination of doubt, but its imaginative suppression, a suppression that is necessarily temporary and incomplete. Just as one cannot doubt everything, one cannot despair of, or be confident about, everything. Nor does one enter life in a state of confidence or despair. Rather, one *acquires* confidence and *sinks* into despair, having first been in a state of doubt. We are naturally disposed to having doubts, which is why doubting should not be presented as a philosophical achievement. And since doubt comes first, and this again is the crucial point, it can in fact never be fully removed.

This is so not just because we are affected in a great many ways. Doubt cannot be annihilated even with regard to a single singular thing. For in imagining something "excluding the existence of the things that put [one] in doubt," one must also have a memory trace of the previous imagination of having that doubt. If one had really completely forgotten that conflicting idea, one would have been essentially indistinguishable from the child imagining a winged horse, who cannot properly be said to be in a state of confidence or despair. As we have seen, in a state of confidence or despair, one does not spontaneously imagine "other things, excluding the existence of the things that put [one] in doubt," but one does so intentionally; that is, actively and purposefully. It is not just any random thing that will have the effect of excluding the cause of doubt, but only certain things. One must therefore imagine something specific that will have this effect. However, in order to do this, one must somehow be aware of the reason for doubt; that is, one must have at least an inkling of what one wants to ignore. To put it differently, if one merely randomly thinks of "something else," one may just as well find oneself thinking of the same thing again. Inasmuch as the imaginative effort is specifically exerted with the purpose of ignoring a specific cause of doubt, this very effort registers the existence of that doubt. To put it even more strongly, it is impossible to ignore anything entirely inasmuch as ignoring is an intentional act. Thus, confidence and despair are not only contingent on the general possibility of having doubt; they also require that the specific cause of doubt be remembered, if only partly, while it is ignored.

To sum up this point, just as certainty concerning clear and distinct ideas cannot give way to doubt, doubt concerning contingencies can never be eliminated; it can only be ignored. Being "capable of perceiving a great many things" and "affected in a great many ways," humans are "necessarily always subject to passions."[68] This equally implies that humans necessarily lack certainty concerning the things they perceive and by which they are affected, and hence that they have reasons to doubt whatever they perceive and feel. Sometimes these reasons must be ignored, imaginatively suppressed. However, this is only done for the purpose of maintaining self-integrity; that is, when circumstances are overwhelming and action requires some arbitrary closure (which is always the case to some extent). This does not result in the elimination of doubt, but only in its partial and temporary suspension. If circumstances allow it again and one regains the capacity to thrive, to increase one's power of acting and force of existing, one ceases to imagine the thing excluding the cause of doubt and thus re-encounters finite reality properly; that is, one re-experiences it as inherently contingent as it actually is. The moment one is able to be more realistic in such a way—that is, to stop imagining as a psychological means of coping—one returns to the more basic state of doubting and of having hopes and fears. Furthermore, since doubt cannot be eliminated, one is necessarily hopeful and fearful and cannot be absolutely confident or desperate, not even about a single singular thing. For as we have seen, confidence and despair demand the effective investment of an imaginative effort specifically targeting the cause of doubt. This implies that the cause of doubt is somehow known and that in disregarding this cause, the thing in relation to which the confidence or despair is experienced is not adequately captured in its contingency; that is, realistically. Thus, one cannot experience these inherently object-related emotions with regard to any object without falsifying its nature.

We can therefore put our conclusion as follows. For Spinoza, doubt is not a means to be used or a goal to be achieved, but a fact of life which cannot be avoided. Doubt is affective to the same extent that it is imaginative, perceptual, mnemonic, and motivated, but only once the affective dimension is clarified (i.e., in the third part of the *Ethics*) is its description completed. Hence, it can properly be labelled affective doubt. Coming last in the order of demonstration—that is, after perception,

---

68 E2p14; E4p4c.

memory, imagination, and striving—the affective dimension subsumes all the others. Without regard to this dimension, all the others appear as strange abstractions—as exemplified by the child imagining a winged horse or the one seeing Paul, Simon, Peter, etc. We have further established that being affective, Spinoza's doubt is an ongoing and continuous process, one which always already takes place and which can only be suspended, never permanently stopped. With regard to contingencies, it is all-encompassing; with regard to clear and distinct ideas, it is inapplicable. This argument, that doubt always already takes place or always already does not take place, cannot be dismissed as trivial or circular or as an argument that leads to an infinite regress, for it does not imply that the taking-place or not-taking-place of doubt in the mode of always already is an earlier, brute repetition of the doubt exercised at the present moment. Rather, it implies that doubt is repeated not just diachronically, in the past, present, and future, but also synchronically, appearing at once in all the various indissolubly related and yet distinguishable dimensions of finite existence. Corresponding to such a conception of doubt, Spinozist scepticism is more radical than Cartesian scepticism. While the former is real and multidimensional, the latter is just a thinking exercise.

7

We are still left with a concern that we have mentioned only briefly: if doubt is so important for Spinoza, why does he appear to devote such little attention to it? In order to address this concern, it is sufficient to point to the centrality of the place of the affects in Spinoza's philosophy, evidenced by the fact that the better part of the *Ethics* is devoted to their explanation. For being affected is indissolubly related to being in a state of doubt, as all the temporal affects, including confidence and despair, originate in hope and fear, which are defined as modes of doubting. Accordingly, the human affects and affective doubt are an integral part of Spinoza's metaphysics, and hope and fear and the other affects are accordingly put on a par with God and other matters that are typically considered the subject of first philosophy. In this regard, the difference from Descartes is particularly telling, for Descartes, as we know, devises his systematic classification of the affects in a separate book, *The Passions of the Soul*, which is also the last one he wrote. By way of conclusion, I would like to show that Descartes treats the passions and affective doubt only as a kind of afterthought, and, more than that, in a manner that is in fact incongruent with his earlier meditations on first philosophy. In this regard too, Spinoza's scepticism appears more radical: not only is it affective, but this dimension is also ascribed metaphysical significance.

The argument against Cartesian scepticism on which I have focused thus far targets its lack of affective dimension. My suggestion has been that the lack of affect corresponds to the lack of intention that Spinoza identifies as a problem when he says that he is concerned with "true doubt in the mind, and not [with] what we commonly see happen when someone says in words that he doubts although his mind does not doubt."[69] But qualifying Spinoza's scepticism as affective means more than that; that is, more than it is sincere. It also means that for Spinoza, scepticism itself—in the sense of both doubt and inquiry—can and should be exercised with respect to the affects themselves. In other words, in Spinoza's account, not only are the passions expressions of doubt and not only is doubt necessarily affective, but the passions themselves, being a kind of contingent thing, are necessarily subject to doubt. By contrast, Descartes does not only overlook the affective dimension of scepticism; he also, as we shall presently see, explicitly excludes the affects from the domain of applicability of his method of doubt and thereby relegates them to secondary significance. In other words, by excluding the affects from the domain of applicability of his method of doubt, Descartes shows that unlike Spinoza, he does not consider the affects a proper metaphysical or epistemological concern.

This makes Spinoza's scepticism all the more radical, but also less intuitive, for we are accustomed to thinking of feelings in accordance with Descartes as things that are beyond doubt. My fear may be unreasonable in that it is based on a misperception or misunderstanding, but no one is in a position to question the fact that I am afraid. For Spinoza, the opposite is the case. Not only can we not be certain of any finite mode due to the inherently contingent nature of finite existence, but it must be even less possible to be certain of the passions, for the passions are even more finite and contingent than finite modes, being the affections *of* finite modes, contingencies of the second order. If the I cannot be known with certainty on the grounds that it is a finite mode, it follows *a fortiori* that the *affections of* the I cannot be known with certainty, not even by the I itself. Nonetheless, even though we cannot know our particular feelings with certainty, we can know with certainty *that* we are affected in various specific ways, such that the affects can be defined just as the cause of itself or the nature of a triangle can be defined; that is, as part of nature and correspondingly as part of a single metaphysical system.[70]

Cartesian scepticism is, by contrast, limited not only because it is exclusively cognitivist, but also because it is articulated in a dualistic framework. Questioning the reality of an external world, it presupposes the reality of internal experience. Descartes confirms this in *The Passions of the Soul* when he states that unlike "perceptions which refer to objects outside us," the passions are "so close and internal to our soul that it cannot possibly feel them unless they are truly as it feels them to be."[71] This idea is foundational and has often been reiterated in modern philosophical psy-

---

69 This argument has been reiterated more recently by Peirce and others in the pragmatist tradition. As Christopher Hookway has it, the basic contention in this regard has been that "we should not doubt in philosophy what we do not doubt in our hearts." See Christopher Hookway, "Doubt: Affective States and the Regulation of Inquiry," *Canadian Journal of Philosophy*, Supplementary Volume 24 (1998): 205.

70 Cf. E3pref and TP 1.1.

71 CSM 1:338.

chology. According to Franz Brentano, for instance, since the "objects of inner perception" are a matter of "immediate insight," in questioning them, one "would reach a state of absolute doubt, a skepticism which would certainly destroy itself, because it would have destroyed any firm basis upon which it could endeavor to attack knowledge."[72]

While Descartes does not hesitate to doubt clear and distinct ideas, he insists that the passions are felt precisely as they truly are. But what about all the reasons for doubt that he lists in the *Meditations*? Descartes himself goes on to consider one of them immediately after the passage just cited from *The Passions of the Soul*:

> Thus often when we sleep, and sometimes even when we are awake, we imagine certain things so vividly that we think we see them before us, or feel them in our body, although they are not there at all. But even if we are asleep and dreaming, we cannot feel sad, or moved by any other passion, unless the soul truly has this passion within it.

Indeed, we can wake up from a nightmare and be uncertain about what we have just dreamt but nonetheless be certain about having experienced a feeling of anxiety. Often, we continue to experience it. While the connection of ideas is disrupted upon waking up, the feeling persists. But even if this example can indeed be taken to show that we have some sort of immediate awareness of our feelings, what about the possibility that evil demons have manipulated them? Cannot such demons implant false feelings in us just as they can implant false ideas about mathematics and the external world? Tellingly, Descartes does not invoke this argument here. This is arguably so because he rejects the very notion that the passions can be meaningfully qualified as true or false, for in his account, they can only be false if they are taken to represent something in the external world. Given that they are impressions produced purely internally by the "fortuitous" movements of the spirits, they do not properly represent anything—that is, they are not about anything—and accordingly, it is meaningless to doubt them.[73] Thus, the invocation of the argument about sleeping is already redundant. The passions simply cannot be doubted as a matter of principle. Descartes thereby confirms what Brentano makes explicit; namely, that by doubting such "objects of inner perception," one "would reach a state of absolute doubt, a skepticism which would certainly destroy itself, because it would have de-

---

72 Brentano, *Psychology*, 7.

73 The contrast with Spinoza is again instructive. According to Spinoza (E3p56), "there are as many species of joy, sadness, and desire, and consequently of each affect composed of these [...] or derived from them [...] as there are species of objects by which we are affected." That is, for Spinoza, the passions are defined by their intentional objects; they derive their meaning from the specific way in which they situate the individual in relation to the object. Descartes explicitly rejects this view (CSM 1:371): "The objects which stimulate the senses do not excite different passions in us because of differences in the objects, but only because of the various ways in which they may harm or benefit us, or in general have importance for us."

stroyed any firm basis upon which it could endeavor to attack knowledge." Spinoza's scepticism is more radical in that it does not impose this restriction.

To conclude, although Spinoza is often considered an antisceptical thinker, he presents us in effect with a more complete and truthful and therefore more radical form of scepticism than Descartes. While in Descartes's account, the affects cannot be doubted and doubt is not significantly affective, in Spinoza's account, the affects themselves are expressions of doubt and it is, in fact, clear and distinct ideas that cannot be earnestly doubted. Cartesian scepticism is transitory and narrowly cognitive and is hence less radical than Spinozist scepticism, which is all-encompassing, ongoing, and equiprimordially cognitive, embodied, imaginative, mnemonic, motivated, and affective. While Descartes considers both the affects and doubt as states that need to be overcome,[74] for Spinoza, they are a force that needs to be reckoned with, as humans are "necessarily always subject to passions." Also unlike Descartes, Spinoza is committed to the view that human understanding is inherently confused. Accordingly, confusion, wonder, or doubt is not only where inquiry begins or how it proceeds, but also where it stays.

Bibliography

Aquinas, Thomas. *The "Summa Theologica" of St. Thomas Aquinas*. Vol. 6, *Part II* (First Part), *First Number* (*QQ. I.–XLVIII*). Translated by the Fathers of the English Dominican Province. London: Burns Oates and Washbourne, 1927.

Aristotle, *The "Art" of Rhetoric*. Translated by John Henry Freese. Cambridge, MA: Harvard University Press, 1926.

Bloeser, Claudia and Titus Stahl. "Hope." In *The Stanford Encyclopedia of Philosophy*, Spring 2017 Edition, edited by Edward N. Zalta: https://plato.stanford.edu/archives/spr2017/entries/hope/.

Brentano, Franz. *Psychology from an Empirical Standpoint*. Edited by Linda L. McAlister. Translated by Antos C. Rancurello, D.B. Terrell, and Linda L. McAlister. 2nd ed. London: Routledge, 1995.

Cairns, Douglas. "Metaphors for Hope in Archaic and Classical Greek Poetry." In *Hope, Joy, and Affection in the Classical World*, edited by Ruth R. Caston and Robert A. Kaster, 13–44. New York: Oxford University Press, 2016.

Caston, Ruth R., and Robert A. Kaster, eds. *Hope, Joy, and Affection in the Classical World*. New York: Oxford University Press, 2016.

Cottingham, John. "The Intellect, the Will, and the Passions: Spinoza's Critique of Descartes." *Journal of the History of Philosophy* 26, no. 2 (1988): 239–57.

Crane, Tim. "Intentionality as the Mark of the Mental." *Royal Institute of Philosophy Supplements* 43 (1998): 229–51.

Deleuze, Gilles. *Expressionism in Philosophy: Spinoza*. Translated by Martin Joughin. New York: Zone Books, 1990.

Della Rocca, Michael. *Spinoza*. London: Routledge, 2008.

74 CSM 1:348: "There is no soul so weak that it cannot, if well directed, acquire an absolute power over its passions."

Descartes, René. *The Philosophical Writings of Descartes*. Translated by John Cottingham, Robert Stoothoff, and Dugald Murdoch. 2 vols. Cambridge: Cambridge University Press, 1984–85.
Doney, Willis. "Spinoza on Philosophical Skepticism." *The Monist* 55, no. 4 (1971): 617–35.
Fulkerson, Laurel. "'Torn between Hope and Despair': Narrative Foreshadowing and Suspense in the Greek Novel." In *Hope, Joy, and Affection in the Classical World*, edited by Ruth R. Caston and Robert A. Kaster, 75–91. New York: Oxford University Press, 2016.
Hegel, Georg Wilhelm Friedrich. *Lectures on the History of Philosophy*. Translated by Elizabeth Sanderson Haldane and Frances H. Simson. 3 vols. Repr. ed. London: Routledge and Kegan Paul, 1974.
Heidegger, Martin. *Die Grundbegriffe der Metaphysik: Welt—Endlichkeit—Einsamkeit*. Edited by Friedrich-Wilhelm von Herrmann. Vols 29 and 30 of *Gesamtausgabe*. Edited by Vittorio Klostermann. 3rd ed. Frankfurt am Main: Klostermann, 2010. English version: *The Fundamental Concepts of Metaphysics: World, Finitude, Solitude*. Translated by William McNeill and Nicholas Walker. Bloomington: Indiana University Press, 1995.
Heidegger, Martin. *Sein und Zeit*. 19th ed. Tübingen: Niemeyer, 2006. English version: *Being and Time*. Translated by John Macquarrie and Edward Robinson. Repr. ed. Oxford: Blackwell, 2001.
Hobbes, Thomas. *Leviathan*. 3rd ed. London: George Routledge and Sons, 1887.
Hookway, Christopher. "Doubt: Affective States and the Regulation of Inquiry." *Canadian Journal of Philosophy*, Supplementary Volume 24 (1998): 203–26.
Kisiel, Theodore. "Heideggers Dankesschuld an Emil Lask: Sein Weg vom Neufichteanismus zu einer Hermeneutik der Faktizität." *Studia Phaenomenologica* 1, nos. 3/4 (2001): 221–47.
Lachterman, David R. "The Physics of Spinoza's *Ethics*." *The Southwestern Journal of Philosophy* 8, no. 3 (1977): 71–111.
Lysemose, Kasper. "The Force of Existence: Looking for Spinoza in Heidegger." *Sophia* (2019).
Melamed, Yitzhak. "Why Spinoza Is Not an Eleatic Monist (Or Why Diversity Exists)." In *Spinoza on Monism*, edited by Philip Goff, 206–22. Houndmills: Palgrave Macmillan, 2011.
Nelis, Damien. "Emotion in Vergil's *Georgics*: Framing and the Politics of Hope." In *Hope, Joy, and Affection in the Classical World*, edited by Ruth R. Caston and Robert A. Kaster, 45–74. New York: Oxford University Press, 2016.
Nietzsche, Friedrich. *Jenseits von Gut und Böse: Vorspiel einer Philosophie der Zukunft*. Leipzig: Naumann, 1886. English version: *Beyond Good and Evil: Prelude to a Philosophy of the Future*. Translated by Walter Kaufmann. New ed. New York: Vintage, 1989.
Parmenides. *Parmenides*. Translated by Leonardo Tarán. Princeton: Princeton University Press, 1965.
Rot, Avraham. "The Ontological Status of the Affects in Spinoza's Metaphysics: 'Being in,' 'Affection of,' and the Affirmation of Finitude." *The Review of Metaphysics* 71, no. 4 (2018): 667–701.
Schmid, Stephan. "Spinoza on the Unity of Will and Intellect." In *Partitioning the Soul: Debates from Plato to Leibniz*, edited by Klaus Corcilius and Dominik Perler, 245–70. Berlin: De Gruyter, 2014.
Seneca. *Letters from a Stoic: Epistulae Morales ad Lucilium*. Translated by Robin Campbell. New York: Penguin, 1969.
Spinoza, Benedict de. *Spinoza: Complete Works*. Edited by Michael L. Morgan. Translated by Samuel Shirley. Indianapolis: Hackett, 2002.
Spinoza, Benedict de. *Spinoza Opera*. Edited by Carl Gebhardt. 4 vols. Heidelberg: Carl Winters, 1925.
Spinoza, Benedict de. *The Collected Works of Spinoza*. Translated by Edwin Curley. 2 vols. Princeton: Princeton University Press, 1985–2016.

Timothy Franz

What was Maimon After? His Metaphysical Foundations of Natural Science in 1790

This essay is an outline of part of a future commentary on the philosophy of Solomon Maimon (?1753–1800).[1] It covers his article entitled "Über die Weltseele (*Entelechia Universi*)," published in the *Berlinisches Journal für Aufklärung* in July 1790, and how it directly follows on from the problems of his larger work, the *Versuch über die Transzendentalphilosophie*, published earlier that same year. First, of course, these problems must be specified. As a mere outline, this essay works towards a larger commentary on Maimon's philosophy from 1790 to 1800. Such a commentary is necessary first of all because of the *status quaestionis*. From at least 1848, historians of philosophy have been discussing Maimon's work. However, even given appreciative and enlightening books by Ernst Cassirer (1920), Richard Kroner (1921), Martial Gueroult (1929), Samuel Atlas (1964), Samuel H. Bergmann (1967), Frederick Beiser (1987), and Aachim Engstler (1990) (to mention only some), it is nevertheless true that the seven volumes of his works which we currently have remain an unknown quantity.[2] The same may be said about the many important articles on the subject. One can read the relevant books and articles, understand and learn from them, and still reasonably ask: "What precisely do Maimon's works argue, and why do they argue whatever it is that they do?" At the same time, we know that Kant, Fichte, and some of Maimon's commentators held him in high regard. Therefore, the *Stanford Encyclopedia of Philosophy* reports Manfred Frank's tantalising suggestion that "Maimon is the 'last great philosopher' about to be discovered."[3]

1 Solomon Maimon, *Gesammelte Werke*, ed. Valerio Verra. 7 vols. (Hildsheim: Olms, 1965–76), abbreviated here as GW. Translations from the German are mine except when from Maimon, *Essay on Transcendental Philosophy*, trans. Nick Midgley, Henry Somers-Hall, Alastair Welchman, and Merten Reglitz (New York: Continuum, 2010; original version: *Versuch über die Transzendentalphilosophie* [Berlin: Bei Christian Friedrich Voss und Sohn, 1790]), and then cited as ETP. This translation also gives the GW pagination.

2 Martial Gueroult, *La Philosophie transcendantale de Salomon Maïmon* (Paris: Libraire Félix Alcan, 1929); Ernst Cassirer, *Das Erkenntnisproblem in der Philosophie und Wissenschaft der Neueren Zeit, Dritter Band: Die Nachkantischen Systeme* (Hamburg: Felix Meiner Verlag, 2000; original version Verlag Bruno Cassirer [1929]); Richard Kroner, *Von Kant bis Hegel*, 2nd edition (Tübingen: J. C. B Mohr/Paul Siebeck, 1961 [1921–24]); Samuel Atlas, *From Critical to Speculative Idealism, The Philosophy of Solomon Maimon* (The Hague: Martinus Nijhoff, 1964); Samuel H. Bergmann, *The Philosophy of Solomon Maimon*. Translated by Noah J. Jacobs (Jerusalem: Magnes Press, 1967); Frederick C. Beiser, *The Fate of Reason: German Philosophy from Kant to Fichte* (Cambridge, MA: Harvard University Press, 1987); Achim Engstler, *Untersuchungen zum Idealismus Salomon Maimons* (Stuttgart-Bad Cannstatt: Frommann-Holzboog, 1990).

3 Peter Thielke and Yitzhak Melamed, "Salomon Maimon," *The Stanford Encyclopedia of Philosophy* (Fall 2015 edition), ed. Edward N. Zalta, https://plato.stanford.edu/archives/fall2015/entries/maimon/.

https://doi.org/10.1515/9783110618839-011

Thus, there is a paradox: a large amount of good work has at the same time increasingly obscured Maimon's argumentation. Here are four (certainly non-exhaustive) reasons for this situation. First, Maimon led what has been described as a "doubly alienated" life.[4] In Polish Lithuania, as a poor but nevertheless privileged Talmudic prodigy within his own community, he remained alienated from the artistic and scientific culture which he knew existed, but was just beyond his grasp. He left for Germany, where he intended, as straightforwardly as possible, to discover the truth in philosophy. There, however, he was inescapably identified with the foreignness of the place he had left, and so he was again alienated from what he wanted to achieve and was not known as he wanted to be known.[5] This fact, related to a fear of actual and potential unappreciative reviews of his works, resulted in a Talmudic style that was even more cryptic and allusive than it would instead perhaps otherwise have been.[6] Secondly, Maimon's alienated life in Germany took its unfortunate course in that "Maimon's appearance on the philosophical scene was like that of a meteor: sudden, of short duration and apparently leaving no trace."[7] This must be true. When Maimon published his most sophisticated work, the *Versuch einer neuen Logik* (henceforth: *Logik*), in 1794, he disappeared from public recognition just as Fichte acceded to it with his *Wissenschaftslehre* and his appointment to the University of Jena.[8] Thirdly, in terms of Maimon's reception, many commentators have voiced certain preconceptions about him based on his personality and cultural situation, but have not further justified (or abandoned) those preconceptions through the effort of close reading.[9] And fourthly, the close reading that is called

4 Abraham P. Socher, *The Radical Enlightenment of Solomon Maimon: Judaism, Heresy, and Philosophy* (Stanford: Stanford University Press, 2006), 112ff.

5 In a letter to Reinhold of March 28, 1794, after a remark about Jews typically gaining "an air of importance for themselves at someone else's expense," Kant stated: "I have never really understood what he is after." See Immanuel Kant, *Correspondence*, trans. and ed. A. Zweig (Cambridge: Cambridge University Press, 1999; original vols. 10–13 of *Gesammelte Schriften* [Berlin: Georg Reimer, 1900–1922]), 476.

6 In the *Versuch* (1790), one finds: "Incidentally, to what extent I am a Kantian, an anti-Kantian, both at the same time, or neither, I leave to the judgment of the thinking reader" (GW 2:9–10). And, in the *Logik* (1794): "I do not always express my thoughts in a sufficiently determined or systematic way, and occasionally leave holes in [their] execution, whose filling in I cannot expect from every reader" (GW 5:26). The *Weltseele* is conducted in the style of a Talmudist who would "position his two legs on the two opposite sides of a Talmudic distinction and ask his disputants where he belonged." For this citation, see Yitzhak Melamed, "Salomon Maimon and the Failure of Modern Jewish Philosophy" (2012), http://www.theapj.com/wp-content/uploads/2013/03/Melamed-on-Maimon.doc, 13.

7 Atlas, *From Critical to Speculative Idealism*, 7.

8 This is not to say that Maimon was not read; he was likely eagerly read by certain interested parties. And so, like the traces of a meteor which surface years after the impact, mixed with local elements, Maimon appears integrally yet anonymously in German Idealism, neo-Kantianism, logical positivism, and other traditions. Reading them aids in reading Maimon, and vice versa.

9 Friedrich Schlegel's fragments include "Maimon ein *entgeisterter* Kant," cited in Engstler, *Untersuchungen zum Idealismus Salomon Maimons*, 95. See also Franz Rosenzweig's remarks to the effect that

for is difficult, and not only because of Maimon's style. Maimon eagerly read the works of Immanuel Kant and claimed to have anticipated their major themes.[10] He also adopted the claims and method of transcendental or critical reflection, though he carried it out via an idiosyncratic sceptical method.[11]

A commentary on Maimon is therefore also necessary because of his apparently quite difficult and complex relationship to Kantian philosophy. One of the foremost contemporary Maimon scholars, Prof. Gideon Freudenthal, has put forth the thesis that Maimon not only found nothing impressive in modern philosophy (as opposed to modern science), but also that in his first philosophical book in German, the *Versuch über die Transzendentalphilosophie* of 1790, he "denies Kant's revolution in philosophy and claims to further improve on him on the basis of Leibniz and Hume."[12] According to this interpretation, Maimon denied Kant's revolution because he had disproved the existence of synthetic *a priori* propositions in mathematics and physics and was therefore able to treat Kant's argument for the categories as merely a limited variation of the age-old ontological question of the relation or community between

---

Maimon left a self-enclosed culture in favour of barbarism and a philosophy of totality destructive of individual value, cited in Socher, *The Radical Enlightenment of Solomon Maimon*, 152–53.

10 In chapter 3 of the *Lebensgeschichte*, Maimon says that memories of studying the Bible with his father caused him to write an essay (around 1785–68) that he sent to Professor Garve concerning how dogmatic commitment to the principle of sufficient reason leads to a contradiction, an idea "which, though at the time I knew nothing of the Kantian philosophy, still constitutes its [dialectical] foundation." (This was the Professor Garve to whom Kant sent the letter of September 21, 1798, confessing that it was the "antinomy of pure reason" which "aroused me from my dogmatic slumber"; Kant, *Correspondence*, 551–52.) In chapter 13, he writes that he derived a merely human use from the ten Sephiroth of the Kabbalah, treating them as logical categories which have ontological application in experience; this angered the Kabbalists, who asserted a divine origin for the categories instead of a human one. In chapter 16, Maimon's encounter with mystical excesses of the Hasidim and others led him to develop an ethics which demands consideration of actions estimated intrinsically "in relation to some end" prior to their adoption as means. Furthermore, the ten chapters on Maimonides continually associate Maimonides's thoughts on God with Kant's transcendental dialectic. Overall, the *Lebensgeschichte* is filled with Maimon's anticipations of Kantian ideas.

11 For examples of Maimon's description of himself as a "critical sceptic," see (among many examples) GW 3:429n.; GW 4:210–11, 225–26, 220, 260; GW 5:351 ff. In each, he asserts his willingness to defend critical philosophy against all comers and proclaims that it should not be dogmatic (as it would be for Reinhold) but have an internal scepticism, whether of the *quid juris* or the *quid facti*. In my reading, Maimon's challenge was one of purifying critical reflection from unjustified hypotheses (the *quid facti*) and at the same time coming to a critical *a priori* logic of physics (the *quid juris*).

12 Gideon Freudenthal, "Maimon's Philosophical Itinerary," afterword to Solomon Maimon, *The Autobiography of Solomon Maimon*, ed. Yitzhak Melamed and Abraham P. Socher, trans. Paul Reitter (Princeton: Princeton University Press, 2018; original version: *Lebensgeschichte* [Berlin: Friedrich Vieweg dem ältern, 1792–93]), 246, 261. However, at the end of chapter 12 of the *Lebensgeschichte*, Maimon argues that it would be just as unfair to condemn Mendelssohn's philosophy as non-Kantian as it would be to condemn Maimonides's Ptolemian astronomy as non-Newtonian, for both had nevertheless achieved great things. This admits a parity between scientific (Newtonian) and philosophical (Kantian) development. There are many further examples of this way of thinking.

the soul and the body.[13] However, an interpretation of the further course of Maimon's philosophy, including not only the *Versuch*, but also the *Weltseele*, the *Logik*, the *Kritische Untersuchungen über den menschlichen Geist* (1797), and the smaller works that subserve these major ones, can convincingly (and, in my opinion, correctly) advance the precise antithesis: that Maimon did not deny Kant's revolution, but in fact intended to perfect it and succeeded—in the *Logik*—in giving an exemplary version of its perfection. A commentary on Maimon may therefore become a locus for renewed discussion of the nature, feasibility, and desirability of the philosophy of transcendental reflection in general.

But, then, how exactly might Maimon's work perfect or complete Kantian philosophy? To see this, one must recall that this very possibility was built into Kant's major works. They are each entitled *Critique*, so that their achievement is to be measured against the idea of *transcendental philosophy.* Kant presents the distinction between critique and transcendental philosophy most fully in section 7 of the introduction to the B edition of the first *Critique.*[14] When transcendental reflection is completed in open-ended critique of the special sciences—i. e., when these sciences are systematically unified in experience (*Erfahrung*) so that the nature of their syntheses is identified and the most basic principles of experience are shown to be constitutive of those sciences—then critique reaches its immanent goal of transcendental philosophy. Maimon, from the beginning of his engagement with the *Critique of Pure Reason,* realised that its reflection did not justify the cognition claims of the science of

---

13 Freudenthal, 249 ff. However, in my opinion, Maimon's *Versuch* did not disprove the existence of synthetic *a priori* propositions in mathematics. Rather, it indirectly proved them by attempting to make them analytic, failing, and thereby discovering "brute facts of intuition" in a way that agrees with Kant's distinction between mathematical and philosophical reason (as in the *Critique*'s "Transcendental Doctrine of Method"). As for physics, see Gideon Freudenthal, "Maimon's Subversion of Kant's *Critique of Pure Reason:* There Are No Synthetic *a priori* Judgments in Physics," in *Salomon Maimon: Rational Dogmatist, Empirical Skeptic. Critical Assessments*, edited by Gideon Freudenthal (Dordrecht: Kluwer, 2003): 144–75. This article is powerful and educative and I agree with its analysis of Maimon's argument (in the 1793 commentary to Pemberton's book on Newton and the 1797 *Kritische Untersuchungen*); i. e., that in his reflection on physics, Kant simply had no right to analytically substitute concepts of dynamics for those of pure mathematics. I note that in this vein, Maimon's commentary to Bertholdy's translation of Bacon's *New Organon*, GW 4:357 ff., argues that critical reflection on experience, even on scientific experience, does not license *a priori* commitment to any *specific* physical theories or hypotheses, e. g., "forces." Maimon is far less committed than Kant to the dynamical paradigm. But this need not mean that Maimon's "critical scepticism" denies the philosophy of transcendental reflection. Rather, Maimon's main works are ultimately devoted to making critique more critical through scepticism—which is to move towards transcendental philosophy.

14 Immanuel Kant, *Critique of Pure Reason*, ed. and trans. Paul Guyer and Allen W. Wood (Cambridge: Cambridge University Press, 1998). Original work: *Kritik der reinen Vernunft* (Riga: Hartknoch, 1781; 2nd ed., 1787). Also, Werner Flach, "Transzendentalphilosophie und Kritik. Zur Bestimmung des Verältnisses der Titelbegriff der Kantischen Philosophie," in *Tradition und Kritik: Festschrift für Rudolf Zocher zum 80. Geburtstag*, ed. Wilhelm Arnold und Hermann Zeltner (Stuttgart and Bad Cannstatt: Frommann-Holzboog, 1967): 69–83.

physics.[15] Without a proper critique of the science of physics, there could be no transcendental philosophy. Maimon attempted to provide that critique. Kant later came to the same realisation and began working on what we know as the *Opus Postumum* sometime in or after 1796.[16]

The reasons for the deficiency of the *Critique* vis-à-vis physics are perhaps obvious in retrospect. That work simply did not give the conditions of the possibility of the science of physics. It was instead propped up on the dynamical hypothesis that matter (as a condition of experience to which any actual cognition must be connected [A216/B266]) is a product of two reciprocally interacting yet independent forces, repulsion and attraction.[17] Furthermore, critical reflection on physics (in the *Metaphysical Foundations of Natural Science*) not only indicated that the experience of the truths of physics was different from the experience specified in the first *Critique*, but its analysis also revealed to interested readers, and to Kant himself, a circularity in the supposedly independent forces. The force of attraction depends upon certain values (the distance between bodies, since repulsive force is in inverse proportion to the cube of that distance) which only it can give (because it is uniquely universally penetrative).[18] Thus, in the earlier parts of the *Opus Postumum*, Kant en-

---

15 Its development in the *Critique of Judgment* could not do so either. One aspect of the justification of the cognition claims of physics must be outlining a type of symbolic cognition that refers to objects either indirectly given to perception or not given at all. Maimon argues: "So we cannot make such concepts cognizable [*kennbar*] by means of the objects in which they are found, but only by means of signs" (GW 2:273, ETP 143). See the entire *Anhang über die symbolische Erkenntnis* to the *Versuch*. A response to the thesis that symbolic cognition can entirely transcend and organise intuition *a priori* is in Kant, KU § 59: "The more recent logicians have come to use the word symbolic in another sense that is wrong and runs counter to the meaning of the word. They use it to contrast symbolic with intuitive presentation" (Immanuel Kant, *Critique of Judgment*, trans. Werner S. Pluhar [Indianapolis and Cambridge: Hackett, 1987]; original work: *Kritik der Urteilskraft* [Berlin and Libau: Bey Lagarde und Friederich, 1790], 227).

16 He perhaps realised this earlier than 1790, and had mostly completed the *Opus* by the middle of 1801. See Förster's introduction to Immanuel Kant, *Opus Postumum*, trans. Eckhart Förster and Michael Rosen, ed. Eckhart Förster (Cambridge: Cambridge University Press, 1993; original work published in Berlin: De Gruyter, 1938), p. xxxvii, where he finds that "early in the 1790s, Kant's thinking on the philosophy of nature went through a transitional period."

17 "Hypothesis" here means an adjunct to a theory that helps explain that theory but is "higher" than it in the sense that it cannot itself be falsified in experiment. In this sense, one could argue that the "ether," so important for Kant, was retained by scientists as a hypothesis because the theoretical phenomena of light and magnetism, for example, seemed to require a substratum. What Michelson's attempt of 1887 did, however, was to contradict the conjecture that the speed of light must change relative to the earth. It did not contradict the ether hypothesis, however; that could simply be dropped. I rely here on Hans Wagner, *Philosophie und Reflexion*, 2nd ed. (Munich and Basel: Ernst Reinhardt Verlag, 1967; first published 1959), 199, who references Hugo Dingler's work on hypotheses.

18 See Förster's introduction to *Opus Postumum*, p. xxxvi, and also Kant's letter to Beck of October 16, 1791, in *Correspondence*, 434 ff. Also Kant, *Metaphysical Foundations of Natural Science*, ed. and trans. M. Friedman (Cambridge: Cambridge University Press, 2004; Original work: *Metaphysische Anfangsgründe der Natur-wissenschaft* [Riga: Hartknoch, 1786]).

deavoured to solve the circle problem by adopting the absolute circularity of the etheric force as a ground for repulsion and attraction. As such, the ether is supposed to be an absolute and unconditioned force, but there is one problem with that supposal. The problem is that it is still a hypothesis, and any hypothesis is conditioned by the theory for which it is a hypothesis (in this case, the Newtonian theory). If Kant is to have an *a priori* logic of physics, then what is needed is not a conditioned (and therefore *a posteriori*) hypothesis, but an *a priori* theory of theories. He comes closest to this in the suggestion that the "object of physics" to be outlined *a priori* is that of the "appearance of appearances"[19]—the ways in which appearances may appear to us, such that they can be described by physics.

This is the thesis I would propose for a larger commentary on Maimon's works: that the *Logik* is primarily an *a priori* logic of the appearances of appearances,[20] which was preceded by a succession of fundamentally different metaphysical attempts to solve the same problem of physics for criticism, such that the failure of each informed and gave rise to the following attempt. The *Versuch*, first of all, was a hybrid transcendental rationalist metaphysics of identity. It was incoherent (and we will see why) and was immediately followed by the speculative dialectical *Weltseele*. The world-soul as a theme occupied Maimon for at least three years, up to the 1793 article "Über die Schwärmerei," the second part of which comments on extracts from Giordano Bruno's *Cause, Principle, and Unity.*[21] There, Maimon pronounces the

---

19 This suggestion is from the tenth and eleventh fascicles, to which Förster gives the heading "How Is Physics Possible? How Is the Transition to Physics Possible?", describing it as going in a different direction from the ether proofs.

20 I cannot give an account of this work here, but it involves at least these two argumentative components: 1) reflection, prior to experience, on the logical principles of the thought of any object which can be given in experience, including scientific experience (Maimon calls this the "object of logic"), and 2) reflection on the ways in which such thinking can systematically approach the complete determination of how that which appears does appear. GW 5: §§ 1–7, 12–13 accomplish the first, while §§ 8–11 accomplish the second. Maimon announces his main thesis of the *Logik* in GW 5:67–68: "The laws and original forms of thinking can be determined *a priori* (through reflection) and made complete, and in view of them logic can be made, as a whole, complete."

21 Maimon's source for Bruno was Jacobi's emended translations in his *Über die Lehre des Spinoza, in Briefen an den Herrn Moses Mendelssohn* (Breslau: Gottlieb Löwe, 1789). Yitzhak Melamed, "Salomon Maimon and the Rise of Spinozism in German Idealism," *Journal of the History of Philosophy* 42, no. 1 (2004): 72, notes that chapter 69 of Maimon's *Give'at ha-Moreh* (1791), his commentary on Maimonides' *Guide* commissioned by the Berlin Enlightenment, includes Hebrew translations of Bruno. He argues that Maimon only included Bruno as a way of esoterically admitting Spinozism into philosophy. I disagree, insofar as Maimon had an evident interest in Bruno's thesis of the absolute as a dynamic mediation of opposition. Melamed writes that part of what allowed Maimon to introduce Spinoza via Bruno is that Maimon had a historically conditioned (mis)understanding of Spinoza's parallel attributes of thought and extension. However, this may not be true, since in the *Weltseele*, Maimon specifically says he is not Spinozistic because he does not admit that "God and the world are one and the same substance to which two properties can be attributed, namely infinite extension (matter) and infinite understanding (form)." Rather, like Bruno, he proposes the world-soul as one force that dynamically unifies what is opposed. Maimon advances along his own path.

conception of the world-soul to be a piece of unphilosophical *Schwärmerei*, or enthusiasm. It is by commenting on Bruno that he intends to "separate the well-grounded from the enthusiastic therein as if through a chemical operation" (GW 4:616). As he was at that time also writing a commentary on Aristotle's *Categories* together with a *Propadeutik zu einer neuen Theorie des Denkens* which accompanied the *Logik* of 1794, here is a picture of the whole: Maimon's philosophy has as its high points the *Versuch*, the *Weltseele*, and the *Logik*.[22] The *Logik* is neither a metaphysics of identity nor of dialectic, but primarily one of analogy, though I cannot go into a description of that work here.[23] But importantly, if this theory of Maimon's development is correct, then the *Weltseele* not only meaningfully followed on from the *Versuch*, but its failure also meaningfully contributed to the *Logik*. How is this the case? If this essay can begin to show this much, if only in outline, then it will make a partial yet essential contribution to the central argument of the larger commentary.

The *status quaestionis* regarding the *Weltseele* is as follows. In a 2004 article, Melamed wrote that Maimon's conception of the world-soul is "relevant" but "of secondary importance," at least to Maimon's relationship to Spinoza and the reception of Spinoza in Western philosophy. For according to that article, Maimon's primary philosophical contribution was in bringing Spinozistic monism to Kant in order to justify the categories, thereby explicitly introducing Spinozism into the philosophical tradition. In a 2006 dissertation, Ehrensperger outlined the breadth of the historical background that one may assume is behind Maimon's notion of the world-soul. The

---

22 Maimon also wrote the *Kritische Untersuchungen über den menschlichen Geist* in 1797; this work consists mainly of dialogues, but its ideas largely follow from the *Logik*. Indeed, the *Prolegomena zu Kritik einer praktischen Vernunft* included therein specifies the metaphysics implied by the cognising subject as it is given in the *Logik*. One aspect left out of consideration here (which would also be necessary for the *Logik*) is Maimon's development of an intentional theory of thinking in contradistinction to Reinhold's *Satz des Bewusstseins*. For this, see primarily his letters to Reinhold in the *Streifereien* and also the letters appended to the *Logik*.

23 See GW 5:18–20 for Maimon's analogical conception. The basic thought is, I think, that he no longer gives a metaphysics in which the absolute entirely transcends the cognising subject (as is the unintended result of the *Versuch*), nor does he give one which identifies the absolute with the cognising subject (as in the *Weltseele*), but one in which the absolute transcends and yet is immanent to the subject, and this is nevertheless a metaphysical view which Maimon critically justifies. This justification involves the claim that the thinking subject, in transcendental reflection, completely or absolutely accounts for the principles of its thinking of being. As absolutely justified in this reflection, the subject's thinking can also rightfully claim to be the thinking of the absolute—but since being transcends the subject which thinks it, the absolute (which is common to the thinking subject and to being) transcends the subject too, and so the subject in its thinking is also entirely contingent to the absolute. In this way, Maimon can justify an analogical metaphysics such that metaphysics is no longer *prima philosophia*, but *philosophia ultima*. Incidentally, the sequence of metaphysics of identity, dialectic, and analogy is described as the true path towards a post-critical metaphysics in Erich Przywara, *Analogia Entis: Metaphysics, Original Structure and Universal Rhythm*, trans. John R. Betz and David B. Hart (Grand Rapids, MI: William B. Eerdman's Publishing Co., 2014); original version: *Analogia Entis: Metaphysik* (Munich: Kösel und Pustet, 1932).

question behind that outline is the following: given that Maimon published the *Versuch* in 1790 and the article “Baco und Kant”[24] soon afterwards (which was closely related to the concerns of the *Versuch*), and wrote to Kant recommending the latter article on May 9 of that year, how was it that he came only six days later, on May 15, to ask Kant to read his just-published 45-page article on the world-soul—a work with apparently divergent, even incompatible, themes and arguments? Ehrensperger cannot save the *Weltseele* from relegation to “secondary importance” either, however, since he finds that it contains some basic thematic similarities to the *Versuch*, but only enough to indicate that Maimon's major effort was to carry over a medieval *Weltanschauung* into Enlightenment philosophy.[25] Like Engstler, then, he sees Maimon mainly as a problem thinker whose “argument” was not much more than the form, practice, and ethos of commentary.

In contrast to these commentators, I would urge that Maimon's brilliance lay in immediately seizing upon the basic underlying problem and task: that critical philosophy was not reconciled with the science and claims of physics, yet if it could be so reconciled, then critique could advance to transcendental philosophy. If he never phrased it quite so explicitly, that is because of the difficulties inherent in his texts (mentioned above), difficulties which include the fact that this transcendental reflection on physics, which would constitute the philosophical meaning of the science, had not only not been conducted before, but also required a confrontation with ontology of the sort which Kant had attempted to ban from philosophy. To see the requirement of this confrontation with ontology, one may note that the parallel courses that Maimon and Kant (in the *Opus Postumum*) took in this endeavour point to a certain paradox. As the philosophers abandoned their uncritical hypotheses about empirically conditioned forces or entities (whether these were the ether or the world-soul as a force, or even differentials), they indicated—and Maimon, specifically, realised—that what the task called for was an unconditioned logic, and therefore an absolute logic, a logic of the absolute. This seems by definition to be something that falls outside the bounds of critical philosophy.[26] So, the solution to this paradox be-

---

24 Solomon Maimon, “Baco und Kant,” *Berlinisches Journal für Aufklärung* 7 (1790): 99–122.

25 Florian Ehrensperger, “Weltseele und unendlicher Verstand, Das Problem von Individualität und Subjektivität in der Philosophie Salomon Maimons,” PhD diss., Ludwig-Maximilians-Universität, 2006, traces the history of monopsychism as Maimon might have understood it from commentaries on Aristotle up through Avicenna, Averroes, and Maimonides, finding that just as the medievals thought of humanity in collective terms while moderns think of individuals as self-made men, in the *Weltseele*, Maimon urges for a biology of “generic preformation” instead of “individual preformation.” However, I do not believe that Maimon was arguing for any preformation at all, but rather for an absolute force which would give structure to what is originally completely abstract and unstructured.

26 In the first *Critique*, Kant attempted to give a logic of the absolute that was regulative of experience, and necessarily projected by the faculty of reason prior to experience, so as to demand conceptual generalisation, specification, and continuity. However, to argue for this, he had to leave the bounds of experience and conduct a transcendental investigation: “Thus the first step we take becomes Maimon's task over his three works, and one with which he deals in detail in the *Logik*: how to develop a critical work, a work of transcendental reflection immanent to human scientific knowledge, which is also a logic of the absolute which transcends that knowledge. The *Weltseele* represents a fundamental moment in this development, since it solves certain definite problems of the *Versuch* by uncovering, for the first time, a compatibility between transcendental reflection and ontology.

In order to give the requisite outline of a commentary on the *Weltseele*, this essay must begin with an interpretation of the *Versuch* from the point of the overarching problem of the critique of physics. Once the *Versuch* has been so interpreted, an account of the problem which the *Weltseele* confronts will follow, then an account of its central argument, and finally a conclusion regarding the status of that argument in terms of the desired *a priori* logic of physics.

1 The *Versuch*

The *Weltseele* is a speculative dialectical portrayal of the absolute as a living whole, while the *Versuch* is a transcendental rationalist metaphysics of identity.[27] Only the interpretation of the *Versuch* according to the central problem of physics will bring out what its argument is and how it was that Maimon could transform it into the metaphysically different *Weltseele* so very quickly after its failure. The central problem is, again, that the *Critique* had failed to integrate the science and claims of physics into itself and that only the solution to this problem could advance it to its goal of tran-

yond the sensible world compels us, in acquiring new knowledge, to begin with the investigation of the absolutely necessary being, and to derive from the concepts of it the concepts of all things insofar as they are merely intelligible; we will set about this attempt in the following chapter" [A567/B595]. Kant's *Prolegomena* §§ 57–59 attempts to account for this stepping beyond experience by arguing that this aspect of cognition relates not to the things in themselves of the soul, world, and God (to which appearances it seems must surely relate), but to "such concepts as express the relation of those beings to the sensible world." See the *Prolegomena to Any Future Metaphysics, With Selections from the Critique of Pure Reason*, ed. and trans. Gary Hatfield, 2nd ed. (Cambridge: Cambridge University Press, 2004; original work: *Prolegomena zu einer jeden künftigen Metaphysik, die als Wissenschaft wird auftreten können* [Riga: Hartknoch, 1783]), 102–12. However, this does not work as either a logic of the absolute or a justified critical cognition, for on the one hand, one cannot legitimately predicate an experiential predicate of the absolute (Kant admits this), and on the other, cognition of the relation of the absolute to experience can only be legitimately meaningful "if an uncritical, transcendent use of the categories can be provided" (Erich Heintel, *Hegel und die Analogia Entis* [Bonn: Bouvier, 1958], 17)—and Kant cannot have given this in either the *Critique* or the *Prolegomena*. The solution to this very problem becomes Maimon's task, since a critical logic of physics requires a logic of the absolute (i.e., of the formal conditions of a theoretically complete explanation of the sort Kant intended with his ether theory) that is constitutive of experience. Incidentally, I do not claim that Maimon forcefully anticipated Kant's *specific* circle problem with forces, but he anticipated the basic nature of the problem nonetheless.

27 The use of these terms will be addressed below, after a full account has been given.

scendental philosophy. Hence, Maimon called his work an essay, a *Versuch*, regarding transcendental philosophy.[28]

1.1 The Central Problem of the *Versuch*, and of the Other Main Works

The commentary tradition has not yet recognised Maimon's central problematic. It has instead consistently assumed that Maimon attacked Kant's *quid juris* question directly, in order to replace it without remainder. (The *quid juris* question [A84/B116] asks by what right can the critical philosopher make objectively valid judgments about what appears to us.) But did Maimon have this intention? For an example of this interpretation, here is Cassirer on how Maimon reframed the categories as applying not to appearances, but instead to appearances as reduced to their infinitesimal elements: "The understanding here does not subsume something given *a posteriori* to an *a priori* rule, but it instead lets [this something] arise *a priori* according to that rule, which is the only way to answer the *quid juris* question in a fully satisfying way."[29] However, if completely replacing Kant's basic argument was Maimon's underlying endeavour, what is one to make of Maimon's remark to Kant in his first letter indicating that his question in the *Versuch* was both a *different* question from Kant's and also an expansion of it? He asks:

> The question *quid juris?* Because of its importance this question was worthy of a Kant; and if one gives it the extension [*Ausdehnung*] you yourself give it, one asks: How can something *a priori* be justifiably applied [*lässt sich*] with certainty to something *a posteriori?* The answer or deduction that you gave us in your writings is completely satisfying, as [only] the answer of a Kant can be. But if one wants to expand [*ausdehnen*] the question further, one asks: How can an *a priori* concept apply to an intuition if it be an *a priori* intuition? Now the question must await the master once more in order to be answered satisfactorily. (GW 2:424)[30]

---

28 See GW 2:8–9, ETP 8–9: "The great Kant supplies a *complete idea* of transcendental philosophy (although not the whole science itself) [...]. My aim in this enquiry is to bring out the *most important truths* of this science." See also GW 5:383 ff. for how, in the *Logik*, Maimon's scepticism answers the *quid juris* question for physics and "sets the solution [...] at the foundation of my critical philosophy."
29 Cassirer, *Das Erkenntnisproblem*, 100.
30 I depart from the English translation of this letter (ETP 229), which reads: "How can an *a priori* concept apply to an intuition even to an *a priori* intuition?" This does not make much sense, since Maimon has just admitted that Kant definitively shows how an *a priori* concept may be applied to "something *a posteriori*," which must imply empirical intuition. The German is "Wie lässt sich ein Begriff *a priori* auf eine Anschauung ob schon auf eine Anschauung *a priori*, applizieren?" The correct stress must be something like "an intuition, albeit an *a priori* intuition," or "an intuition, if it should happen to be an *a priori* intuition," in which case the intuition would be determined *a priori* by processes to be described by physical law.

What is the nature of this question? If one takes it as a direct attack on Kant's *quid juris* and his deduction of the categories, then that line of interpretation demands an investigation of how Maimon must have found that deduction unsuccessful on its own terms (even though he evidently admits that he found it very successful and did not attack it, but expanded it). This investigation has been carried out by the tradition that has come down to us from Cassirer to the works of Engstler and Thielke.[31] According to Engstler, Maimon attacks Kant's deduction by bringing a Humean objection against the Second Analogy; i.e., Kant's principle according to which we are justified in assuming every given action or event to be an effect of some cause, such that this assumption makes it possible for us to discriminate objective causal events from subjective sequences of perceptions. Thus, Maimon must have argued that this principle does not allow us to proceed from subjective perception to objective causal judgments, because one cannot perceive the categories as one would an image; one cannot directly perceive a characteristic of objective causality. But, as Engstler correctly says, this argument is not an argument so much as it is a misunderstanding. Maimon's scepticism only results from a confusion of the transcendental and empirical levels. He does not understand the nature of "transcendental schemata"; namely, that they are transcendental time determinations methodically intrinsic to the *a priori* collective and regulative constitutive application of the categories to experience, according to the Analogies.[32] For in truth, instead of going from the subjective to the objective, Kant argues (A193/B238) that we must project the categories *a priori* in such a manner that we are first enabled to derive even subjective apprehension from the objective sense of experience which it must presuppose.

Thielke agrees with Engstler that such a Humean objection would be an obvious error, but disagrees insofar as he finds that Maimon actually makes a much better argument. That argument is that if the cognising subject does apply the objective category of causality *a priori* and in such a manner, then Maimon does not see what "constrains" the subject to use this application to distinguish any particular event or set of events as irreversible in the first place. If we apply the category *a priori*, it is still wholly incapable of determining individual perceptions. There is, at the transcendental level, a failure to specify how it is that the category of causality should be applied in order to discriminate it from subjective perception or from other types of causality. Therefore, Maimon is correct that the Second Analogy

31 Engstler, *Untersuchungen zum Idealismus Salomon Maimons*; Peter Thielke, "Discursivity and Causality: Maimon's Challenge to the Second Analogy," *Kant-Studien* 92 (2001): 440–63.

32 It is worth noting that the *Critique* is not simply divided into the constitutive understanding and the regulative reason. Instead, the first two Synthetic Principles of experience (the Axioms and Anticipations) are constitutive of intuition (all intuition is of certain extensive and intensive magnitudes, which we can order in basic ways). The second two Synthetic Principles (the Analogies and Postulates) are regulative of intuition and constitutive of experience. The Principles of Reason (which are, finally, the three demands to make experience more general, more specific, and more continuous) are regulative of experience.

does not work. But Thielke's interpretation cannot work as an argument either, and for two reasons. First, for Kant, the ability to apply the categories is epigenetically obtained.[33] Because of the nature of this epigenesis, any experience as such—i.e., about which judgments can be made—already entails that the categories have already been applied *altogether*, including all three categories of relation, and this makes possible the corresponding Analogies (of substance, one-directional causality, and reciprocal causality). Thus, prior synthesis makes it possible for judgment to analyse what has been synthesised and to come to inductive empirical cognition about what is substantial and causal in experience and how it is so. The virtue of an experiential cognition is therefore simply that it is objectively valid; i.e., it can be meaningfully said to be true or false.[34] Secondly, Maimon is quite aware that this is Kant's argument, and he finds it successful on its own terms.[35] (Thankfully this is so, otherwise his philosophy would have had an unpromising beginning.)

---

33 The process of epigenesis (B167) involves the concepts of reflection and the three "levels" of synthesis (apprehension, imagination, and recognition) and generates the schemata along with the basic structure of the "concept tree" described in the "guiding thread" chapter. In this way, it makes it possible to apply the categories in judgment according to the synthetic principles of experience. The deduction of the categories proves that experience is both synthesised and analysed by the same categories and that there cannot be any meaningful analysis beyond the bounds of such syntheses. For a strong reading of this epigenesis, see, of course, Béatrice Longuenesse, *Kant and the Capacity to Judge: Sensibility and Discursivity in the Transcendental Analytic of the Critique of Pure Reason*, trans. Charles T. Wolfe (Princeton and Oxford: Princeton University Press, 1998), which advances this interpretation.

34 And so Maimon does not make a "third man" argument either; i.e., the complaint that another judgment is needed to certify a previous one, and so on. Kant is not interested in indubitability of judgment, but in objective validity, and Maimon understands this.

35 For example, in GW 2:5–8; ETP 6–9, Maimon gives an appreciative account of Kant's argument which *rules out* possible objections to Kant (as in Engstler's interpretation) in which the categories, through the Synthetic Principles of experience, would already be applied with *a priori* empirical markers for certain objects or events (where "experience, that is an objective connection" would nevertheless be only "a *particular* proposition" that we could intuit) and possible objections (as in Thielke's interpretation) in which the categories would, by some constraint, only apply *a priori* to some objects ("*a few* objects are so constituted that if the one is posited, the other must necessarily also be posited"). Also, in GW 2:51–54; ETP 31–33, he again asks how to justify the category of cause and gives Kant's answer as something that "removes all difficulties at a stroke [*auf einmal*]." However, he indicates, he can still ask *his own* question about scientific cognition (and the following lines, likely added after Kant's review of the work, doubt the success of his own effort). At the end of chapter 2, Maimon argues that Kant has evoked a type of necessity in cognition—one that says something scientific about matter—but has (so far) unjustly related that scientific cognition to ordinary experience, and, of course, in the full sense of "experience" necessary for transcendental philosophy, the two must be related. This is the meaning of the statement (GW 2:73; ETP 42–43) that Kant has argued successfully, but has as yet only made *a priori* authorisation of associative experiential propositions: "It is just what in animals we call the expectation of similar cases; and even if *Kant* has proved that we cannot have abstracted these forms from experience because experience only becomes possible in the first place by means of them, *David Hume* (or his representative) can gladly admit this. He will say: the concept of cause is not in the nature of our thought in general such that it would also occur in

Maimon's different *quid juris* does not direct itself against Kant's own problem, but rather its expanded scope would preserve what Kant had achieved. The point of that expansion is the discovery of how the categories have *a priori* application to how intuition is determined *a priori* by the physical world; i.e., to the ways in which sensibility in general can be affected, as an object itself, by the physical world. This discovery would (if correctly achieved) give the conditions of the possibility of physics—i.e., natural scientific cognition—and secure the relation of ordinary (pre-scientific) cognition to that cognition. With this in mind, let us see how Maimon asks the main question of the *Versuch*. That question is: "How can an *a priori* concept apply to an intuition if it is an *a priori* intuition?" Can it mean anything but "How can we come to *a priori* cognition of how intuition is determined *a priori* according to physical law?"

In chapter 1 (GW 2:20; ETP 15), Maimon recognises one fundamental presupposition of physics: that what is given to us is *already* necessitated; i.e., such that it can be cognised in terms of antecedent conditions and general laws. For if "a synthesis in general is unity in the manifold," then there are necessary, arbitrary, and spontaneous syntheses. The first is "given (*reale* in sensation)," the second is a product of unguided psychology or imaginative association, and the third is an effort of the understanding that "spontaneously" (*Freiwillig*) attempts to discover the ground of the syntheses which would accord with the necessitated given reality. The truly *Freiwillig* understanding that comes to cognition "counts as an object only a synthesis that has an objective ground (of the determinable and the determination) and that must have consequences [*Folge*], but no others." According to such a synthesis, if a physical object is cognised as a determinable and is given a unique determination as a condition, there then follows another determination as a unique consequence. Maimon indicates what he means by a synthesis indicating certain consequences and no others in the middle of chapter 2. There (GW 2:42; ETP 27), he argues that in geometry, one can come to complete explanation of a given geometrical object. One proves the "possibility of a circle [...] analytically": a nominal explanation of a circle ("all lines that can be drawn from a given point within it to its limits are equal") may be considered completely explained by a real definition of it (i.e., a rule of construction: "a line moved about one of its endpoints is identical to itself in every possible position"), since each merely intuited component can be accounted for (even if *ad infinitum*) by the real definition and an underlying identity can be discovered between the nominal relation and the real relation. And so, the same explanation of

symbolic cognition, and is also not grounded in experience (in the sense in which Kant uses this word)." Maimon attempts to answer this further iteration of Hume in order to account for *scientific* objective validity. Thus, it is not so much that Maimon criticises Kant's hypothetical judgment as unjustified in itself (as the tradition of commentary might have it), but rather as unjustified from Maimon's own question about an *a priori* logic of physics. Finally, as he admits in a note to the introduction (GW 2:338; ETP 175): "I maintain that Kant's *Critique of Pure Reason* is, in its way, just as classical as the work of Euclid's and just as irrefutable."

underlying physical reality can be given in terms of general laws and antecedent conditions.[36] Thus, Maimon's question becomes:

> [W]hat determines the faculty of judgment to think the rule-governed succession as corresponding to the rule of the understanding itself (so that, if a comes first and b follows, but not the reverse, then the faculty of judgment thinks the relation of cause and effect between them) and to think each particular member of this sequence as corresponding to each particular member of the rule of understanding (the preceding corresponding to the cause and the succeeding to the effect)? (GW 2:54; ETP 33)

That is, how can a nominal relation of appearances be completely explained by a real relation (which is the work of theory) so that its particular elements are precisely determined and how is the subject motivated to come to such an explanation?[37]

This question breaks down into further questions. Physical reality that is to be explained by theory is not directly given to the subject as are, for example, geometrical shapes. When Maimon grapples with the application of Kant's category of cause according to the Second Analogy of experience, his ultimate argument is that without stepping outside the bounds of that principle (and also outside of the bounds of the accompanying principles of substance–accident and reciprocal causality; i.e., outside appearances), there can be no theoretical explanation of reality.[38] True, Kant is right that we can recognise a distinction of cause and effect in ordinary experience (and that those concepts make experience possible), but for actual theoretical scientific cognition of causal relations "within an action, we must be acquainted with the nature of things outside the action. So, we cannot have any direct cognition of it [as cause or effect] within the action, but only indirect cognition" (GW 2:222; ETP 117). Thus, we are given the ultimate elements of reality within experience only (if at all) indirectly. We see the illumination, but are not given whatever light

---

36 "Experience is perception of the very same thing connected with different determinations that are changing in time" (GW 2:137; ETP 76). Scientific cognition intends to discover this "very same" identity between the nominal and the real.

37 In the *Critique*, Kant entirely denies the possibility of the scientific/physical cognition which Maimon has in mind. See A178/B222, the introduction to the Analogies of Experience: "Now the way in which something is apprehended in appearance can be determined *a priori* so that the rule of its synthesis at the same time yields this intuition *a priori* in every empirical example, i.e., can bring the former about from the latter. Yet the existence of appearances cannot be cognized *a priori*, and even if we could succeed on this path in inferring to some existence or other, we still would not be able to cognize it determinately, i.e., be able to anticipate that through which its empirical intuition is differentiated from others." Maimon is asking about this very application of *a priori* concepts to *a priori* intuitions, and has confidence that the answer can actually determinately cognise existences.

38 In the *Opus* (eleventh fascicle, 22:479; p. 135), Kant admits that in this regard, Newton had stepped beyond the limits of experience, since his discovery transcended sensation in general in order to determine it *a priori:* "It was *Newton* who first introduced this concept [of gravitational attraction at a distance]; not as an empirical proposition (for how can one experience an effect which does not occur on the senses, but only on the object of pure intuition?)."

is for physics; or we feel gravitation or observe magnetism, but are not given what gravitational or magnetic fields are for physics. What is it to cognise such indirectly given entities or elements? What is it to even recognise that such cognition is possible or demanded? First, it is intrinsic to such cognition that it determines more than is given in appearances.[39] It determines the lawful nature of *all such* appearances, and given antecedent conditions, it can also determine them precisely. In order to do so more accurately than fictionally, such cognition requires an intrinsic "drive" to theoretical completeness. Thus, the most universal theory-concept should also give the most precise determination of the underlying reality. Maimon recognises this: "In the end all concepts must be reduced to one concept, and all truths to a single truth," so that we bring to light "what kind of connection holds between the claim that air is elastic and the claim that the magnet attracts iron, and between the latter and the Pythagorean theorem for example" (GW 2:429; ETP 218). Secondly, because such cognition refers to what can only be given indirectly or mediately in appearances, and because it must do so in ways that demand and accommodate the approach to theoretical completeness, it must possess a certain symbolic conceptual form.

Thus, Maimon's exploration discovers that his *quid juris* requires the following parts for its solution: indirectly given real elements of appearances which are to be recognised as determining how appearances appear, theoretical cognition of those appearances which contains the demand for its own completeness, and a symbolic form of that cognition. The difficulty with Kant's answer to his *quid juris*—i. e., his deduction of the categories—was its synthetic nature. Maimon's *quid juris*, by contrast, can be answered via the discovery of the conditions *within* Kantian synthetic cognition for complete analyticity of theoretical cognition. The desired outcome is that the question then "falls away," so that Kant's deduction does not have to be asked anew or modified.[40] How does this work?

39 The experience of such scientific cognition is "perception of something that would determine what is left undetermined by the subjective laws of my mode of representation" (GW 2:127; ETP 72). Moreover, "it must not be the case that any possible experience follows every possible experience [i. e., as in the case of Kant's Second Analogy], but rather one out of all the other possible appearances must necessarily follow."

40 Maimon describes his *quid juris* in GW 2:48 (can the symbolic be made intuitive?); GW 2:55 (he recognises the *Versuch*'s failure: it tried to make an analytic deduction instead of critically explaining the possibility of the synthetic facts of intuition; i. e., to make scientific cognition into an *a priori*, but not pure cognition); GW 2:59 (he again recognises the *Versuch*'s failure; the *quid juris* is impossible to answer analytically in the 'straight line is the shortest' case, and so one cannot see how a subjective rule is met with in the given object); GW 2:60–61 (with symbolic analytic cognition the question falls completely aside); GW 2:62 (we achieve the deduction by transforming it into the question of how a world arises from an intelligence); GW 2:355 (it is answered because the categories apply to elements of appearances); GW 2:363–64 (it is really the *quid rationis*, asking what hypothesis must be adopted for both certain and possible experiential facts); GW 2:128 (it asks how one can justify *formaliter* how one thing follows with objective necessity from another, in terms of a particular determination of the

1.2 The Argument of the *Versuch*

Maimon attempts to construct the conditions for the valid theoretical cognition of physics from within the Kantian *Critique*. His approach takes two lines of argument. We will see that these two arguments correspond to the conditions for the possibility of the subject S and the predicate P, respectively, of the basic categorical proposition ("S is P") of physics. The first line extends over the course of the book, though mainly in chapters 1, 2, and 8 and in the Short Overview. It is as follows. First, consider Kant's transcendental reflection on experience. This reflection discovers experience (*Erfahrung*) as not only the medium which makes all the "special sciences" (mathematics, natural science, biology, even aesthetics, etc.) possible, but also as the product of the cognising subject. This is the subject which for the most part does not cognise its experiences in objective terms (i.e., according to the categories), but is in principle always able to do so. The structure of what is in principle always possible for the subject is the "original synthetic unity of transcendental apperception," and what is possible, as a condition of objectivity, is the consciousness "that I am" and that I am the same I, in each possible synthesis of different representations which the I that thinks determines. The effort of objectivity is always related to the effort and possibility of self-consciousness, the structure of which is discovered in transcendental reflection.

It is necessary to consider Kant's argument because Maimon essentially reverses it. Here is the sense of this reversal. If Kant says that we are always at least implicitly aware of our transcendental unity and that coming to objectivity means also coming to and maintaining the sense of the objective unity of the self, then Maimon replies that if so, then this implicit awareness must *also* be a primordial self-intuition, of the self as uniquely single. Instead of Kant's view that the representation of the "I" is "empty of content," Maimon says that it has a definite *formal* content—"By contrast, I maintain that the I is a pure *a priori* intuition"—and that this is so "because it is the condition for all thought in general" (GW 2:209; ETP 111).[41] How is the pure intuition of the "I" a formal condition for thought? The answer is this: for this intuition to even be possible for consciousness, it could not be consciousness of complete undifferentiated unity, because then "there is therefore no comparison, and conse-

preceding and succeeding and without looking to perception); GW 2:135; ETP 75 (the question falls away because the relation between the forms and matter is justified, i.e., "the forms of intuition [time and space] presuppose forms of thinking and these in turn presuppose something objective [matter]"); GW 2:187 (he doubts the fact of experience, but proves the possibility of applying concepts to intuitions in general, and so the *quid juris* falls away); GW 2:192–93 (the *quid juris* is answered because concepts apply to the elements).

41 We should note the marked difference between chapter 10 and the Short Overview (GW 2:208–10; ETP 110–112) on this issue, since the former is in agreement with Kant's Paralogisms, while the latter calls the subject a simple substance—so that the *quid juris* falls away or "does not arise." This contradiction is inherent to the argument, as is shown below.

quently no consciousness (and also no consciousness of identity)" (GW 2:15; ETP 13). However, it could not be of complete and total difference, because "then there is no unity and, once again, no comparison, and consequently also no consciousness" (GW 2:15; ETP 13). In other words, this primordial intuition has a certain form, and indeed projects this form so that what is given to it is given in terms of that form.[42] This form is that of space and time, because it is only space and time which provide the simultaneity and succession which make it possible for the subject to both relate (and so identify) and distinguish (and so differentiate) what is given to it.

To echo Kant's "Refutation of Idealism," this primordial intuition does not take place without something material given to it. But the given is *given as* a singular intuition of a *simultaneous* unity (i. e., spatiality) of the *preceding-succeeding* (i. e., temporality) that is a spatiotemporal "point" or differential relation (dy/dx).[43] The conditions of givenness correspond to the conditions of thought; subsequent relating and distinguishing of the material given is nothing other than "thought in general." This intuition, which immediately implies thinking, can be explicated as follows (GW 2:16–17): thinking posits time in what it identifies as an individual object, and negates space in that same object, but in doing so, it posits space in general, so that the negation of that object becomes a possibility. It simultaneously posits space in an individual object and thereby negates time in it, but in doing so, it posits time in general, so that again, the negation of that object becomes possible. Thinking therefore begins by the projecting of an intuitional form on what is given—i. e., an intuitional form in which conceptual spatiotemporal relations of identity and difference are implicit—such that thinking these concepts also generates spatiotemporal representations in general. This makes imagination possible, which, as both passive and active, represents (as "fictions") universal spatiotemporal regularities based on what it is given, and so also makes possible the understanding's conceptual recognition of unique objects which are opposed to other objects and can be completely

---

42 I use the term "primordial intuition" here because it describes a fundamental "categorial intuition" of reason, or the pure thinking I. Maimon does not use "primordial intuition," but in chapter 2, GW 2:30; ETP 20, he calls it "primitive consciousness [*primitiven Bewußtseyn*]." This primordial or primitive consciousness does not have representations, but rather presentations. "The world 'representation' [*Vorstellung*], used of primitive consciousness, here leads us astray; for in fact this is not representation, i. e. a mere making present of what is not [now] present, but rather presentation [*Darstellung*], i. e. the representation of what was previously not as [now] existing."

43 See GW 2:32; ETP 21. "These differentials are objects of the so-called *noumena*; but the objects themselves arising from them are the *phenomena*. With respect to intuition = 0, the differential of any such object in itself is dx = 0, dy = 0 etc.; however, their relations are not = 0, but can rather be given determinately [*bestimmt angegeben werden können*] in the intuitions arising from them." As the translator remarks, this " = 0" of intuition—i. e. of what appears—means that "we cannot imagine or be conscious of" these differentials, but we do have "primitive consciousness" of them, as presentations, and we can, it is argued, come to spontaneous and symbolic-conceptual cognition of them.

differentiated from them.[44] Thus, Maimon agrees with Kant that space and time are forms of sensibility, but he "only add[s] that these particular forms of our sensibility have their ground in the universal forms of our thought in general, because the condition of our thought (consciousness) in general is unity in the manifold" (GW 2:15; ETP 13); i.e., the unity of self-intuition.

What does Maimon want to achieve by this argument? First of all, if self-intuition projects space and time as forms of its intuition, such that thinking is given the task of differentiating and relating all that is given to it in space and time to preserve and concretely develop the consciousness of self-identity, then the effort of cognition becomes the beginning of an infinite task. This infinite task demands cognition of the identity of the self in precise terms which indicate what everything else is in relation to it.[45] Cognising self-identity becomes the demand to cognise absolute identity, and the finite subject demands its expansion into an infinite subject. Secondly, since the forms of space and time are continuous, such that what is given in them is so given in a continuity of difference, they can be reduced to their smallest possible relations; i.e., their differentials. Maimon therefore calls what we perceive in appearances—i.e., the possible visual markers of the differentials which we cannot perceive, but are nevertheless real—relations of "maximum identity" and "minimum difference."[46] The task of the understanding becomes the conceptual recognition of objects in appearances as they are advertised by these proximate relations and then reduced to the very differentials which are primordially given, so that they are explained by the corresponding conceptual expressions. For physics, as Maimon knew, states truths about indirectly given objects in terms of differential relations (they express, for example, rates of change; they can be objects of theory). The subject is therefore not only rightfully permitted, but in its self-intuition, it is driven to reduce what is given in the forms of sensibility to their smallest possible relations and to cognise them as such, because it is driven to achieve absolute identity of cognition. It is so

44 Again, as in GW 2:21, there are three types of possible syntheses: of the given and necessary, the imagined and arbitrary, and the spontaneous and understanding.

45 Individual objects should have their relatively individual grounds, "but the sufficient ground [*zureichender Grund*] is merely an idea of reason, which we can approach more closely (and in doing so extend the use of reason), but can never reach" (GW 2:106; ETP 61).

46 See GW 2:372. If "b" follows "a" in perception, then "b" and "a" must stand under a rule of "the relation [*Verhältnisse*] of maximum identity [*Einerleyheit*]," and this apparent relation is therefore what requires theoretical explanation; GW 2:216, which states that if the difference in perception were not "minimal," then one would refer only to "something completely different"; also, GW 2:218, where Maimon's rule or principle of experience determines only "the relation of objects to each other (maximum of sameness), not however the objects themselves in relation to those relations," and indeed, we do not directly perceive cause and effect, but determine it symbolically and conceptually by theoretical explanation of the differentials which are indicated by the relations of maximum of identity and minimum of difference. (Of course, since Maimon is monistic, there is no ontological difference between the differentials and the perceptible relations; instead, in chapter 3 and in the notes to it [GW 2:394 ff.], he argues for a fundamental qualitative difference.)

driven thanks to the demand of self-consciousness, and its aim is ever-greater theoretical completeness. This argument therefore addresses some of the desiderata mentioned above: how the indirectly given objects of physics, which determine how things directly appear, can be given to us at all, and how there is a drive to theoretical completeness.

The above argument addresses all that is necessary for the possibility of the subject "S" in the categorical proposition of natural science, "S is P." The propositional subject, in scientific experience, can first be taken as a nominal relation; i. e., whatever is observed to be closely associated. And because it can be reduced to differentials, it is susceptible to theoretical explanation. But what is the nature of this explanation, which contains what is needed to reduce the "S" from its nominal to its real relations and which must be given by the relation expressed by the predicate "P"?[47] Maimon deals with this type of predicate in his second basic line of argument, which also extends over the course of the book, though it is mainly in chapters 1 and 4 and in the appended monograph on symbolic cognition. Here, he considers the fact that the truths of physics are theoretical concepts expressed symbolically which possess a certain logical structure. Maimon calls this structure "determinability." Any concept that has objective reference must consist of a determinable which refers to a "substance," or what persists over a period of time, and a determination or "accident," which uniquely determines how that persisting object acts.[48] Determinations must uniquely determine the determinable over time, because this relation is an explanation of reality and because it intends to find strict identity between the nominal and the real, and this of course cannot be a merely partial identity.[49] Therefore, the relation between the determinable and the determination must be functional.[50] Functions can uniquely designate physical entities, according to their differential elements, in time and space, since as they are determinable, they already have the form of particular syntheses and so can refer to lasting types or patterns of reality on their own. With the predication of determinations, such functions can uniquely

---

47 A succinct expression of this is in GW 2:87n; ETP 50n: "However, for the finite subject, the subject is not what is *thought* in itself but only what is *given* in itself; and the predicate is what is thought merely in relation to it, as object."

48 GW 2:95–96; ETP 54: "Applied to objects of experience, the concepts of subject and predicate provide us with the concepts of substance and accident." The substance-concept can be disjunctively expressed in many different syntheses, so that the different determinations are its accidents, and since "time is the form of intuition, different representations cannot be thought at the same time."

49 Again, GW 2:20, ETP 15: "[T]he understanding counts as an object only a synthesis that has an objective ground (of the determinable and of the determination) and that must thus have consequences, and no others."

50 As evidence of this claim, see GW 2:373–74; ETP 192: differentials are, on the one hand, real objects, and on the other, they are not consciously represented, but "expressed merely by means of a function-relation that, as a numerical relation, is continuously variable," and, if "a determination cannot be thought without a determinable," then "it follows of itself that a determination with respect to our consciousness can only be a relation" (GW 2:94; ETP 54).

refer to particular physical entities in time and space. And, because of the demand for theoretical completeness, Maimon admits a possible hierarchy of determinable functions, so that, as he says (GW 2:429), all concepts can be reduced to one concept and all truths to a single truth.[51] Finally, it must be repeated that if Maimon has completely established the conditions of the possibility of the theoretical scientific proposition "S is P" from within the Kantian critique and its categories in the *Versuch*, then he has answered his own expanded *quid juris* question—and again, the question itself should "fall away," since there is a universally possible analytic connection, which is also demanded, between the subject relation and the predicate relation, so that the symbolic can be made intuitive (GW 2:48, 60–61, 128, 135, 187, 192–93, 355).

---

51 Maimon gives three proofs of the determinable relation which unfold in sequence and address each other. The first is GW 2:88–89, the second is GW 2:90, and the third is GW 2:142–44. The first shows that assuming that objective syntheses have unique and new consequences, then two determinables cannot have the same determination, because the assumed unique and new consequences could not take place. The consequences could not follow from the same determination (since it is the same), nor from the different determinables (because then we would not be referring to unique new consequences), nor even from the determinable–determination relations (because these would only be combinations of what had already been ruled out, and so are ruled out too). Maimon perhaps means by this last argument that such syntheses would not be entirely unique and new, but partially identical. Thus, there must be a unique determination for a determinable in any objective synthesis. The second proof shows that assuming (from the first proof) that different grounds (determinables) must have completely different consequences (determinations), then they cannot have partially similar consequences. If such seems to be the case, then the correct interpretation must be that the common element, "c," is the determinable with two determinations, "a" and "b." This argument seems to clarify what Maimon meant by his last argument in the first proof. The third proof then addresses the case of a determinable with two determinations. Again, one assumes that there is an object that is a unique and new reference or meaning of a synthesis. Given "Abc," then either "A" has determination "b" and "b" has determination "c," or "A" has two simultaneous determinations. If it has two simultaneous determinations, then (since they are determinations, not determinables) they do not have any reason in themselves for their simultaneity, but they do have reasons in themselves for being determinations of "A." Given this, then "Abc" is arbitrary and not necessitated by an object. Thus, "A" has "b" for its determination, and "b" has "c" for its determination. These proofs seem questionable, but no doubt they depend on certain definite assumptions (they seem to assume all they wish to prove). One troubling thought is that given hierarchies of determinables (which Maimon must legitimate, given the demand of theoretical completeness), then in "Abc," "b" is both a determinable and a determination. However, Maimon has strongly differentiated the determinable from the determination by saying (GW 2:378; ETP 194) that the subject of a synthesis "constitutes a synthesis in itself," while the predicate in it "does not constitute a synthesis in itself." This would seem to make "b" impossible. It should be noted that the *Logik* makes an entirely different use of the principle of determinability, where it is a central principle of reflective thinking, but not of the object in itself (see GW 5:250).

1.3 The Failure of the *Versuch*

The ambitious project of the *Versuch* fails, but it is certainly a productive failure, and it is redeemed in the *Weltseele* and the *Logik*. The central flaw is that the two main lines of argument cannot connect as they should. On the one hand, the self-intuition of the finite subject demands theoretical completeness in terms of the manifold of space and time which it projects onto the given: this provides the meaning of the subject "S." On the other hand, the conceptual structure of determinability provides the nature of the predicate "P" which should explain "S." Natural scientific cognition is, conceptually speaking, a more precise form of ordinary cognition: identifying the very same thing in its nominal relation that is identified again in its real relations.[52] But why precisely must the absolute identity demanded by thinking's self-intuition take the form of determinability? And what is the structure of the determining cognition which intends this goal? This argument seems to be missing. Either Maimon's proofs for determinability are tautological (determinability should itself be the condition of "an objective possibility of a synthesis in general," but this condition is grounded "in the objects" [GW 2:85; ETP 49]), or the requirement for this conceptual form comes from Maimon's argument in chapter 8 for the *a priori* principle of the continuity of the perceptible manifold (GW 2:136 ff.). This argument is that the spatiotemporal medium, if it is in principle able to be completely synthesised in experience, must be composed of determinations that are by definition different from each other and so reducible to differentials. That is, because they are different, the determinations must exclude each other. But in order to overcome this destructive contradiction, they must be separated from each other by the smallest possible break; i.e., they must also be differential relations. Because of this *a priori* and absolute continuity, we are enabled to perceive, and thus experience, what persists over time and what changes in this medium, which must therefore be expressed by determinable concepts: the "S" for the persisting, the "P" for the changing. But this direct correlation of the terms of determinability with differentials makes little sense, for the argument equates (real) differentials with (conceptual) determinations and also (appearing) substances with (conceptual) determinables, destroying any meaningful difference between physical object, appearance, and symbolic expression and also destroying the idea of any hierarchy of functional determinable propositions. Indeed, Maimon generally stipulates that in the "S is P" proposition, the determinable "S" should independently refer to some general reality—but to consider such a subject as a function would, under this argument, be to abstractly divorce it from the determinations amongst which it was found and for which it is uniquely the "S" and then to reapply it. However, given the continuous flux of the immediate spatiotem-

52 This is the definition of what it is to know given by Schlick in his *General Theory of Knowledge*, trans. Albert E. Blumberg (La Salle, IL: Open Court Classics, 1985; original version: *Allgemeine Erkenntnislehre* [Berlin: Springer, 1918]); see, for example, his chapters 4 and 5.

poral medium, it can no longer apply to the same determination values in experience. Hence, one must go back and revise the proposition to read "S is not P."[53]

This contradiction inherent in the scientific proposition indicates a problem with the meaning of its copula: the "is" in "S is P." What does this "is" mean? It was to indicate the underlying identity between nominal and real relations, but according to the type of theoretical completeness demanded (GW 2:103; ETP 59: "The concept of a thing can be distinguished from the thing itself only with respect to completeness, either material or formal";[54] GW 2:209; ETP 111: "I hold the representation or concept of the thing to be one and the same as the thing, and that they can only be distinguished though the completeness of the latter with respect to the former"[55]), this identity "between" rather becomes an identity "of." Thus, this problem is really about the cognising subject. Does it, in its primordially reflexive self-intuition, project the mere *demand* for an ideal thoroughgoing identity of cognition, along with the ideal manifold that it must think? Or does this thoroughgoing identity of cognition actually *produce*, by its own absolute or divine thinking, the differentials

---

53 For Hegel, Maimon's approach would be a perfect example of the dogmatic fixation on thinking which he subjected to critique, since here indeed are philosophical propositions which attempt to express the absolute in terms of "S is P"; i.e., here is "the opinion that here is the usual relation of subject and predicate and the usual attitude of knowing." However, the "philosophical content of the proposition destroys this usual attitude and the opinion which comes with it. Opinion [*Meinung*] discovers that things are meant [*gemeint*] differently than in the way it has opined [*meinte*]; and this correction of its opinion requires of Knowing that it return to the proposition and apprehend it differently." "The above can be expressed formally by saying that the nature of a judgment or a proposition in general, which contains the difference between subject and predicate, is destroyed by the speculative proposition." That is to say, it would be possible to take the very movement of the contradiction "S is P" and "S is not P" as an intelligible dialectical movement, which, taken speculatively as a whole, is cognition of the absolute. See George Wilhem Friedrich Hegel, *Preface to the Phenomenology of Spirit*, trans. and comm. Yirmiyahu Yovel (Princeton and Oxford: Princeton University Press, 2005; original work: *Phänomenologie des Geistes* [Bamberg and Würzburg: Joseph Anton Goebhardt, 1807]), 182, 185; 61–63 in original.

54 The material incompleteness can be completed *ad infinitum*; the formal incompleteness can be completed by the correct "insight," whether obtained gradually or right away.

55 I would here argue against the usual interpretation of Maimon as an anti-Kantian sceptic (as in both Freudenthal and Melamed) who argues that Kant's concepts and intuitions are heterogeneous. Maimon does not argue this, because it is not true (concepts and intuitions in Kant may be discursively related, but they are both transcendentally ideal and therefore not at all heterogeneous). Rather, the issue for Maimon is that since according to his *own* problem, 1) intuitions may be *a priori*, and so they are ontologically heterogeneous with concepts of the understanding, and 2) there may be an infinite intellect similarly heterogeneous to sensible intuitions; and Kant's solution does not help us in either case. The translation of GW 2:64 at ETP 38 can perhaps aid in this misunderstanding. It reads "we have already shown that even if they are *a priori*, intuitions are still heterogeneous with concepts of the understanding." However, the German, "Da aber, wie schon gezeigt worden, Anschauungen, sie mögen auch *a priori* seyn, doch mit Verstandsbegriffen heterogen sind," may be given the emphasis: "But since, as was already shown, intuitions, should they also be *a priori*, are indeed heterogeneous with concepts of the understanding." One notes that Maimon adopts the Leibnizian infinite intellect and its relation to intuitions as a solution to his own problem, not to Kant's.

as real objects which the finite subject is given and is supposed to think? In the first case, the finite subject can only claim to project the transcendentally ideal manifold,[56] but not to be the cause of the real objects which are given to it according to its form of intuition. But in the second case, in which the subject *produces*, the apparent fact that there is any cognitive mediation by a finite subject at all cannot be explained, nor in fact is any mediation possible, since "the infinite thinking being thinks all possible concepts all at once with the greatest completeness without any admixture of sensibility" (GW 2:183; ETP 98).[57] To put it simply: if the finite subject claims some reality, then the infinite subject is ideal; but if the infinite subject is itself real, then because it is absolutely infinite[58] the finite subject cannot be real.

This absolute problem—or contradiction—of the copula and its meaning accounts for other evident irreconcilabilities in the *Versuch*. The first is that the *quid juris* question cannot be answered. The categories cannot be justified via the complete analytic use of the scientific proposition within synthetic cognition, for if the finite subject does project its intuition of given objects as a demand for absolute theoretical identity, then there can of course follow a genesis of the categories from that projection. For example, in any intuition at all, the "persisting and the changing" determinables and determinations will be "substance" and "accident" respectively, while "the necessary succession of determinations one after the other" will be "cause and effect" (GW 2:137; ETP 76). However, these will be formal relational concepts,[59] such that the discovery of actual categorial relations will depend on the discovery of the differentials of whatever is given as objects of scientific experience.

---

56 One should note that if the subject produces, in the sense of creates, this manifold, this is a transgression of the bounds of critique, which never says *whence* the particular ontological quality of transcendental ideality itself arises. However, it can be argued Kant already had this problem of transgressing into the ontological—see note 26 above.

57 See the models of finite/subjective and infinite/objective orders of cognition given in chapter 3 (GW 2:81–82; ETP 48). If the objective order has no sensibility, it is difficult to see how it would have any ideas of the understanding at all, since these presuppose sensibility and are "the infinitely small of every sensible intuition and their forms, which provides the material for explanation of the ways objects arise." To give a further example of the priority of the reality of the finite subject, Maimon argues that it "projects" itself as infinite in GW 2:418, where (for example) the triangle is not a real object that can precede the thinking of it (so that its possibility depends on its actual thinking by the finite subject). However, he argues for the priority of the infinite subject again in GW 2:415; ETP 216, where he says it "thinks all possible objects [...] always according to an analytic rule," and so, for a finite subject of cognition, the possibility of an object would precede its thinking.

58 Maimon is indeed dealing with the absolutely infinite. See, for example, GW 2:417; ETP 214, where he defines the "absolutely *a priori*" as final-grounding cognition that can ground the principles of cognition because it is unconditioned, or rather its "condition" is that it is independent of any "particular determination of the subject (because it presupposes an infinite series)." The absolute *a priori* is that which transcends any "infinite series" of worldly determinations; hence, it is the absolute, i.e., the good or true infinite.

59 In GW 2:39–40 and GW 2:86, such concepts are described as being like functional relations of variables (x and y) such that the determination of one gives the determination of the other.

But in this case, there is no guarantee *a priori* that objects will be given in such a way as to grant theoretical completeness. In fact, there is a contradiction in the very idea of theoretical completeness.[60] Therefore, chapter 1 of the *Versuch* argues that the categories have as much reality as spatiotemporal representations: insofar as they are generalised, they are products not of understanding, but of imagination, and they are condemned to remain so.[61]

On the other hand, should Maimon give an *a priori* guarantee that objects are given in such a way as to grant theoretical completeness, then he must resort to the productive power of the absolute subject's thinking. In such a case, the categories cannot be derived from the finite subject's primordial reflexive intuition (since one can consistently have either an active finite subject or an active infinite subject, but not both), but only from the absolute *a priori* itself, which has as its sole principle "the principle of contradiction (or identity)" (GW 2:169; ETP 91). This leads to an attempt to derive the categories from absolute identity in chapter 6, which is a nascent attempt at an objective logic of being. In speaking about absolute identity, Maimon finds something strange. Identity is not contradictory opposition, of course. It excludes it by definition. They are opposed to each other, and their relation is one of opposition. However, this also means that they are not opposed to each other, for identity signifies absolute (or "transcendental," as Maimon says, after Kant) reality and opposition signifies absolute ("transcendental") negation, but since they transcend finite cognition, "these [...] teach us nothing about the matter or the content of judgments [...] but merely express their form, or the way they relate to one another." In expressing their relation, "one does not cancel out [*aufheben*] the other, but they are instead defined by one another." They are defined such that "opposition is what the two correlates reality and negation have in common" and "the one is instead a different positing than the other" (GW 2:113–15; ETP 64–66). Thus, Maimon realises the odd idea that the absolute is a contradiction. But because it is a contradiction of being and nothing, or reality and negation, it is also a defining relation between the two, such that their relation becomes a third thing, a medium of difference. This, of course—the assertion that "being is not nothing" and "being is nothing," i. e., that they are opposed but not opposed, and so productively preserve themselves in the opposition or nothingness they equally share—is an argument which will echo in the opening of Hegel's *Wissenschaft der Logik*. It will also be an under-

---

60 GW 2:164–65; ETP 89. The completion of what is approached "is not merely an idea, but contains a contradiction, since it is both an object and not an object at the same time," and this is analogous to how if an angle becomes 90 degrees, then the sine function has a symbolic meaning, but no real meaning. The real problem here is that Maimon demands the identity of thought with being, but at this point, thought cannot express being, and so it ceases to be thought.

61 See the argument of chapter 1 from GW 2:22–26; ETP 16–18. "I can maintain that representations of space and time have just the same degree of reality as the pure concepts of the understanding or categories [...]. Space and time can therefore only be called empirical intuitions (as predicates of intuitions) and not pure intuitions."

lying argument of the *Weltseele*. That is to say, this attempt to derive the categories does not relate itself to the finite subject's cognition at all in the *Versuch*, and so it also cannot answer the *quid juris* for that work. But it is (at least) an alternate argument for a logic of being, or of the real and objective relations of things, and one that could replace the logic of determinability to provide an explanation or theory of real relations.

There is one more important irreconcilability to point out. If, again, self-intuition sets up for thinking the task of theoretically complete cognition in absolute identity, then it can be assumed *a priori* that this will be cognition of nature according to the laws of mechanical causality. Determinability relations that do the work of theoretical explanation are such that the determinable is always the ground of its determination as consequence, such that determinations are in a cause–effect sequence. The cause is external and prior to the effect, and so it is in what Kant calls the "descending order" of *the nexus effectivus*, as opposed to the order of the *nexus finalis*, which would "carry with it dependence both as it ascends and as it descends [that is, the cause would also depend on the effect]."[62] Thus, anything that appears in sensibility must have a cause that is external to it—i.e., an efficient cause—and Maimon demands cognition of theoretical completeness in terms of this causal order for everything that falls under the absolute subject, including the absolute subject itself. But it follows that the biologically embodied process of cognition that posits absolute identity as its purpose or final cause also falls under the absolute subject and also demands explanation, although it is irreducibly teleological.[63] Therefore, since Maimon proposes a monism, the task of cognising mechanism must be equally the task of cognising teleology and thus of reconciling the apparent contradiction between them in the absolute.

The effort to present the absolute as a contradiction, but also as that which expresses and resolves the contradiction between mechanical and teleological causation in the real and living human "body" which cognises, is a good first definition of the *Weltseele*.

---

62 Kant, *Critique of Judgment*, § 65.

63 This problem arises seemingly unexpectedly in chapter 8, GW 2:140–42; ETP 78–79. There, it comes to mind that there is a distinction of "cause in itself or outside itself," and the continuity between the two "must be sought [...] in the analogy between bodily movements and sensations [...] and rests on the question *de commercio animi et corporis*." Thus, this is an unanswered question that has to do with relation between living movement and materiality. Yet this problem is already obliquely referenced in chapter 2, GW 2:62–63; ETP 37, where Maimon changes the *quid juris* to ask instead about "the world's arising (with respect to its matter) from an intelligence," and in the notes to this idea in GW 2:362; ETP 187, where he says that this is rather a "*quid rationis*" that asks "what sort of hypothesis must I adopt for it to be comprehensible" that there are not only possible, but objectively necessary and certain facts of experience. Thus, the *Weltseele* is already referred to as answering the same basic question about how *a priori* concepts relate *to a priori* intuitions in a different way; sc., this question is now taken as asking how objective reality in itself can account for the transcendental subject.

2 The *Weltseele*

"Über die Weltseele (*Entelechia Universi*)" is a 45-page article that was published in 1790, after the publication of the *Versuch* (and "Baco und Kant") in that same year. A description of its contents does not at all reveal the nature of its argument, but for the sake of the clarity of further discussion about it, here—very briefly—is what it contains.

First (47–52), there is a defence that need not concern us too much. Maimon recalls the idea of monopsychism from the distant past and wishes to analyse and develop it anew in comparison with the contemporary preference for the essential individuality of souls and with the Leibnizian effort to ground that preference. He quickly defends its harmlessness to the religious concern with the soul's personal immortality and asserts its essentially moral nature (citing Malachi 2:10: "Do we not all have one father [Origin (*Ursprung*)]? Has not one God created us? Why should each of us fall asunder [*zerfallen*] with his brother?" [52]).[64] He argues that it is entirely compatible with Kantianism and in sympathy with its basic convictions.[65] He also argues that he does not advance Spinozism, because Spinoza upheld parallel attributes (of extension and intellect), while the world-soul is the ground of the three forms of objective reality—i.e., matter, life, and understanding—and it is itself "assumed *a posteriori*" rather than by ontological proof connected to the *causa sui*.[66]

Second (52–63), since Maimon had apparently only just recently skimmed Kant's *Critique of Judgment* and seized upon Johann Friedrich Blumenbach's idea of a force of biological growth or *Bildungstrieb*, he proposes to analyse that force as a reality that was quite compatible with the idea of a world-soul. He then gives a history of the origin of the *Bildungstrieb* as a theory of the "epigenesis" of life from matter which emerged in opposition to two contemporary theories of "evolution"; namely those of sperm and of egg preformation. Epigenesis promises a mundane and therefore more scientific explanation of how a living organisation arises as a structure where there was previously no structure at all.[67] Since it is a "force" in the Newtonian

64 This article is in the *Berlinisches Journal für Aufklärung* 8 (1790): 47–92. It did not make it into the collected works, but one can find a truncated version under "W" in the *Philosophisches Wörterbuch*, GW 3:203–32.

65 His opponent is not at all Kant, he says, but the pre-critical dogmatism which, "since it holds that mere thinking is sufficient to determine an object by its means, has no reason to reject this notion" (51). I hold this to mean that he develops an objective logic of being, or of ontology, about how the world "comes to be" which the finite or transcendental subject must presuppose.

66 This is the literal truth; the world-soul is a hypothesis whose justification is in its explanatory ability. However, Maimon would eventually reject it because it is an uncritical hypothesis; it does not result from reflection on cognition.

67 Blumenbach made this hypothesis after cutting off the limbs of fresh-water polyps. They always grew back in such a way that the body of the organism was diminished in size according to a definite and measurable proportion. No preformation could account for such spontaneous change and measured development of structure, but perhaps a force could. For an account of how measurement pre-

sense of the word, however, it requires philosophical grounding, or rather critique: what is the relation between the cognition of the force and the force itself? But because it is a force that would explain even the origin of the life in which human cognition is embodied, Maimon engages in this critique in a rather unique way. He asserts again that the world-soul grounds the three real forms of matter, life, and understanding. Then, he opposes what he calls an Aristotelian view of the world-soul to a Leibnizian view of it. The former assumes an ontological opposition of one transcendent Being to all other beings—i. e., the three real forms—which grounds those beings. The latter assumes instead that thinking beings ground themselves, relating to yet differentiating themselves from an infinite thinking being, and that the real forms of the world become merely phenomenal or illusory. Neither viewpoint can justify the nature of the world-soul which it presupposes. From this Aristotelian-Leibnizian opposition, Maimon makes four seemingly arbitrary conclusions about the world-soul (which is, upon first reading, in no way specified at this point); namely, (1) that thanks to it, Leibniz's *petites perceptions* can be explained materially instead of ideally; (2) that the true judgments which the transcendental subjective forms of intuition and understanding make in conjunction prove that they must have this common condition; (3) that it confirms a long-held belief in the preservation of movement or force; and (4) that there is a coincidence between teleology and mechanism in terms of the three characteristics of it given by Blumenbach (and Kant in § 64 of the *Critique of Judgment*). These three characteristics are (1) that an organism generates itself according to its species, so that it is cause and effect at the same time; (2) that an organism generates itself as a particular organism; and (3) that it generates itself as an individual, since its parts also generate themselves (in regeneration and healing).

Third (63–74), the opposition of the Aristotelian view to the Leibnizian view also has as consequence that the world-soul, whatever it is, is a universal nature that is in analogy with human nature; i. e., with embodied human understanding. This means: (1) that the world-soul is a substance that results from a pure intelligence that must be thought, yet the world-soul is also an intelligent subject limited in its effect to the world as an organised form, so that (2) it is the formal and final cause of all objects, (3) the individual human and animal consciousnesses as "souls" are not true but appearing substances and the unity of consciousness of individuals is indeed formally specified, but not real, and (4) the psychological law of association is mechanical, as it is a law of bodily ideas (nerve traces) and not of proper ideas of the soul (which are ideas of real living purposes in the world).

---

cedes hypothesis or fiction in Newtonian science, see Alexandre Koyré, *Newtonian Studies* (Chicago: Chicago University Press, 1965), especially chap. 2. And for an account of the *Bildungstrieb* as a structure-giving force, see Robert J. Richards, "Kant and Blumenbach on the *Bildungstrieb*: A Historical Misunderstanding," *Studies in History and Philosophy of Biological and Biomedical Sciences* 31, no. 1 (2000): 11–32.

Fourth (74–80), in order to develop the details of the hypothesis of the world-soul, Maimon also analyses the debate between Locke and Leibniz—regarding the soul either as a *tabula rasa* or as substantially possessing innate ideas—in relation to the world-soul. Leibniz had been conciliatory in this debate: he had argued that some of Locke's "ideas," or objects of thought, cannot come from the senses, but from reflection on what comes from the senses, while he admitted that his innate ideas were dispositions that become attained as ideas in reflection thanks to the actions of *petites perceptions* that must accompany all consciousness, no matter how minimal. Leibniz then offered *a posteriori* and *a priori* arguments for such *petites perceptions*, which are ideal and infinitesimally small thought-elements that are necessary for psychological explanation, just as corpuscles are necessary for mechanical explanation.

Fifth (80–88), Maimon reduces the Locke–Leibniz debate to rather simplified Aristotelian–Platonic terms: it is really about the nature of the relation of the body to the soul, in which the former (Aristotle and Locke) claim that the soul is a mere epiphenomenon (as it were) of the body and the latter (Plato and Leibniz) claim that the body is instead a mere epiphenomenon of the soul. Maimon conciliates these broad views by arguing that on the one hand, Leibniz does not maintain a pre-established harmony between the consciousnesses of the monads and the events of their shared phenomenal and perceived world grounded in God as a third being, but rather an "essentially inner harmony" which Maimon does not strictly specify except for the fact that it is monistic, not dualistic, and "has its ground in the nature of the objects themselves" (82). Because of this inner harmony of the world-soul, real forms of understanding may come about in representational relations to some of the world, since their living bodies are their very unconsciousnesses, or composed systems of what Leibniz called *petites perceptions*. But these *petites perceptions* ("dark representations," in Maimon's terminology) can now be held to be of the same stuff as material corpuscles. On the other hand, though Locke only admits the soul as a passive feature developed in the body, the very force which must impress itself on the body in order that the body may develop an intelligence must itself be an intelligent force, a "living and thinking substance" (84). Thus, both Locke and Leibniz become modified in their opinions so that they agree with the conception of a world-soul, according to which the three types of real forms of matter, life, and understanding are materially homogeneous, but formally differentiated such that each is the condition for the higher form (or each form presupposes the lower).

Sixth (88–92) and finally, the idea of a world-soul is up to this point only compatible with the two views of Locke and Leibniz in different ways that remain to be reconciled. That is, for Locke, so far the world-soul inheres in the real forms of the world as the force of their totality, while for Leibniz, it is the "pure thinking" common to all understanding beings. For the former, the world-soul "does not exist for itself outside the world" (89), and for the latter, it "relates itself [...] to the universe, i.e., to the totality of all possible things, and is therefore infinite" (89). Maimon does not explicitly reconcile these two views after these remarks, suffice to say that Leibniz had good reason to count individual beings as substances, though not "self-standing [*selbstständige*]" (89) in themselves, but conditionally standing on their own, for themselves (*für sich bestehende*), since without limiting itself, the "infinite thinking being" "would not be, itself, anything outside itself" (89). And so, again, the world-soul agrees with the *Bildungstrieb* because of the conditional relations between forms. This guarantees that in some measure, the understanding of biological classifications (of species, etc.) is an arbitrary act of the understanding, but in some measure, it is also grounded in and determined by nature itself; it also explains the fact that the development of species displays regularity, but that there can also be irregularities, hybrids, and "bastards," and it furthermore explains how there can be spontaneous generation of life from matter and higher organisations of life from lower. Matter, then, is not the efficient cause of organisation, but it can be a "conditional cause," depending on the organisation of the organism for which it is a condition. "And how can this be denied? Since one cannot admit the arising of the first men and animals in any other way, if one does not want to take refuge in the supernatural, through which nothing is explained, but merely betrays one's ignorance, or assumes an already completed infinite sequence, which contradicts itself" (92).

It is evident from this summary of the *Weltseele* that it does not explicitly advance its thesis, though it surely has one. At the outset, Maimon declares that he only intends to develop the "notion" of the world-soul by comparing it to the "opposed notion," so that "the thinking reader may draw the result of this comparison himself" (48). In truth, however, there are three meaningful oppositions in the article. They are not necessarily between the world-soul and the opposed notion, however, but between 1) the Aristotelian and Leibnizian philosophies in general, 2) the Lockean and Leibnizian philosophies of the understanding or representing soul reduced to Aristotelian and Platonic positions, and 3) the Lockean and Leibnizian positions as reconciled to materialistic and idealistic conceptions of the world-soul. These various oppositions must be thought together, since neither notion in any of the oppositions can by itself give the unified conception of the world-soul upon which Maimon nevertheless relies for the various partial conclusions that he does draw along the course of the argument. Thus, there are three essential steps in each case: the first opposition, the second opposition, and the conclusion that the thinking reader must draw by way of reconciling or otherwise relating the first two. Therefore, this third step is a reflection that is an essential part of the article and its thesis.[68] How can the three sequences of oppositions and what follows from them be understood as a meaningful thesis?

---

68 In this sense, the article is esoteric. See, for comparison, Schelling's definition of philosophy as necessarily esoteric in Friedrich Wilhelm Joseph Schelling, *Bruno, or On the Natural and Divine Principle of Things*, ed. and trans. Michael G. Vater (Albany, NY: SUNY Press, 1984; original version: *Bruno, oder über das göttliche und natürliche Prinzip der Dinge*, 1802), 134; 233–34 in original.

2.1 The Central Problem of the *Weltseele*

The *status quaestionis* demands an account of how the *Weltseele* follows from the *Versuch*—especially as the former appeared immediately thereafter as a completely distinct metaphysical vision. And now, the question has been further specified, since the world-soul unfolds according to a sequence of three oppositions, each of which must entail third moments as definite conclusions for the thinking reader. So, what in the *Versuch*, and its failure, would at least explain the need for such a structure, whatever it happened to be? A clue is found in the fact that Maimon does not intend to give a dogmatically rationalist thesis, as in the *Versuch*, according to which "mere thinking" is held to be "sufficient thereby to determine an object" (51). Instead, the argument is intended to be not only compatible with the nature of transcendental subjectivity, but also to give an account of something that is a necessary and objective presupposition for it. There were arguments and statements in the *Versuch* which pointed in that direction, for in the suggestion in chapter 2 that Maimon should adopt a Leibnizian infinite subject that was to be equated with the finite subject but differ from it only in degree, he also suggested that he give an "explanation of the community between soul and body," by which is meant "the explanation of the world's arising (with respect to its matter) from an intelligence" (GW 2:62; ETP 37). The note to this remark explained that it would be possible to answer the *quid rationis* question; i.e., a version of Maimon's *quid juris* which would justify a hypothesis about the nature of the objective world in order to explain the certainty and possibility of the facts given in experience to the subject (GW 2:362). The way Maimon answered his *quid juris* in that work did not remotely fulfil these suggestions, but the *Weltseele* does.

The major thesis of this outline is that the *Versuch* intended to develop the critique of physics which was missing from Kant's philosophy and that the transition between the two works of Maimon in question can only be grasped in terms of this central problem. Maimon approached this problem as a reflection on the possibility of the scientific proposition "S is P." So how does the question of this proposition account for the transition to the next work? Look again at this proposition and its status. The outline of the *Versuch* above has shown that the finite subject which makes scientific judgments intends to discover the identity of a subject "S" composed of nominal relations— i.e., hypothetically assumed relations between phenomena that are proximate to each other—with a predicate "P" composed of precisely formulated conceptual relations. The subject can therefore designate with ever greater precision how appearing objects actually function in terms of their smallest possible elements and their relations and can thereby approach theoretical completeness under one single truth and concept. However, Maimon also attempts to understand this one single truth and concept and what theoretical completeness actually is.[69] We have seen why and how he attempts to do this and just what contradictions this involves him in. Nevertheless, if he is going to have a theory about this infinite subject—i. e., the absolute—it will follow that if everything can be completely explained by theoretical terms and relations, then the absolute will be the infinite subject which creates by its means from such relations; i. e., by thinking them. Maimon says just this: "The infinite thinking being thinks all possible concepts all at once with the greatest completeness without any admixture of sensibility" (GW 2:183; ETP 98). Therefore, the infinite thinking being does not judge, as finite thinking does, but constructs concepts. The thought process of the infinite thinking being is one of constructing concepts, and that of the finite thinking being is judging. "The understanding's procedure in forming [*Bildung*] concepts is opposed to its procedure in judging. In the former case it acts synthetically, but in the latter analytically" (GW 2:93; ETP 53); i. e., the former implies the complete determination of determinables, and the latter implies the discovering of the determinables for given, or nominal, determinations and so replacing the nominal with the real. Because of this, "the terms subject and predicate must be exchanged in the two cases" (ibid). For the infinite subject, the "S" in the "S is P" stands for determinability relations, while for the finite subject the "S" stands for nominal relations. But given the thesis of total theoretical completeness, "concept and judgment are identical" (ibid), and what occurs in judging is the Platonic "re-membrance [*Wiedererrinerung*] of the concept" (GW 2:94; ETP 53).[70]

However, a fundamental question remains. If there are no nominal relations for the infinite subject to encounter—i. e., no "admixture of sensibility" for it—and conceptual determinability relations comprise the "S" of the "S is P" which it thinks, then what comprises the predicate "P" for it? Ostensibly, this "P" refers to differentials.[71] However, chapter 8 argued that differential relations simply *are* determinability relations; there is no meaningful difference between them. Thus, Maimon begins to give a meaningful account of subjective judgment, but he cannot give a meaningful account of the objective conceptual structure about which the subject judges. The divine or absolute "S is P" has no structure to speak of. Therefore, as recounted above, there is no place for cognitive mediation in and by the infinite subject without the finite subject, and when Maimon equates the two subjects, as he must, he completely loses this possibility as well. The problem is, therefore, that Maimon does not develop what Hegel would famously call the theory or "exposition of God as he is in

69 In his *General Theory of Knowledge*, Schlick argues that this is philosophy's basic error: it tries to say what reality *is*.

70 That concept and judgment are the same and different is, of course, another central contradiction.

71 See notes 43 and 50 above.

his eternal essence before the creation of nature and of a finite spirit"[72] in the *Versuch*. If he could develop a hypothesis which would give a complete structure to the "S is P" construction of concepts, such that this structure would be the objective presupposition for the judgments made by the transcendental subject, then he could answer his own *quid juris*. The judgments made by the finite subject could be argued to be in complete accordance with the concepts of the infinite subject. This, of course, would not be an answer given critically in terms of reflection on the thought of the transcendental subject (which is a possible direction for the question considered by the *Versuch* and fully taken up by the *Logik*), and it would not be an answer that further specifies how it is that the finite subject can make judgments in the science of physics. It would be an answer given *via* hypothesis about how the world arises so as to account for the "certain and possible" judgments the subject can make. It would instead be an answer to his *quid juris* as *quid rationis*; it would be an account of a world-soul. It would also be a strong commitment to the dynamic paradigm. The world-soul, as Maimon hypothesises, is the force of the *Bildungstrieb*, as it explains how it is that life in general can relate both to matter and to the understanding which comes to have cognition of it.

The structure of the absolute or infinite "S is P" which Maimon desires must therefore reveal or indicate itself in the argument of the *Weltseele*. The argument, again, takes the peculiar form of three different oppositions which have a certain pattern. In each case, there are two opposing arguments and the third argument which follows from them for the thinking reader. This third argument, or conclusion, fully characterises the world-soul. Now, the nature of this conclusion must be investigated.

2.2 The Argument of the *Weltseele*

Maimon's argument in this text is profoundly suggestive, and an outline cannot capture every detail or aspect of it. However, here is an attempt to express its core, step by step. First, we may note Maimon's definition of the world-soul: "The world-soul is a force [*Kraft*] residing in the matter in general (the stuff of all real objects) and acting [*wirkende*] on it, whose effects [*Wirkungen*] are different according to the different modifications of the matter." It is the "ground of the particular type of combination [*Zusammensetzung*] in every body"—i.e., in merely material bodies and in organic bodies—including "life in animals, and the understanding and reason in humans" (48–49).[73] The world-soul, therefore, unifies three types of objectively

72 Georg Wilhelm Friedrich Hegel, *Science of Logic*, trans. Arnold V. Miller (Amherst, NY: Humanities Press, 1969; original work: *Wissenschaft der Logik*, 3 vols. [Nuremburg: Johann Leonhard Schrag, 1812–16]), 50; 21.34 in original.

73 Alternatively, one can read the definition of it given to Kant in Maimon's letter about his article: "The world-soul is a force [*Kraft*] inherent in matter in general (the material of all real objects), a force

real forms: forms of matter, forms of life, and forms of understanding.[74] Since Maimon advances a theory of the world as intelligible and as presupposed by the transcendental subject, it seems he has chosen traditionally acceptable forms with which to describe it. First, one can see how matter is a unique real form on its own, since it is evident that matter as such is elemental, such that each element, whether on its own or in combinations, falls into various structures and patterns. Thus, physics may take these elements as its subject-matter. Second, one can also evidently see the form of life which is "higher" than mere matter. A life or life-form such as a bee colony, like matter, consists of elements or parts, but its parts are functional parts of a living whole. An organism is a living whole "because [its] members are different from each other and reciprocally complementary in virtue of their differences."[75] Thus, a different science—for example, biology—may take anything exhibiting organic structure as its subject-matter. (It seems that one may reduce the biological sciences to physical sciences, and perhaps do so completely, but then the subject-matter in question would no longer be an organism or organic system.) Furthermore, beings with understanding manifest an even higher form; namely, the form of that which can autonomously conceptualise the lawfulness of its own organisation and thus come to have its own unique structure and life to the degree to which it does so. And in turn, disciplines such as ethics or political philosophy may take such forms of understanding (ethical persons, republics) as their subject-matter.[76]

But Maimon did not choose these forms, however traditional, as arbitrarily as it might first seem, for each corresponds not only to certain specialised sciences, but also to definite philosophical approaches to reality. Erich Heintel has shown how this is: he has analysed these approaches as the inevitable course of philosophy's "falls from grace [*Sündenfälle*]";[77] i.e., its progressive falls from states of apparent im-

that affects matter in general in different ways according to the various ways that matter is modified. It is the ground of the particular sort of combination in each sort of matter (even in unorganized matter), the ground of the organization in every organized body, the ground of the life in an animal, of the understanding and reason in human beings, etc.; in short, the world-soul confers forms on all things according to the constitution of their matter, in such a way that it adapts matter, enabling it to change from a single form, to take on other forms, forms of a higher order. And since matter can undergo unlimited modification, so this entelechy too can supply an unlimited variety of forms. It is thus the ground of all possible agency" (GW 6: 429–30; *Correspondence*, 352).

74 One must note, of course, that Maimon is here proposing objectively real forms that are also syntheses that can be thought by the subject. They are therefore different from the subjective forms of sensibility and understanding which Maimon defines in chapter 1 of the *Versuch*, though of course they are also related.

75 Michael Beresford Foster, *The Political Philosophies of Plato and Hegel* (Oxford: Clarendon Press, 1935), 2 ff.

76 A standing difficulty is the difference between understanding as demiurge, which is given the pattern of the structure which it is to understand and synthesise, and understanding as spirit, which freely creates its own structure. Foster's book asks which one Hegel describes; one can ask the same of Maimon here.

77 Heintel, *Hegel und die Analogia Entis*, 43 ff., and *passim*.

mediacy to more complicated and alienated states of reflective distance from that immediacy.

First, nascent philosophical reflection, departing from mythology which blurs natural and divine generation, asks after a first principle of the origin of things; i.e., that which gives a universal explanation of particulars. And it is natural that it looks for this principle among the material elements of natural things. One unique being among beings must be the principle of all beings (such as water was for Thales); this is called thinking in terms of the ontic difference. But this reflection reveals that any element thus chosen has been chosen only contingently, since it is a part of processes which could have any other element as their principle or starting point; or, admittedly, the processes in question could have been and might be otherwise. This reflection therefore (and for other reasons, certainly) gives way to another, according to which the principle must not be a being among other beings, but must transcend them.

Second, thinking thus adopts the terminology and problematic of the ontological difference—of the difference between a transcendent Being from beings. This philosophical approach corresponds not to the question of matter, but to that of the form of life; traditionally, philosophers have sought the principle of the unity of the living organisation in terms of the ontological distinction (since a final cause is not ever now, or here). But this approach has also posed an insoluble problem. For reasonable speech about the ontological principle must not simply speak about it positively-ontically as a being among beings (such speech cannot be univocal), since it must be a *principle* of them, nor must it speak about it negatively-ontically as entirely unlike other beings (such speech cannot be equivocal), since, again, it must be a principle *of them*. This dual-negative prescription is that of the doctrine of the *analogia entis*. But the prescription cannot be fulfilled in these terms; any attempt to predicate something of the ontologically distinct principle (whether by making human speech more eminent or by restricting it) can nevertheless only either treat the ontologically different in an ontic manner or entirely and abstractly negate it.

However, third, throughout the course of this reflection, thinking has approached reality in a way that corresponds to the question of the form of the understanding. It has asked what must be presupposed in order to cognise ontic and ontological explanations in the first place, and so (without recognising or reflecting on itself) it has implicitly asked what human or finite thinking is. Therefore, this approach, first made explicit by Kant, recognises the transcendental difference, because the question of the understanding involves a difference between object-directed thinking and thinking about such thinking; i.e., reflection. This difference is peculiar. On the one hand, reflective thinking takes itself as object, because it is finite, and presupposes itself as given, thereby distinguishing itself from the infinite and productive thinking of the ontological principle (which presupposes nothing); on the other hand, it is productive in that its critical reflection autonomously determines itself as objectively valid thinking, and in doing so, it is something that in principle cannot be treated like an object. (The original synthetic unity of transcendental apperception cannot be determined, but only thought, by the categories.) The transcendental difference is therefore really that between beings and thinking. This difference is not the same as that between objects.

However, the transcendental difference also has its problems in that, just as the ontological difference cannot be differentiated from the ontic, it has not quite been differentiated from the ontological. One sees this problem exemplified in Kant. In order to account for itself as a principle (though not for Kant as *the* absolute principle), reflective thinking must independently determine and justify the criteria for the objective validity of thought itself, but in doing so, it has to transform the failed attempt to think the ontological principles (the ideas of the soul, world, and God, unified in the ideal of reason) into regulative principles. However, the attempt to do so (see note 26 above) cannot justify itself. In order to avoid predicating the experiential concepts of the absolute (to avoid univocal speech), Kant proposes predicating them of the relation of the absolute to appearances (to avoid equivocal speech), but this would still require a transcendent use of the categories. The problem of the *analogia entis* returns, and Kant is unable to fully distinguish the principle of the transcendental difference, thinking, from the absolute ontological principle and its insoluble problems. Heintel observes that the transcendental difference signals that it could potentially justify the ontological difference as a philosophical principle—for the first time—since it promises a way to differentiate the ontological from the ontic in a non-ontic manner (since thinking must differ from being in a different way to how beings differ from beings). Thus, the ontological principle is further confirmed as a divine intellect or *intellectus archetypus*. However, the distinction between finite and infinite intellects, and between these two principles (the transcendental and the ontological), is not yet fully worked out. Both succumb to the problem of the *analogia entis*. But to conclude regarding the *Sündenfälle*, thus it is that the three forms of Maimon's world-soul correspond to three types of philosophical reflection, and it is of course notable that Maimon explicitly recognises all three types of thinking and, *pace* Kant, attempts in the *Versuch* to reconcile the principles of the transcendental and ontological differences.

Of course, Maimon's reconciliation of the transcendental and ontological differences in the *Versuch* failed, but how exactly did this occur? This was outlined in section 1.3 above, but can be stated anew. First of all, that work is a theory of the absolute, meaning that the theory attempts to give the ground of our cognition of the absolute (the single true concept) and also to give the absolute ground of our cognising in general (the grounding of our cognising both for itself and in the single true concept).[78] This, as Heintel says, is to unite the subjective genitive (cognition *of the absolute ground*; the absolute's cognition) with the objective genitive (ground *of our cognising*; our cognising's ground) in one foundation. Ever since Parmenides, the "absolute" has referred to precisely this uniting foundation.

78 Heintel, 7 and 15.

In the *Versuch*, Maimon approaches this absolute in two ways. On the one hand, he asks about it in terms of the ontological difference. He proposes an infinite thinking being that is one with the world and with thinking, so that it would think and thus give existence to all relations which follow from its own pure identity, equated with all that is non-contradictory. Thus, being and thinking fall together according to this principle, as do the objective and subjective genitive. But this falling together only occurs by negating all determinacy, so that the absolute identity of being and thinking is an empty Being without difference, which (as Maimon recognises in chapter 8) is indistinguishable from nothing. But since Maimon has only been able to recognise that this being is identical with thinking because of the transcendental difference—of which, as a post-Kantian, he is explicitly aware—he actually denies the immediate and finite thinking which is a presupposition for transcendental reflection and which is considered as an object in reflection. Furthermore, he makes absolute the pure reflexive thinking that cannot be taken as object and is revealed by the transcendental difference. So, what occurs in the *Versuch* is that Maimon's positing of this ontological principle contradicts itself (being is being, being is nothing), but, as it contradicts itself, it reveals the pure form of transcendental reflection (thinking).

On the other hand, Maimon approaches the absolute in terms of the transcendental difference. When he "reverses Kant" and posits the original synthetic unity of transcendental apperception as a primordial intuition, such that any empirical given is given to the subject in nascent conceptual spatiotemporal relations, this is to posit the transcendental distinction in the self. For, since thought-relations are immediately given to the subject, the subject must think about those given thought-relations, taking them as objects. And the product of this thinking is a transcendentally ideal spatiotemporal manifold which, as in Kant, is a thought-thing, to be distinguished from things in themselves. However, there is a problem. On the one hand, Maimon posits this self-grounding of the subject pre-reflexively, primordially, so that it possesses immediate self-certainty, and this is the same immediate self-certainty which it must pursue as its final cause. This immediate self-certainty, as beginning and end, is therefore that which Maimon takes as his justification for identifying it with the ontological principle, absolute being. On the other hand, the reflexive self must, in its account of how it thinks itself, develop a doctrine of its categories and of the structures of thought by which it thinks what is given. It must account for its own validity. But to take the immediate self-certainty of thought's intuition as justification for its identity with absolute being is to ultimately (once again) negate all given and thought determinacy. Thus, there is a denial (once again) of the finite thinking presupposed by the transcendental difference. As recounted in section 1.3, to take the judgments of the finite subject and identify them with the concepts of the infinite subject is to destroy their possibility; there is no structure of propositional truth in this absolute identity, and so no expression of truth as unity in a manifold. Therefore, this beginning with the transcendental difference contradicts itself (being is thinking, being is not thinking), but as it does so, it reveals itself as the pure absolute *a priori* of the ontological principle (being).

The account of the mutual self-contradictions of the ontological and transcendental arguments of the *Versuch* explains just why it is that Maimon's first opposition in the *Weltseele* takes place between the Aristotelian and Leibnizian philosophies of the absolute. For after first describing the nature of the *Bildungstrieb* as a force of generation, growth, and healing and then asserting that the force requires a philosophical critique, Maimon states that the three real forms must be grounded, and there is a question about whether they have their grounds in themselves or in something outside themselves. From that question "sprang two opposed opinions: namely, the Aristotelian school maintained that matter and form are entirely heterogeneous; matter is entirely without all effect, and acts merely in a passive manner, but the form is the ground of all actuality" (65). The Leibnizian–Wolffian school, by contrast, supposes that "the forms of bodies have the ground of their existences in themselves, and they arise and pass away not according to the laws of nature (without, because of that, existing as necessary beings, however)" (66). Both approaches contradict themselves, however. The Aristotelian approach, which proceeds in terms of the ontological difference only, upholds the existence of these three forms grounded in an absolute being transcendent to them, but "this *anima mundi*" is "a form without matter" (66) and therefore a form without determinacy, and so it is instead absolutely nothing. The Leibnizian approach, which proceeds in terms of the transcendental difference only, "makes matter and form into only one substance (insofar as the difference between them only occurs relative to the subject)" and so the world-soul is "entirely banned" and the "infinite being is by no means according to this an *anima mundi*, but an *ens extramundanum*" (66). Therefore, one can now understand how the Aristotelian view reflects the ontological principle in the *Versuch*, which contradicts itself (being is being, being is nothing) and in this contradiction reveals the pure form of transcendental reflection, since what really occurs is the elimination of all determinacy of finite thinking which this reflection presupposes. The Leibnizian view reflects the transcendental principle in the *Versuch*, which contradicts itself (being is thinking, being is not thinking) and so reveals the pure form of the ontological principle, being.

However, in the *Weltseele*, there is a new reconciliation of these two oppositions in the third step; i.e., the recognition of the structure of the *Weltseele* that is left to the thinking reader. The reconciliation must be this: the three real forms (matter, life, understanding) can only be grounded in relation to each other if they have grounds *both* outside themselves *and* in themselves. This would be to propose a foundational identity of contradiction. Maimon's own example should help to explain what this means. For immediately after giving the opposition between the Aristotelian and Leibnizian opinions, he lists four conclusions that follow, and the main conclusion is that mechanical (material) and teleological (living) processes have the same origin. As an indication of this same origin, he recites the three unique properties of organic life described in § 64 of Kant's *Critique of Judgment*, entitled "On the Character Peculiar to Things [Considered] as Natural Purposes." These also correspond to Blumenbach's description of the *Bildungstrieb* according to its functions of "generation, nourishment, and reproduction" (the latter also in the sense of healing).

These three properties of organisms are as follows. "A tree, firstly, generates another tree according to known laws of nature, i. e., it generates itself according to its species, and is therefore cause and effect simultaneously" (69). Secondly,

> A tree also generates itself as a particular tree. This is the growth which is entirely different from every other increase of magnitude according to natural laws, and is to be recognized as a generation, insofar as it first processes the material which it adds to itself into a specifically characteristic quality, which is something that no mere natural mechanism can do, and it builds itself up by mediation [*vermittelt*] of a material, which according to its composition is its own product (69–70).

This process of self-generation is twofold. First, the natural thing severs from its context and consumes an amount of natural resources, which must be seen only as an "educt, according to its properties." This is a first "negation," to use Hegelian terminology. Second, the natural thing transforms this educt so that "there is in the separation and new composition of this raw stuff such an originality that leaves all art far behind it." This is a second negation, i. e., a "negation of the negation," to again use Hegelian terminology. And finally, or thirdly, "a part of this creature also generates itself, something with which everyone aware of nature should be sufficiently familiar." The particular tree as such is then also propagated as a unique singular tree.

There is an important distinction of the universal, the particular, and the singular in this example. The particular tree carries out the essential function of *mediation*. It takes external matter from its medium or environment, then digests and processes that matter into its own substance, and so it grows. However, in its essence, this growth must also be decay. Matter is excreted, and the organism dies. Therefore, the material environment of the tree enacts the very same, yet opposite mediation; it also takes external matter from the creature, and in decay, that matter is transformed (another mediation) back into the elemental. Hence, this mediation of the particular thing is an identity of contradiction, at its heart. This expression and resolution of contradiction gives the organic its organic nature. An organism is brought about mechanically, and yet is an intrinsically purposive thing, thanks to what its inherent contradiction enacts. Thus, each act or event, such as consumption or digestion, is in itself mechanically causal, and may be described as mechanical. Or, as Maimon says, the forms (whether of organism or matter—or even of understanding) are "brought forth according to mechanical laws of nature, but only through a process by which its objects are brought forth in a certain way [*auf eine bestimmte Art*], according to mechanical laws of nature, and acquired" (68–9). The process of mediation therefore propagates the tree as a *universal* genus, while at the same time, the *particular* tree as such generates and heals its own parts, specifically differentiating it over time as an irreducibly unique *singular* tree. And, at the same time, the opposite process of mediation also brings about matter in particularly generic and specific ways. Maimon argues that this contradictory mediation is the force of the world-soul, universalised. It explains matter's relationship to the organism, and the organism's relationship to matter. The force of the *Bildungstrieb* is the force of the world-soul in this movement of negation of negation at the heart of the differentiating and relating of forms. Taken as a whole, as absolute, it must explain the relation of both matter and organism, as itself, to its own understanding.

The second and third oppositions can also be taken as examples and as elucidating the first foundational opposition. The second opposition considers Leibniz's debate with Locke about the nature of the understanding's representations and the former's conciliation, which nevertheless entails an abstract dualism of material corpuscles and psychical *petites perceptions*. The intentionally unspoken resolution of this dualism must be that again, the forms are grounded both in and outside of themselves. Therefore, a representing organism will represent in the same manner that an organism consumes its food. It will negate some given matter, which tradition describes in terms of "primary qualities," and negate it again, transforming it via this mediation into psychological "secondary qualities" which it may then perceive and think. And so, thanks to the limitation of itself to a body, each particular representing being represents only a limited portion of the world-soul. In this way, the representing organism may also be susceptible to mechanical explanation, in all its individual processes. Instead of a dualism, Maimon can maintain "that it is yet these [material] elements and their modifications that serve just as well for explanation in the soul as they do for the doctrine of nature, as I have just sufficiently shown" (87). The third opposition, finally, questions the very nature of this material; i.e., the form of matter and the force of the world-soul that matter becomes. Is it something entirely being-like, or thought-like? It is, of course, both, but conditionally one or the other, as the second opposition has shown. And so, this opposition leads the reader back, in a circle, to the first opposition, which was the opposition of ontological and transcendental principles.

What, therefore, does the reconciliation of the evidently fundamental opposition of the Aristotelian/ontological and Leibnizian/transcendental viewpoints mean? It must mean, first of all, that the two are reconciled so that the Aristotelians' absolute being *is* the Leibnizians' transcendental principle of thinking. But at the same time, absolute being is *not* thinking, because it is different from it. Because of this sameness and difference, the absolute being is *also* nascent reflexive thinking. Its determinations, which (according to the problem of the *analogia entis*) are ontic and so necessarily contradictory, beginning with "being is being" and "being is nothing," are no longer only externally connected predications and qualities; they are also internally connected in a movement of predication in a thinking of being which expresses itself. The understanding which was held entirely apart from being and took its contradictions for ultimate impediments to speech about it is now the immanent life and dialectical mediation of itself. Thus it is that the world-soul can be, conditionally, matter, life, or understanding (or not): the mediation by negation of negation of matter can become, under certain circumstances, living matter (and vice versa), and the dialectical mediation of this living matter by negation of negation can become, under certain circumstances, representation of the understanding (and vice versa). And this justifies the three types of philosophical reflection. The ontic difference can be distinguished from the ontological difference because there is now a way of expressing the ontological principle in a way that distinguishes it from all mere beings. Furthermore, the transcendental difference can be distinguished from the ontological because it is the expression of the ontological principle (of Being) that comes about in and for it; that is, it specifies how the world-soul is the absolute ground of cognition of it and how it also absolutely founds our cognising in general.

To see *in detail* how the world-soul is the absolute and therefore joins the subjective and objective genitive of cognition, recall section 2.1 on the problem of the *Weltseele*. There, Maimon considered how concepts and judgments are different and yet the same. The concept is constructed by the objective or absolute intellect, and it proceeds with the "S" and (it is stipulated) the "P" follows from that "S" according to an analytic rule. However, judgments are made by finite intellects, so that the "S" is nominal and the "P" determines it, in a way that approaches completeness. In this completeness, the subject and predicate of the infinite intellect's concept are respectively the predicate and subject of the finite intellect's judgment. But the *Versuch* could not carry out this conception, since the infinite subject could only speak tautologies; i.e., nothing at all. However, the *Weltseele* lays out a possible solution that can be extrapolated from its oppositions, which is also found in Hegel. According to the fundamental opposition, there are Aristotelian/ontological and Leibnizian/transcendental approaches to the absolute. The first approach would begin with the determinable of absolute being or identity. This is the subject "S," and (as in chapter 8 of the *Versuch*) it is indistinguishable from nothing. "S is S" is then "S is not S." But this contradiction must be productive and yield a medium of difference, "P," which is a determination in which the original determinable would (unreflectively) disappear.[79] In this way, Maimon would, like Hegel, set up a "Doctrine of Being." Its successive predicates would be predicates of any being whatsoever, which could therefore include quality, quantity, and measure,[80] and its result would be the revealing of the form of pure transcendental reflection, because all along, this doctrine has been the reflective thinking of any being whatsoever to the exclusion of the determinations of the finite thinking which it must nevertheless presuppose. Thus, the first approach gives way to the second.

The second approach, as explicitly reflective, would not proceed by sequential determination of a determinable which disappears in its determinations. Rather, it

79 Hegel's *Science of Logic* includes these three doctrines of being, essence, and concept. See *Encyclopedia Logic* § 161 for description of the corresponding types of dialectical processes and their transitions into each other.

80 A study of the similarities between Maimon and Hegel would have to reference how in Maimon, quality (as a differential) *becomes* quantity (GW 2:29n, 395) and how this *becomes* measure as well (GW 2:29n), anticipating the basic content and movement of the "Doctrine of Being" in Hegel's *Logic*.

would proceed by specifying the ways of reflection which are necessary in order for thinking to discover a determinable "S" for a given determination "P," and via this procedure, it would begin to recall, in mediated ways, the progressively more universal determinables of the "Doctrine of Being" which had disappeared into their determinations. This would yield further terms applicable either to beings or to thoughts, which could therefore be used as subjects or predicates. Via this procedure, the terms that are generated contradict each other by being similar (and appearing in each other) and yet different. One example of this would be "ground" and "condition";[81] others would be "substance and accident" and "cause and effect." This second approach could be called, as in Hegel, the "Doctrine of Essence," and its reflective procedure would result in the reconciliation of the ontological and transcendental approaches to the absolute, such that any determination of being at all, even being that appeared via representation to reflective thought, could be conceptualised, or placed in conceptual form such that a being could be said to be determined in mechanical ways (mere being), in living ways (beings that have reflexive structures), or in understanding ways (beings that speak beings). Consequently, the world would have its genesis in such a way that the judging subject would be enabled to represent any aspect of that reality—even including the judging subject itself which would exist in it, or even that reality as a whole—and to take that representation as a nominal subject of its judgment, and would be enabled to predicate something real of it. The possibilities of any such predication would have already been accounted for in the genesis of the world-soul. The ways that the judging subject could judge in ways that would be, as Maimon says, a "re-membrance [*Wiedererrinerung*] of the concept" (GW 2:94; ETP 53) would comprise what Hegel would call the "Doctrine of the Concept." In this way, the *Weltseele* may be called a dialectical speculative metaphysics, as indicated above: it calls for certain dialectical movements whose unification is absolute cognition.

---

81 In Maimon, grounds are generally determinables which necessitate certain determinations, but then he in turn counts those determinations as necessitating their determinables (GW 2:91; ETP 52: "The necessity [...] rests on the other part [...] that cannot be thought without the first"). Thus, ground becomes condition and condition ground, as in Hegel's "Doctrine of Essence." The substance–accident and cause–effect relations display a similar movement. Oded Schechter, "The Logic of Speculative Philosophy and Skepticism in Maimon's Philosophy: *Satz der Bestimmbarkeit* and the Role of Synthesis," in Freudenthal, *Salomon Maimon*, 18–53, argues that Maimon's "determinability" can be taken as the core of Hegelian speculative logic. If so, it must also come to comprise these three types of dialectical movements. That is, if it strikes the reader that Hegel's *Logic* is indeed a strange catalogue or classificatory scheme of different "logical" sequences, then the nature of Maimon's problematic can show why and how that should be.

2.3 The Failure of the *Weltseele*

Of course, Maimon did not complete the extrapolation of the ideas in the *Weltseele* which has just been suggested. He did not explicitly develop a "logic of the concept" that would objectively account for the possibility of the subjective judgment of a transcendental subject. The work in question left that task to the "thinking reader."[82] But regardless of the particular details of how such a task would be carried out, the article does make one thing clear, which is that any transcendental philosophy at all, especially one which would be reconciled with the nature and claims of physics about the objective world, must reconcile and justify the transcendental and ontological differences, and this means that it must have a logic of the absolute, specifically such that it can claim that its logical speech is a mediation of being. Thus, Maimon learns this lesson and retains it for the working out of his *Logik*. However, at the same time, he did explicitly abandon the conception of the world-soul, as we know from his article on *Schwärmerei* and its commentary on Bruno. Without giving a close reading of that article, what are some reasons for retracting the claims of the *Weltseele*, and which particular claims are retracted?

First, Maimon's explanation of how a world arises in order to answer the *quid rationis* and to show how it is objectively comprehensible that the subject can have true and certain judgments replaces the more evidently scientific use for conceptual determinability relations that he had proposed in the *Versuch*. There, he had come to a model, which he would improve upon in the *Logik*, of how these concepts are used to cognise, theoretically and by symbolic means, precisely how it is that the physical elements of appearances determine those appearances. This is no longer possible, since cognitive representation in the *Weltseele* is a higher type of negation of negation. It is a form of life, and so it comes to be and passes away as life does, by consumption and transformation of the material of representation. But theoretical or scientific cognition, which the *Versuch* proposes and the *Logik* answers, is not an eating of its object, but the opposite: it asks after the conditions of the possibility of symbolically presenting objects as they are in themselves. It aims at theoretical fidelity to the object, and because of this, it demands a definite *Gelassenheit* rather than heteronomous consumption. However, the philosophical speculation of the *Weltseele* offers no route to this type of theoretical cognition of the physical world.

Second, consider the apparently perfect way in which the *Weltseele* indicates that the ontic, ontological, and transcendental differences can be reconciled. Given this perfect coinciding of principles, how could one abandon it? But then

---

82 In this suggestion, I attempt an approximation to Hegel's *Science of Logic*, since I think Maimon in his own way is dealing with the nature of Hegel's speculative proposition. Like this proposition, which cannot as such be stated, but only shown in regular propositions which are to be taken speculatively, Maimon shows but does not state the truth of the world-soul to the thinking reader who reflects on the product of his three oppositions.

again, the nature of ontic processes is that they could always be otherwise. One must, in fairness, admit this. And if they could be otherwise, then perhaps they might not have existed at all, or at least the type of processes which resulted in or made possible understanding might not have existed. This thought evokes the question of creation, and so is also an ontological thought.[83] The *Weltseele*, which (like Hegel's *Science of Logic*) offers a distinct ontic theory about force, is thrown into doubt along with its ontological hypotheses. At this point, the entire "battlefield of endless controversies" of metaphysics may return. The only response which will answer philosophical questions with scientific dignity in the face of these controversies is the Kantian one, which autonomously establishes the possibility of the objective validity of thought. And so, the question for Maimon's *Logik* is, again, can he carry out such a critical reflection while also giving the demanded logic of the absolute?

3 Conclusion

This outline will have been successful if it contributes to a future commentary on Maimon. It should have proved that Maimon articulated a certain basic problem of the critique of physics that anticipates Kant's *Opus Postumum*. It should also have shown that the interpretation which takes Maimon as half-"rational dogmatist" and half-"empirical sceptic" is only partially correct, and ultimately misleading. Though Maimon gave himself this designation at the end of the *Versuch* (GW 2:432 [436]; ETP 222), it is a designation which is a symptom of a failure to solve the basic problem.[84] Thus, to take Maimon as such would be to take a symptom of the failure of a work to be that work's cause and motivation. Finally, this essay should have shown that the *Weltseele* follows from the *Versuch* (if it has not done so in every detail) and that it represents one of three major high points of Maimon's phi-

83 See Wagner, *Philosophie und Reflexion*, 414. There, he recalls that it was the pre-Kantian metaphysics which prescribed the laws of thinking to beings, to things in themselves (and those who misunderstand Kant accuse him of doing just that). Thus it was that Aristotle wrongly denied the possibility of an actual infinite because it was not thinkable, and thus it is that even those who currently ground their philosophies on the question "why is there Being at all, and not nothing?" do so fallaciously, because they uncritically exclude "the possibility of a necessity of Being" simply because "they believe themselves able to think the Not-being of Being."

84 In GW 3:455 (the year 1792), Maimon replied to the Swiss philosopher Jacob Hermann Obereit, who had coined the term "nihilism" as a description of Kant's apparent isolation of the subject from God, that he too had "shuddered before the [Kantian] nothing" and had taken a *salto mortale* into a combination of Spinozist rationalist dogmatism and Kantian transcendental philosophy in the *Versuch*. His recognition of the failure of that leap, however, saw him instead attempt to combine "Kantian philosophy with Humean scepticism." In the interpretation of this article, the latter combination is a heightened criticism, one that proceeds to solve the central problematic physics poses for criticism, and one sees now how this comes to pass.

losophy. The third high point, the *Logik*, would have as its task an apparent paradox: a critically justified philosophy of reflection that is also a philosophy of the absolute, in order to finally bring criticism to transcendental philosophy.

Bibliography

Atlas, Samuel. *From Critical to Speculative Idealism: The Philosophy of Solomon Maimon*. The Hague: Martinus Nijhoff, 1964.

Beiser, Frederick C. *The Fate of Reason: German Philosophy from Kant to Fichte*. Cambridge, MA: Harvard University Press, 1987.

Bergmann, Samuel H. *The Philosophy of Solomon Maimon*. Translated by Noah J. Jacobs. Jerusalem: Magnes Press, 1967.

Cassirer, Ernst. *Das Erkenntnisproblem in der Philosophie und Wissenschaft der Neueren Zeit, Dritter Band: Die Nachkantischen Systeme*. Hamburg: Felix Meiner Verlag, 2000. Originally published by Verlag Bruno Cassirer, 1929.

Ehrensperger, Florian. "Weltseele und unendlicher Verstand, Das Problem von Individualität und Subjektivität in der Philosophie Salomon Maimons." PhD diss., Ludwig-Maximilians-Universität, 2006.

Engstler, Achim. *Untersuchungen zum Idealismus Salomon Maimons*. Stuttgart and Bad Cannstatt: Frommann-Holzboog, 1990.

Flach, Werner. "Transzendentalphilosophie und Kritik. Zur Bestimmung des Verältnisses der Titelbegriff der Kantischen Philosophie." In *Tradition und Kritik: Festschrift für Rudolf Zocher zum 80. Geburtstag*, edited by Wilhelm Arnold und Hermann Zeltner, 69–83. Stuttgart and Bad Cannstatt: Frommann-Holzboog, 1967.

Foster, Michael Beresford. *The Political Philosophies of Plato and Hegel*. Oxford: Clarendon Press, 1935.

Freudenthal, Gideon, ed. *Salomon Maimon: Rational Dogmatist, Empirical Skeptic. Critical Assessments*. Dordrecht: Kluwer, 2003.

Gueroult, Martial. *La Philosophie transcendantale de Salomon Maïmon*. Paris: Libraire Félix Alcan, 1929.

Hegel, Georg Wilhelm Friedrich. *Hegel's Preface to the Phenomenology of Spirit*. Translation and commentary by Yirmiyahu Yovel. Princeton and Oxford: Princeton University Press, 2005. Original work: *Phänomenologie des Geistes*. Bamberg and Würzburg: Joseph Anton Goebhardt, 1807.

Hegel, Georg Wilhelm Friedrich. *Science of Logic*. Translated by Arnold V. Miller. Amherst, NY: Humanities Press, 1969. Original work: *Wissenschaft der Logik*. 3 vols. Nuremburg: Johann Leonhard Schrag, 1812–16.

Heintel, Erich. *Hegel und die Analogia Entis*. Bonn: Bouvier, 1958.

Kant, Immanuel. *Correspondence*. Edited and translated by Arnulf Zweig. Cambridge: Cambridge University Press, 1999. Original work published in Volumes 10–13 of *Gesammelte Schriften*. Berlin: Georg Reimer, 1900–1922.

Kant, Immanuel. *Critique of Judgment*. Translated by Werner S. Pluhar. Indianapolis and Cambridge: Hackett, 1987. Original work: *Kritik der Urteilskraft*. Berlin and Libau: Bey Lagarde und Friederich, 1790.

Kant, Immanuel. *Critique of Pure Reason*. Edited and translated by Paul Guyer and Allen W. Wood. Cambridge: Cambridge University Press, 1998. Original work: *Kritik der reinen Vernunft*. Riga: Hartknoch, 1781; 2nd ed., 1787.

Kant, Immanuel. *Metaphysical Foundations of Natural Science*. Edited and translated by Michael Friedman. Cambridge: Cambridge University Press, 2004. Original work: *Metaphysische Anfangsgründe der Naturwissenschaft*. Riga: Hartknoch, 1786.

Kant, Immanuel. *Opus Postumum*. Translated by Eckhart Förster and Michael Rosen, edited by Eckhart Förster. Cambridge: Cambridge University Press, 1993. Original work published in Berlin: De Gruyter, 1938.

Kant, Immanuel. *Prolegomena to Any Future Metaphysics, With Selections from the Critique of Pure Reason*. Edited and translated by Gary Hatfield. 2nd ed. Cambridge: Cambridge University Press, 2004. Original work: *Prolegomena zu einer jeden künftigen Metaphysik, die als Wissenschaft wird auftreten können*. Riga: Hartknoch, 1783.

Koyré, Alexandre. *Newtonian Studies*. Chicago: Chicago University Press, 1965.

Kroner, Richard. *Von Kant bis Hegel*. 2nd ed. 2 vols. Tübingen: Mohr/Siebeck, 1961; first published 1921–24.

Longuenesse, Béatrice. *Kant and the Capacity to Judge: Sensibility and Discursivity in the Transcendental Analytic of the Critique of Pure Reason*. Translated by Charles T. Wolfe. Princeton and Oxford: Princeton University Press, 1998.

Maimon, Solomon. "Baco und Kant." *Berlinisches Journal für Aufklärung* 7 (1790): 99–122.

Maimon, Solomon. *Essay on Transcendental Philosophy*. Translated by Nick Midgley, Henry Somers-Hall, Alastair Welchman, and Merten Reglitz. New York: Continuum, 2010. Original version: *Versuch über die Transzendentalphilosophie*. Berlin: Bei Christian Friedrich Voss und Sohn, 1790.

Maimon, Solomon. *Gesammelte Werke*. Edited by Valerio Verra. 7 vols. Hildsheim: Olms, 1965–76. Originally published 1789–1801.

Maimon, Solomon. *The Autobiography of Solomon Maimon: The Complete Translation*. Edited by Yitzhak Melamed and Abraham P. Socher. Translated by Paul Reitter. Princeton: Princeton University Press, 2018. Original version: *Lebensgeschichte*. Berlin: Friedrich Vieweg dem ältern, 1792–93.

Maimon, Solomon. "Über die Weltseele (*Entelechia universi*)." *Berlinisches Journal für Aufklärung* 8 (1790): 47–92.

Melamed, Yitzhak. "Salomon Maimon and the Failure of Modern Jewish Philosophy." September 2012. Available online: http://www.theapj.com/wp-content/uploads/2013/03/Melamed-on-Maimon.doc.

Melamed, Yitzhak. "Salomon Maimon and the Rise of Spinozism in German Idealism." *Journal of the History of Philosophy* 42, no. 1 (2004): 67–96.

Przywara, Erich. *Analogia Entis: Metaphysics, Original Structure and Universal Rhythm*. Translated by John R. Betz and David Bentley Hart. Grand Rapids, MI: William B. Eerdman's Publishing Co., 2014. Original version: *Analogia Entis: Metaphysik*. Munich: Kösel und Pustet, 1932.

Richards, Robert J. "Kant and Blumenbach on the *Bildungstrieb*: A Historical Misunderstanding." *Studies in History and Philosophy of Biological and Biomedical Sciences* 31, no. 1 (2000): 11–32.

Schechter, Oded. "The Logic of Speculative Philosophy and Skepticism in Maimon's Philosophy: *Satz der Bestimmbarkeit* and the Role of Synthesis." In *Salomon Maimon: Rational Dogmatist, Empirical Skeptic. Critical Assessments*, edited by Gideon Freudenthal, 18–52. Dordrecht: Kluwer, 2003.

Schelling, Friedrich Wilhelm Joseph. *Bruno, or On the Natural and Divine Principle of Things*. Edited and translated by Michael G. Vater. Albany, NY: SUNY Press, 1984. Original version: *Bruno, oder über das göttliche und natürliche Prinzip der Dinge*. Berlin: Unger, 1802.

Schlick, Moritz. *General Theory of Knowledge*. Translated by Albert E. Blumberg. La Salle, IL: Open Court Classics, 1985. Original version: *Allgemeine Erkenntnislehre*. Berlin: Springer, 1918.

Socher, Abraham P. *The Radical Enlightenment of Solomon Maimon: Judaism, Heresy, and Philosophy*. Stanford: Stanford University Press, 2006.

Thielke, Peter. "Discursivity and Causality: Maimon's Challenge to the Second Analogy." *Kant-Studien* 92 (2001): 440–63.

Thielke, Peter, and Yitzhak Melamed. "Salomon Maimon." *The Stanford Encyclopedia of Philosophy* (Fall 2015 edition), edited by Edward N. Zalta. Available online: https://plato.stanford.edu/archives/fall2015/entries/maimon/.

Wagner, Hans. *Philosophie und Reflexion*. 2nd ed. Munich and Basel: Ernst Reinhardt Verlag, 1967; first published 1959.

Michah Gottlieb

Does Judaism Have Dogma? Moses Mendelssohn and a Pivotal Nineteenth-Century Debate

Dogmas serve important functions in religious communities. R. Marston Speight identifies several such functions, including (a) providing a basis for membership in the community; (b) testing for orthodoxy; (c) expressing communal self-understanding; and (d) affirming communal unity.[1]

Since the Middle Ages, Jewish thinkers have debated the question whether or not Judaism has dogma.[2] Moses Mendelssohn launched the modern Jewish discussion of this question with what many interpreted to be a negative response to this question. In the first part of this essay, I will show how Mendelssohn's eighteenth-century discussion is informed by *interreligious* concerns; namely, seeking to express a distinct Jewish communal self-understanding in relation to Christianity. Mendelssohn rejects dogma in a Christian sense while affirming that there are principles that define Judaism analytically. In the second part of the essay, I will explore a pivotal nineteenth-century debate over dogma in Judaism, in which the problem is addressed from an *intrareligious* perspective showing how it is informed by the problem of Jewish communal unity.[3]

My discussion will focus on four thinkers, each of whom was an outstanding representative of a major school of German Judaism: Mendelssohn, the leading figure of the Haskalah; Samson Raphael Hirsch, the ideological leader of German Neo-Ortho-

1 See R. Marston Speight, "Creeds: An Overview," in *Encyclopedia of Religion*, 2nd ed., ed. Lindsay Jones (Detroit: Macmillan 2004), 3:2053. Speight writes specifically about the function of creeds, which he defines as "confession[s] of faith, put into concise form, endowed with authority, and intended for general use in religious rites." See Speight, 2052. Given that creeds have confessional, liturgical roles, their functions are broader than dogmas. Nevertheless, there is a great deal of overlap between the functions of dogma and creed. In the nineteenth-century debate over dogma in Judaism that will be explored, some writers focus on creed.

2 Menachem Kellner has done more than anyone else to explore this topic. See Menachem Kellner, *Dogma in Medieval Jewish Thought: From Maimonides to Abravanel* (Oxford: Oxford University Press, 1986); Kellner, *Must a Jew Believe Anything?* 2nd ed. (Oxford and Portland, OR: Littman Library of Jewish Civilization, 2006). On Maimonides's account of dogma, see Marc B. Shapiro, *The Limits of Orthodox Theology: Maimonides' Thirteen Principles Reappraised* (Oxford and Portland, OR: Littman Library of Jewish Civilization, 2004); Louis Jacobs, *Principles of the Jewish Faith: An Analytical Study* (London: Vallentine Mitchell, 1964).

3 While scholars have largely forgotten the nineteenth-century debate over dogma in Judaism, the one exception is Kerstin von der Krone. See her brief treatment in von der Krone, "Jüdische Wissenschaft und modernes Judentum: eine Dogmendebatte," in *Die "Wissenschaft des Judentums": Eine Bestandsaufnahme*, ed. Thomas Meyer and Andreas Kilcher (Paderborn: Wilhelm Fink Verlag, 2015), 118–20.

https://doi.org/10.1515/9783110618839-012

doxy; Manuel Joël, a leading figure in Positive-Historical Judaism, and Abraham Geiger, the intellectual leader of Reform Judaism. I will show how each author's discussion of dogma is informed by his concept of Judaism.

Mendelssohn's Haskalah: Judaism as Non-Dogmatic Religion

In an important 1926 article titled "Besitzt das überlieferte Judentum Dogmen?" ("Does Traditional Judaism Possess Dogmas?"), Leo Baeck refers to the "long and widely held belief" that for Mendelssohn, "Judaism was a religion without dogmas."[4] The problem with this claim, Baeck notes, is that the term "dogma" is notoriously undefined. He observes that in the original Greek, "dogma" (*dokein*) has several meanings. In the Septuagint and the New Testament, it means "order" or "decree" and usually refers to governmental regulations. In ancient philosophy, *dokein* is used in a derivative sense to refer to intellectual axioms that were regarded as "decrees of cognition." According to Baeck, it was only when ancient philosophy joined Christianity that "dogma" came to refer to "statements of faith."[5] Catholicism enumerated revealed doctrines found in sacred texts or traditions that were confirmed by an authoritative religious body as dogma.[6] Affirming dogma was regarded as a divine commandment and a necessary condition for salvation, while denying it was deemed a heresy that would consign a person to eternal perdition. Baeck notes that Protestant theologians largely accepted their Catholic counterparts' understanding of dogma,[7] and recent scholars have tended to affirm the general contours of Baeck's genealogy.[8]

---

4 Leo Baeck, "Does Traditional Judaism Possess Dogmas?" in *Studies in Jewish Thought: An Anthology of German Jewish Scholarship*, ed. Alfred Jospe (Detroit: Wayne State University Press, 1981), 41.
5 Baeck, "Does Traditional Judaism Possess Dogmas?", 45. M. E. Williams notes that in ancient Greek, "dogma" (*dokein*) can also mean "private opinion" and that in early Christianity, it could refer both to particular teachings (e. g. the immortality of the soul) and to an entire system of teachings (e. g. Christian dogma as opposed to pagan dogma). See M. E. Williams, "Dogma," in *New Catholic Encyclopedia*, 2nd ed., ed. Thomas Carson (Detroit: Thomson/Gale, 2002), 1:810–11.
6 Baeck, "Does Traditional Judaism Possess Dogmas?", 46–47.
7 Baeck notes that some Protestant theologians denied that Christianity had dogma because they denied the existence of infallible religious institutions to confirm it. See Baeck, 46. The relationship between Jewish thinkers such as Mendelssohn, Geiger, and Baeck who denied that Judaism had dogma and Protestant thinkers who denied that Christianity had dogma is an important subject that exceeds the bounds of this essay.
8 See Williams, "Dogma"; Adolf Darlap and Karl Rahner, "Dogma," in *Encyclopedia of Religion*, 2nd ed., ed. Lindsay Jones (Detroit: Macmillan, 2004), 4:2387–90.

Despite the widely held view that Mendelssohn denied that Judaism has dogma, he never uses the term "dogma" in connection with Judaism.[9] Mendelssohn's denial of dogma in Judaism is most commonly inferred from his distinction between *revealed religion* and *revealed legislation.* For Mendelssohn, *revealed religion* refers to revealed theological doctrines, while *revealed legislation* refers to revealed law. For Mendelssohn, Christianity is grounded in revealed religion, while Judaism only recognises revealed legislation.[10]

In Menachem Kellner's influential book *Must a Jew Believe Anything?* (first edition 1999; second edition 2006), he interprets Mendelssohn's distinction between revealed religion and revealed legislation as "orthopraxy." Kellner does not claim that Mendelssohn deems doctrines unimportant in Judaism. Rather, he contends that Mendelssohn interprets Judaism as comprising two elements: principles of universal natural religion known through reason (including God's existence and the immortality of the soul) and ritual laws specific to Jews known through revelation. Kellner criticises Mendelssohn's philosophy of Judaism as schizophrenic, splitting the Jew into a public human being and a private Jew.[11] Kellner's interpretation of Mendelssohn has a long history. In 1840, the Reform theologian Solomon Ludwig Steinheim charged Mendelssohn with being "a heathen in his brain and a Jew in his body."[12]

This interpretation seriously misunderstands Mendelssohn for three reasons. First, Mendelssohn does not limit the doctrines at the core of Judaism to universal teachings of natural religion. In his 1783 *Jerusalem*, he asserts that Judaism comprises three elements: (a) universal eternal truths of natural religion, including God's existence, divine providence, the immortality of the soul, and natural morality, which

---

9 However, shortly after Mendelssohn's death in 1786, his disciple David Friedländer attributed this view to his teacher. In 1788, Friedländer wrote: "Among our great men, Mendelssohn was the first, not indeed to discover, but to emphasise and teach with full clarity that the distinguishing principle of Judaism is that *we have no dogmas* [*wir keinen Dogmen*], no articles of faith, no teachings about salvation that are indispensable and necessary for blessedness" (my emphasis). See David Friedländer, *Der Prediger* (Berlin: Friedrich Maurer, 1788), 23. I thank Uta Lohmann for calling my attention to this source.

10 See Moses Mendelssohn, *Gesammelte Schriften: Jubiläumsausgabe*, ed. Alexander Altmann et al. (Stuttgart and Bad Cannstatt: Frommann-Holzboog, 1971–), 8:164, 96 (henceforth: *JubA*); Mendelssohn, *Jerusalem, or On Religious Power and Judaism*, trans. Allan Arkush, ed. Alexander Altmann (Hanover, NH: University Press of New England, 1983), 97, 248; Mendelssohn, *JubA*, 7:90–102; Mendelssohn, *Moses Mendelssohn: Writings on Judaism, Christianity, and the Bible*, trans. Curtis Bowman, Elias Sacks, and Allan Arkush, ed. Michah Gottlieb (Hanover, NH: University Press of New England, 2011), 16–27.

11 See Kellner, *Must a Jew Believe Anything?*, 112, note 3. In claiming that Mendelssohn splits the Jew into a public human being and a private Jew, Kellner cites Yehuda Leib Gordon's famous expression (often mistakenly attributed to Mendelssohn): "Behave like a human being on the street, and like a Jew in your tent."

12 See Salomon Steinheim, *Moses Mendelssohn und seine Schule* (Hamburg: Hoffmann und Campe, 1840), 37, cited in Michael Meyer, *Response to Modernity: A History of the Reform Movement in Judaism* (New York: Oxford University Press, 1988), 69.

can be known "at all times and in all places" through reason or common sense; (b) historical truths about the Jewish nation, including the miraculous exodus from Egypt and the revelation at Mount Sinai, which are known through tradition and recorded in the Torah; and (3) the revealed laws of the Torah.[13] As the second and third elements include historical doctrines that cannot be known "at all times and in all places," it is evident that Mendelssohn does not identify the doctrines at the core of Judaism solely with universal principles of natural religion.[14]

Second, Mendelssohn does not completely separate universal principles of natural religion from revelation, as he asserts that these universal principles *can* be communicated through revelation. Thus, Mendelssohn interprets Ecclesiastes as a dialogue teaching the immortality of the soul and he asserts that rational truths about God's existence and providence were revealed to Moses and recorded in the Pentateuch.[15]

Finally, Mendelssohn does not completely separate universal principles of natural religion from ritual law, as he asserts that a primary function of ritual law is to bring the principles of natural religion to life and provide an occasion for contemplating their meaning. As he puts it: "All laws refer to, or are based on, eternal truths of reason, or remind us of them, and rouse us to ponder them. Hence, our rabbis rightly say: the laws and doctrines are related to each other, like body and soul."[16]

13 Mendelssohn, *JubA*, 8:191–93; *Jerusalem*, 126–27.

14 Alexander Altmann notes, however, that in his earlier "Counter-Reflections to Bonnet" (1769) and his 1771 letter to Elkan Herz, Mendelssohn seems to posit a more dualistic concept of Judaism comprising eternal truths and revealed law. See Alexander Altmann, "Moses Mendelssohn's Concept of Judaism Reexamined," in *Von der mittelalterlichen zur modernen Aufklärung: Studien zur jüdischen Geistesgeschichte* (Tübingen: Mohr, 1987), 234.

15 Mendelssohn's 1769 Hebrew commentary on Ecclesiastes is found in Mendelssohn, *JubA*, 14:148–207. Edward Breuer and David Sorkin have recently published a translation of a significant portion of this text. See Moses Mendelssohn, *Moses Mendelssohn's Hebrew Writings*, trans. Edward Breuer, ed. Edward Breuer and David Sorkin (New Haven: Yale University Press, 2018), 123–218. Mendelssohn's most famous discussion of the rational content of revelation in Judaism is in his commentary on Exodus 3:14 on God's revelation of the Tetragrammaton to Moses. Translating the Tetragrammaton as "The Eternal" (*Der Ewige*), Mendelssohn notes that this concept includes God's eternality, necessary existence, and providential governance. See Mendelssohn, *JubA*, 16:26–27; *Moses Mendelssohn: Writings*, 217–18. See also *Jerusalem*, where Mendelssohn writes: "Although the divine book that we received through Moses is, strictly speaking, meant to be a book of laws containing ordinances, rules of life, and prescriptions, it also includes, as is well known, *an inexhaustible treasure of rational truths and religious doctrines* that are so intimately connected with the laws that they form but one entity" (emphasis mine). Mendelssohn, *JubA*, 8:165–66; *Jerusalem*, 99.

16 See Mendelssohn, *JubA*, 8:165–66; *Jerusalem*, 99. For an elaboration of Mendelssohn's argument, see Michah Gottlieb, *Faith and Freedom: Moses Mendelssohn's Theological-Political Thought* (New York: Oxford University Press, 2011), ch. 2; Gideon Freudenthal, *No Religion Without Idolatry: Mendelssohn's Jewish Enlightenment* (Notre Dame, IN: University of Notre Dame Press, 2012), ch. 6; and most recently, Elias Sacks, *Moses Mendelssohn's Living Script: Philosophy, Practice, History, Judaism* (Bloomington: Indiana University Press, 2017), ch. 2.

Given Mendelssohn's affirmation that the universal principles of natural religion *can* be revealed and that they are intimately connected to ritual law, in what sense does he deny dogma in Judaism? In the Christian sense, as he asserts that Judaism recognises no divinely commanded irrational doctrines known only through revelation and confirmed by an authoritative religious body the affirmation of which is a necessary condition for salvation. Mendelssohn has three reasons for denying that Judaism possesses dogma in this sense.

First, he claims that Judaism teaches that there are no truths known *exclusively* through revelation that are necessary for salvation. Were there such truths, they would either contradict God's goodness, by implying that God consigns all those to whom God did not reveal salvational truths to eternal suffering, or they would contradict God's omnipotence, by implying that God lacked the power to provide all people with the tools necessary for salvation. As Mendelssohn memorably put it: "Why must the two Indies wait until it pleases Europeans to send them a few comforters to bring them a message without which they can, according to this opinion, live neither virtuously nor happily?"[17]

For Mendelssohn, Judaism teaches that salvation depends on the practice of universal morality, which constitutes the true worship of God.[18] This is reflected in the rabbinic teaching that Gentiles who observe the seven so-called "Noahide laws" merit salvation in the afterlife.[19] Mendelssohn asserts that the Noahide laws "more or less comprise the essential laws of natural right" and that the talmudic rabbis call those who follow them "virtuous men of other nations" and "children of eternal bliss."[20] For Mendelssohn, Judaism teaches that the affirmation of universal principles of natural religion is connected to salvation not because affirming these doctrines bestows salvation *by itself*, but because affirming them provides a crucial mo-

17 See Mendelssohn, *JubA*, 8:161; *Jerusalem*, 94.

18 As he put it in his 1770 "Counter-Reflections to Bonnet": "The divine religion into which I was born teaches me that all the peoples of the earth are saved if they live in accordance with the laws of reason, that is, if they practice virtue." Mendelssohn, *JubA*, 7:90–91; *Moses Mendelssohn: Writings*, 16.

19 The talmudic doctrine is found in Babylonian Talmud, *ʿAvodah Zarah*, 64b.

20 The rabbinic appellation for Gentiles who follow the Noahide laws is "the pious of the nations of the world" (*ḥasidei ummot ha-ʿolam*). However, Mendelssohn translates the term as "virtuous [*tugendhafte*] men of other nations," thereby emphasising that moral virtue is the basis for salvation. The seven Noahide laws as enumerated by Mendelssohn involve prohibitions on (1) idolatry; (2) blasphemy; (3) murder; (4) incest; (5) robbery; and (6) eating live animals, as well as (7) the obligation to establish and administer justice. See Mendelssohn's "Open Letter to Lavater" in Mendelssohn, *JubA*, 7:11; *Moses Mendelssohn: Writings*, 9. There is a large literature on Mendelssohn's treatment of the Noahide laws. The classic study is Steven Schwarzschild, *The Pursuit of the Ideal: Jewish Writings of Steven Schwarzschild*, ed. Menachem Kellner (Albany, NY: State University of New York Press, 1990), 29–60.

tivation for ethical action which merits one salvation.[21] So, while for Mendelssohn, Judaism teaches that principles of natural religion *can* be communicated through revelation, it also teaches that revelation is not *needed* in order to know them. As he puts it: "Judaism boasts of no *exclusive* revelation of eternal truths that are indispensable to salvation, of no revealed religion in the sense in which that term is usually understood" (emphasis Mendelssohn's).[22]

Second, Mendelssohn claims that according to Judaism, belief in religious dogma cannot be commanded because *no* beliefs can be commanded. Mendelssohn asserts that human beings have two distinct capacities: *intellect*, which judges truth, and *will*, which intends actions. Will can be manipulated through force, threats, and bribes, but intellect only affirms belief on the basis of rational considerations. For example, by holding a gun to my head a person can force me to say the words "one plus one equals three." But they can never force me to *believe* it. God therefore cannot command belief in dogma, as "ought" must imply "can."[23] Mendelssohn finds this teaching reflected in the opening of the Decalogue: "I am the Eternal your God who led you from the land of Egypt" (Exodus 20:2). Mendelssohn makes clear that this statement is "not a commandment of belief," since belief cannot be commanded. As he puts it, "a miraculous voice would not have instilled any concepts [...] since [a person] demands *rational proofs*" (emphasis Mendelssohn's).[24]

For Mendelssohn, just as the eternal truths at the heart of Judaism, such as God's existence, must be affirmed through reason, so too must the historical truths, such as the miraculous exodus from Egypt and the revelation of the Torah on Mount Sinai. Mendelssohn defends the rationality of these teachings by claiming that they can be rationally affirmed on the basis of eye-witness reports and trust in a subsequent chain of tradition.[25]

---

21 This is a somewhat complex argument that I will not repeat here. For its elaboration, see Michah Gottlieb, *Faith, Reason, Politics: Essays on the History of Jewish Thought* (Brighton, MA: Academic Studies Press, 2013), 132–38.

22 Mendelssohn, *JubA*, 8:164; *Jerusalem*, 97.

23 See Mendelssohn, *JubA*, 8:129–30; *Jerusalem*, 61–62.

24 See Mendelssohn, *JubA*, 8:164–65; *Jerusalem*, 97–98; *JubA*, 16:186; *Moses Mendelssohn: Writings*, 222 (Commentary on Exodus 20:2). Mendelssohn's approach can be seen in a rendering of Maimonides's Thirteen Principles of Judaism that he composed for students at the first maskilic school, the Jüdische Freischule. Maimonides's principles, outlined in his introduction to the tenth chapter of the Talmudic tractate Sanhedrin, were given creedal formulation in the fifteenth century in the anonymous text *Ani Maʾamin*. In this text, each of Maimonides's principles was prefaced with the phrase "I believe with perfect faith" (*ani maʾamin be-emunah šelemah*). However, in Mendelssohn's rendering of these principles, he replaced this formulation with the words "I know as true and certain" (*Ich erkenne für wahr und gewiß*). Mendelssohn's version of Maimonides's Thirteen Principles can be found in Mayer Kayserling, *Moses Mendelssohn: sein Leben und seine Werke* (Leipzig: H. Mendelssohn, 1862), 565–68.

25 I will not elaborate on Mendelssohn's argument here, but its elements can be found in his commentary on Maimonides's *Treatise on Logic* in Mendelssohn, *JubA*, 14:71–72; *Moses Mendelssohn's Hebrew Writings*, 84–85; Mendelssohn's introduction to his commentary and translation of the Torah,

Third, Mendelssohn claims that Judaism has never demanded that its adherents swear allegiance to "articles of faith" (*Glaubensartikel*), the denial of which made one a heretic. He acknowledges that some medieval Jewish thinkers, most notably Maimonides, attempted to create a list of binding articles of faith and claimed that a Jew who did not affirm his articles was not an "Israelite."[26] However, Mendelssohn deems Maimonides's idea to be contrary to "the spirit of true Judaism" (*Geiste der ächten Judentums*), as evidenced by the fact that his principles were disputed by other medieval Jewish thinkers and the debate never led to sectarian divisions or accusations of heresy.[27]

Unlike Maimonides's principles, the first two elements of Mendelssohn's concept of Judaism are general categories rather than specific teachings. Rather than being a litmus test for whether or not a Jew is included in the community, they represent Mendelssohn's analytical understanding of Judaism. Indeed, one of Mendelssohn's most famous teachings is that Judaism deems excommunication illegitimate.[28]

Mendelssohn's denial that Judaism has dogma served an important interreligious function. Mendelssohn was challenged to convert to Christianity several times. The first challenge came in 1769, when the Swiss pastor Johann Caspar Lavater dedicated a philosophical defence of Christianity to Mendelssohn asking him to either refute the defence or convert. The heart of Mendelssohn's response to Lavater was his assertion that he adhered to Judaism because it was a rational, tolerant religion which held that all who practised universal morality merited salvation, which was why Judaism did not proselytise.[29] Mendelssohn's not-very-subtle implication was that this was not true of Christianity.[30] So, for Mendelssohn, Judaism's lack of dogma distinguished it from Christianity. Mendelssohn repeated and elaborated this point fourteen years later in *Jerusalem, or on Religious Power and Judaism* in response to a second challenge to convert to Christianity, this time from the enlightened Protestant writer August Friedrich Cranz.

There was a clear political subtext to Mendelssohn's denial of dogma in Judaism. While the logic of Enlightenment political theory, with its emphasis on human equality and freedom, naturally entailed that Jews should be granted citizenship, many Christians, including Frederick the Great, the enlightened ruler of Prussia, resisted

---

Mendelssohn, *JubA*, 14:211–28; *Moses Mendelssohn's Hebrew Writings*, 241–70; and his discussion of miracles in Mendelssohn, *JubA*, 7:324–25; *Moses Mendelssohn: Writings* 255–56. For a recent attempt to defend Mendelssohn's position, see Sacks, *Moses Mendelssohn's Living Script*, 122–70.

26 On Maimonides's claim, see Kellner, *Must a Jew Believe Anything?*, 52–56.

27 See Mendelssohn, *JubA*, 8:167–68; *Jerusalem*, 100–101. For the history of the dispute over Maimonides's principles, see the books cited in note 2.

28 See Mendelssohn's "Preface to Menasseh ben Israel" in Mendelssohn, *JubA*, 7:16–25; *Moses Mendelssohn: Writings*, 44–52; *JubA*, 8:103–42; *Jerusalem*, 33–75.

29 See note 20 above.

30 Mendelssohn made this claim explicit in an unpublished work that he wrote at the time. See Mendelssohn's "Counter-Reflections to Bonnet" in Mendelssohn, *JubA*, 7:90–105; *Moses Mendelssohn: Writings*, 16–28.

this logic. Part of their argument was that Jews could not be granted citizenship because they adhered to a tyrannical, irrational religion which was at odds with Enlightenment principles.[31] Mendelssohn's claim that Judaism did not possess dogma must therefore be understood in the context of his plea for Jewish civil rights. For Mendelssohn the *Maskil* and *Aufklärer*, Judaism's status as a rational, tolerant religion showed it to be more closely aligned with Enlightenment principles than was Christianity.

The Nineteenth-Century Debate

While Mendelssohn's discussion of dogma in Judaism was informed by the interreligious need to draw a distinction between Judaism and Christianity and advocate for Jewish civil inclusion, in the nineteenth century, the debate was shaped by intrareligious considerations, most notably the problem of Jewish communal unity.

Until the last third of the eighteenth century, most German Jewish communities were organised into traditional structures called *qehillot* (sing. *qehillah*). The authority of the *qehillah* was recognised by Christian rulers and included the right not only to tax its members and administer communal institutions, but also to enforce Halakhah. With the ascension of enlightened absolutist rulers in the eighteenth century who sought to extend a single, rational state law over all the inhabitants of their countries and sweep away estate-based medieval institutions, the authority of the traditional *qehillah* waned. In this context, new Jewish congregations began to emerge.[32]

Reforming the traditional synagogue service was a central concern for these new Jewish congregations. Between 1810 and 1818, Reform Temples were opened in Seesen, Berlin, and Hamburg. The first reforms introduced included a weekly sermon in German to replace the semi-annual talmudic discourse in Yiddish, some prayers in German rather than all prayers being in Hebrew, and the playing of organ music during the Sabbath prayers. Gradually, substantial changes were introduced into the lit-

---

31 The literature on this is vast. For a sample of the eighteenth-century arguments against Jewish civil rights, see Jacob Katz, *From Prejudice to Destruction: Anti-Semitism, 1700–1933* (Cambridge, MA: Harvard University Press, 1980), 13–62.

32 On the pre-modern *qehillah* and its decline, see Jacob Katz, *Tradition and Crisis: Jewish Society at the End of the Middle Ages*, trans. Bernard Dov Cooperman (Syracuse: Syracuse University Press, 2000), and more recently, Elisheva Carlebach, "La communauté juive et ses institutions au début de l'époque moderne," in *Aux Origines du Judaïsme*, ed. Julien Darmon and Jean Baumgarten (Paris: Actes Sud, 2012), 358–89. In addition to the government stripping the *qehillah* of prerogatives, its authority was weakened by internal disputes over Sabbateanism, Hasidism, and Haskalah, as well as tensions between the wealthy and the poor. See Katz, *Tradition and Crisis*, 181–236; Meyer, *Response to Modernity*, 10–13; Carlebach, "La communauté," 365–66, 387–89.

urgy, most notably eliminating the hope that the Jews would return to Palestine, rebuild the Temple, and restore the sacrificial cult in Jerusalem.[33]

In 1844, Reform rabbis convened the first of several rabbinical assemblies to establish unified standards for Reform Judaism. This assembly, held in Braunschweig, proved contentious. However, a few reforms were agreed upon, the most controversial being the ruling that marriage between a Jew and a Christian was not halakhically prohibited as long as children born into the marriage could also be raised as Jews.[34]

Historians generally understand Orthodox Judaism as a reaction to groups perceived to be tampering with Jewish tradition.[35] Jeffrey Blutinger notes that in the nineteenth century, the term "Orthodox" came to be used specifically to designate Jews opposed to Reform.[36] As such, the beginnings of nineteenth-century Orthodoxy are often dated to the 1819 publication of a collection of rabbinic responsa opposing the Hamburg Temple titled *Eleh Divrei ha-Brit* (*These Are the Words of the Covenant*).[37] However, it was opposition to the Reform rabbinical assemblies of the 1840s that galvanised and consolidated Orthodoxy. In response to the Braunschweig Assembly, seventy-seven Orthodox rabbis signed a manifesto denouncing it, and a collection of thirty-seven condemnatory rabbinic responsa was published.[38]

The most important of these responsa was that of Samson Raphael Hirsch, the leading exponent of German Jewish Neo-Orthodoxy at the time. Hirsch harshly criticised Reformers for deviating from Halakhah and expressed concern that the actions of the Reformers were precipitating schism among German Jews. As he put it: "You

---

33 See Meyer, *Response to Modernity*, 28–61.

34 Meyer, *Response to Modernity*, 131–35; David Philipson, *The Reform Movement in Judaism* (New York: Macmillan, 1907), 211–12; Katz, *Divine Law in Human Hands* (Jerusalem: Magnes Press), 234–35, 373. Other reforms included eliminating recital of the *Kol Nidrei* prayer annulling vows at the beginning of Yom Kippur and eliminating the practice of *meṣiṣah be-feh*, which involved the person performing the rite of circumcision sucking the penile blood of the baby being circumcised. For a discussion of this ritual and the controversy surrounding it in the nineteenth century, see Katz, *Divine Law in Human Hands*, 357–402.

35 See Jacob Katz, "Orthodoxy in Historical Perspective," in *Studies in Contemporary Jewry, Volume 2*, ed. Peter Medding (Bloomington: Indiana University Press, 1986), 3–4; Michael Silber, "Orthodoxy," in *The Yivo Encyclopedia of Jews in Eastern Europe*, ed. Gershon Hundert (New Haven: Yale University Press, 2008), 2:1292–97.

36 Jeffrey Blutinger, "'So-Called Orthodoxy': The History of an Unwanted Label," *Modern Judaism* 27, no. 3 (2007): 320.

37 See W. Gunther Plaut, *The Rise of Reform Judaism: A Sourcebook of Its European Origins* (New York: World Union for Progressive Judaism, 1963), 34. A partial translation of the text can be found in Paul R. Mendes-Flohr and Jehuda Reinharz, *The Jew in the Modern World: A Documentary History*, 3rd ed. (New York: Oxford University Press, 2011), 187–89.

38 Signatories continued to be gathered and eventually three hundred Orthodox rabbis signed the protest. See Judith Bleich, "Jacob Ettlinger, His Life and Works: The Emergence of Modern Orthodoxy in Germany" (PhD diss., New York University, 1974), 186–91; Philipson, *The Reform Movement*, 225–29; Katz, *Divine Law in Human Hands*, 236–43.

children of the nation, whose assembly has caused all this [protest] [...]. Do your eyes not see that evil and bitterness will follow the path on which you have begun? Do you not know and understand that if your words bear fruit this time, the house of Israel will be torn into two?"[39]

Undeterred by Orthodox opponents like Hirsch, Reformers held a second rabbinical assembly in Frankfurt the following year in 1845. At that assembly, a major controversy erupted over whether Hebrew should remain the language of Jewish prayer. Abraham Geiger, the chief rabbi of Breslau and leading theoretician of Reform Judaism, argued that all Hebrew prayers should be eliminated from the synagogue service, since German was the mother tongue of German Jews and, as he put it, "a German prayer strikes a deeper chord than a Hebrew prayer."[40]

Geiger's push to eliminate Hebrew from the liturgy was opposed by the leading moderate Reformer, the chief rabbi of Dresden, Zacharias Frankel. Frankel argued that Hebrew must retain a central place in the Jewish liturgy since it was historically the language of Jewish prayer and Scripture, and he dismissed Geiger's contention that German prayers struck a deeper chord than Hebrew prayers as "purely subjective."[41] Ultimately, the majority of the assembly sided with Geiger, and an outraged Frankel withdrew from the conference, severed his ties with Reform, and founded a separate denomination he called "Positive-Historical Judaism" that charted a middle path between Orthodoxy and Reform.[42]

It would, however, be a mistake to see nineteenth-century German Jewry as marching inexorably towards fragmentation. A powerful impetus for Jewish communal unity came from the Christian governmental authorities. In 1847, Prussia passed a law making it mandatory to belong to the local Jewish community. Jews were required to pay taxes to their local Jewish community, and these taxes were partially used by the supreme Jewish council or *Oberat* to fund the community's religious institutions, including the local synagogue. According to the governmental regulations,

---

39 Samson Raphael Hirsch, *Sefer Šemeš Marpe᾿: še᾿elot u-tešuvot, ḥiddušei ha-Šas, iggerot u-mikhtavim*, ed. Eliyahu Meir Klugman (New York: Mesorah, 1992), 195–96. See Katz, *Divine Law in Human Hands*, 242–44.

40 Geiger also opposed retaining Hebrew for prayer because he saw it as a marker of Jewish nationhood. See Philipson, *The Reform Movement*, 239–40. For a partial translation of Geiger's arguments, see Mendes-Flohr and Reinharz, *The Jew in the Modern World*, 203. On the Frankfurt Assembly more generally, see Meyer, *Response to Modernity*, 136–38; Philipson, *The Reform Movement*, 231–72.

41 Frankel also rejected Geiger's contention that reciting prayers in Hebrew prevented Jews from being patriotic Germans. For a partial translation of Frankel's response to Geiger, see Mendes-Flohr and Reinharz, *The Jew in the Modern World*, 203–4.

42 On Frankel, see Ismar Schorsch, *From Text to Context: The Turn to History in Modern Judaism* (Hanover, NH: University Press of New England, 1994), 255–65; Andreas Brämer, *Rabbiner Zacharias Frankel: Wissenschaft des Judentums und konservative Reform im 19. Jahrhundert* (Hildesheim: Olms, 2000).

a Jew could only be exempted from paying these taxes by withdrawing from the Jewish community by converting to Christianity.[43]

The *Oberat* did not have a uniform denominational affiliation. In some communities, it was controlled by Reformers, while in others, the Orthodox controlled it. This created tension, since the local community synagogue generally followed the religious orientation of the majority of the *Oberat*. In a few communities, Jews belonging to a different denomination from the *Oberat*, whether Reform or Orthodox, received a special governmental dispensation to form an independent congregation and conduct separate services. However, they still had to be members of the official community and pay taxes to it, and were therefore left with a double tax burden, forced to fund religious institutions of which they did not approve.[44]

S.R. Hirsch's Neo-Orthodoxy: Dogma as the Foundation of Denominational Sectarianism

In 1869, there was a new development in Karlsruhe, the state capital of Baden. The previous year, the Reform-controlled *Oberat* had introduced reforms into the community synagogue that included the playing of an organ on the Sabbath. The Karlsruhe Orthodox protested and decided not only to form a separate congregation, but also to refuse to pay their communal taxes. This amounted to seceding from the official Jewish community, and the *Oberat* promptly sued the Orthodox secessionists. However, Baden had a tradition of progressive liberalism, and the State Supreme Court acquitted the Orthodox Jews, accepting their argument that they had a right to secede for reasons of liberty of conscience. In response, the *Oberat* issued a harsh public condemnation of the Orthodox secessionists. At the heart of the *Oberat*'s attack was a claim about the theological tenets of Judaism.[45]

---

43 See Salo Baron, "Freedom and Constraint in the Jewish Community: An Historic Episode," in *Essays and Studies in Memory of Linda R. Miller*, ed. Israel Davidson (New York: Jewish Theological Seminary, 1938), 11–12; Robert Liberles, *Religious Conflict in Social Context: The Resurgence of Orthodox Judaism in Frankfurt Am Main, 1838–1877* (Westport, CT: Greenwood Press, 1985), 167–69; Jacob Katz, *A House Divided: Orthodoxy and Schism in Nineteenth-Century Central European Jewry*, trans. Ziporah Brody (Hanover, NH: Brandeis University Press, 1998), 9.

44 Reformers formed a separate congregation in Berlin from 1847 to 1858 and in Hamburg from 1819 until 1865, when the official Jewish community was dissolved after emancipation was granted. The Orthodox formed separate congregations in Frankfurt, Mainz, Cologne, Darmstadt, Karlsruhe, and Wiesbaden. A different type of arrangement existed in Breslau from 1846 to 1856, where Orthodoxy and Reform had equal status and a person could choose which congregation to join. Robert Liberles notes, however, that in 1858, members of the Reform congregation in Berlin were exempted from paying taxes to the official Jewish community, despite remaining members of it. Given that the Hamburg official community was dissolved in 1865, by that year, no Reformers in German lands suffered from double taxation, only Orthodox Jews did. See Liberles, *Religious Conflict*, 172–73.

45 See Katz, *A House Divided*, 237–38.

The *Oberat* charged the Orthodox with "religious indifference" and with lacking a proper "sense of community" (*Gemeinsinn*), since for generations, Jews had regarded membership in the local Jewish community as an inviolable religious obligation.[46] The *Oberat* understood that the Orthodox saw secession as a matter of religious principle. However, it rejected this argument, claiming that the dispute between Orthodoxy and Reform amounted to "a minor difference of opinion over certain provisions in religious services which has not the least bearing on principles of religion [*Religionsgrundsatz*]."[47] According to the *Oberat*, Orthodox and Reform Jews affirmed the same basic religious dogmas. For the Orthodox to tear the community apart over minor disagreements concerning the synagogue service was petty and unconscionable.

In response, the Orthodox faction in Karlsruhe appealed to Samson Raphael Hirsch, the leader of a separatist Orthodox congregation in Frankfurt, for support. In the late 1860s, as Orthodox secession was debated in Hungary, Hirsch nursed the hope that his own Orthodox congregation in Frankfurt would be able to secede from the official Reform-controlled community there.[48] As such, he was eager to support the Orthodox faction in Karlsruhe. In April 1870, he published a blistering reply to the Karlsruhe *Oberat* titled "Die Trennungsfrage in Karlsruhe" ("The Secession Question in Karlsruhe").[49]

For Hirsch, the dispute between Orthodoxy and Reform was no matter of minor differences over the synagogue service, but rather a "symptom" of a much deeper disagreement between them.[50] Hirsch had long battled to affirm the principle of the timeless, unchanging nature of the Torah whose laws were binding on all Jews.[51] According to Hirsch, the Orthodox followed the traditional Jewish view that "the first and most fundamental principle of Judaism" was the acceptance that all Jewish life—including not only the synagogue service, but also all domestic and public life—should be governed by Halakhah as found in the Bible, Talmud, and later rabbinic rulings and customs (*minhagim*) whose final authoritative codification was in the sixteenth-century *Šulḥan ʿArukh*.[52] By contrast, Reformers regarded Juda-

---

46 Samson Raphael Hirsch quotes the *Oberat*'s argument. See Samson Raphael Hirsch, *Gesammelte Schriften*, ed. Mendel Hirsch, 6 vols. (Frankfurt am Main: J. Kauffmann, 1902–12), 5:532 (henceforth: *GS*).

47 Hirsch, GS, 5:532.

48 Katz notes that in the 1850s, Hirsch sought exemption from paying taxes to the community that would be used to support Reform institutions, but did not yet conceive of full Orthodox secession. See Katz, *A House Divided*, 27. On the effect of events in Hungary on Hirsch, see Katz, *A House Divided*, 108–10, 237.

49 The text is reprinted in Hirsch, *GS*, 5:532–42.

50 Hirsch, *GS*, 5:535. Hirsch used the term "Neologie" to refer to Reformers.

51 For a discussion of Hirsch's arguments, see Michah Gottlieb, "Oral Letter and Written Trace: Samson Raphael Hirsch on the Bible and Talmud," *Jewish Quarterly Review* 106, no. 3 (2016): 316–51.

52 Hirsch, *GS*, 5:534. In his 1836 *Neunzehn Briefe über Judentum*, Hirsch wrote that while Maimonides's Thirteen Principles "perhaps" (*vielleicht*) serve as a basis for Judaism, they do not constitute the

ism as comprising a few simple theological beliefs and the universal moral law, while Halakhah was largely seen as antiquated and hence no longer obligatory.[53] As such, Hirsch contended that the dispute between Orthodoxy and Reform was no minor spat, but rather reflected differences "far greater [...] than the opposition between Catholic and Protestant, Lutheran and Reformed, or any other different Christian ecclesiastical confessions."[54]

For Hirsch, it was Reformers, not the Orthodox, who had precipitated religious schism by changing Judaism's traditional religious principles. While the *Oberat* claimed that it was fighting to preserve historical Judaism, since pre-modern Jews regarded membership in the local *qehillah* as a religious obligation, Hirsch argued that this was a distortion of history because the basis of membership in the Jewish community had changed. Hirsch asserted that pre-modern Jews freely consented to be members of the *qehillah* because the community defined itself by the dogma that obedience to divinely revealed Halakhah was mandatory.[55] By contrast, the *Oberat* defined communal membership as the payment of taxes, enforced by non-Jewish governmental authorities. For Hirsch, this artificial, oppressive basis of Jewish communal unity had nothing to do with historical Judaism, and so the Orthodox were entirely justified in seceding.[56] As he put it, "it is not a common tax box, but a *com-*

totality of it. See Samson Raphael Hirsch, *Neunzehn Briefe*, 2nd ed. (Frankfurt: J. Kauffmann, 1889), brief 17, 94. The English translation edited by Joseph Elias omits Hirsch's "vielleicht." See Samson Raphael Hirsch, *The Nineteen Letters about Judaism*, trans. Karin Paritzky, ed. Joseph Elias (New York: Feldheim, 1995), letter 17, 244. In his 1854 essay "Religion Allied to Progress," Hirsch was more forceful in rejecting Maimonides's Thirteen Principles, claiming that there could be no principles of Judaism since every commandment is fundamental. As he put it: "The rabbinical authorities opposed the proclamation of Jewish articles of faith [*Glaubensartikeln*] not because Judaism has no basic tenets, but because they regarded every aspect of Judaism as basic. Judaism, they held, knows neither thirteen articles of faith, nor three. In Judaism, every commandment and every prohibition constitutes a fundamental religious principle and *the denial of even one of the 613 commandments and prohibitions thereby places oneself outside Judaism.*" Hirsch, *GS*, 3:492; Hirsch, *The Collected Writings*, ed. Marc Breuer, Jacob Breuer, Meta Bechhofer, and Elliot Bondi (New York: Feldheim, 1984–2012), 6:112. Mendelssohn made a similar point in *Jerusalem* which he attributed to "Luria and the latter-day Kabbalists." Zev Harvey notes that Mendelssohn was actually referring to the view of the sixteenth-century rabbi David ibn Zimra (Radbaz), who was Isaac Luria's teacher. See Warren Zev Harvey, "Hasdai Crescas and Moses Mendelssohn on Beliefs and Commandments," in *Moses Mendelssohn: Enlightenment, Religion, Politics, Nationalism*, ed. Michah Gottlieb and Charles H. Manekin (Bethesda: University Press of Maryland, 2015), 85–86. Mendelssohn made his sympathy for this position evident, writing that "laws cannot be abridged. In them everything is fundamental, and in this regard we may rightly say: to us, all words of Scripture, all of God's commandments are fundamental." See Mendelssohn, *JubA*, 8:167–68; *Jerusalem*, 100–101. The originator of this view was Maimonides's fifteenth-century critic Isaac Abravanel in his 1494 work *Roš Amanah*. See Kellner, *Dogma in Medieval Jewish Thought*, 179–95.

53 Hirsch, *GS*, 5:535.

54 Hirsch, *GS*, 5:539.

55 See Katz, *A House Divided*, 238–39.

56 Hirsch, *GS*, 5:536–38.

mon creed [*das gemeinsame religiöse Bekenntnis*] that makes a community a community" (emphasis mine).[57]

The dogma that Hirsch affirmed as central to Judaism was very similar to Mendelssohn's third element of Judaism, and Hirsch followed Mendelssohn in rejecting dogma in the sense of a commandment to believe in revealed doctrines, instead asserting that the revealed law was to be accepted on the basis of rational considerations.[58] Hirsch's sense of affinity with Mendelssohn can be seen in an article attacking the Reformers that he published sixteen years before the events in Karlsruhe. While the Reformers lionised Mendelssohn as their pioneer, Hirsch asserted that "the thesis of all of Mendelssohn's writings is that Judaism is essentially a law." Hirsch then asked the Reformers: "Do you seriously believe that if [...] Mendelssohn were to come into your midst today he would accept you as one of his own?"[59]

For Hirsch, the obligation to fulfill all the laws of the revealed Torah was the basic dogma of Judaism. Accepting it constituted the basis for membership in the Jewish community and served as a test of Orthodoxy. If the leadership of the official, government-recognised Jewish community broke with this traditional Jewish dogma, then those who remained loyal to it had no choice but to form a separate Jewish community.

Manuel Joël's Positive-Historical Judaism: Dogma as a Basis of Pan-Communal Unity

As secession was being debated in Karlsruhe and Hungary, many German Jewish leaders were growing increasingly worried that Orthodox separatism would spread, with disastrous results.[60] They worried that without a unified tax base, many smaller Jewish communities would become financially insolvent and disintegrate. Among those most concerned about this was Manuel Joël, the chief rabbi of Breslau and a close associate of Zacharias Frankel who identified with his "Positive-Historical" banner. The idea of Jewish communal unity was central to Frankel and to Positive-Historical Judaism, and Joël's position must be understood in light of Frankel's.

At the Frankfurt Conference, Frankel, who called himself a "moderate [*gemäßigte*] Reformer," presented his view of how reforms should be introduced into Judaism.[61] For Frankel, only Judaism's *laws* were subject to evolution, not its *beliefs*, so any reform of Judaism had to apply only to Halakhah. He noted that in undertaking

57 Hirsch, *GS*, 5:539.

58 See Hirsch, *Neunzehn Briefe*, brief 15, 79; *Nineteen Letters*, letter 15, 200.

59 Hirsch, *GS*, 3:494, 504; *Collected Writings* 6:114, 124.

60 On secession in Hungary, see Katz, *A House Divided*, 31–236.

61 On this label, see Zacharias Frankel, "Über Reformen im Judenthume," *Zeitschrift für die religiösen Interessen des Judenthums* 1, no. 1 (1844): 27.

any reforms, concern had to be directed to the needs of the people *as a whole*, not to the needs of isolated individuals. A primary concern had to be to "prevent any split [*Zerspaltung*]" in Judaism and to avoid creating new factions. Instead, one should seek to reconcile existing groups.[62] Joël's 1869 pamphlet "Zur Orientirung in der Cultusfrage" ("On the Question of Religious Worship") clearly reflected his sympathy with Frankel's commitment to Jewish communal unity and his sense that Jewish beliefs were not subject to reform. At the heart of Joël's argument was his contention that dogma was central to Judaism.[63]

Like the Karlsruhe *Oberat*, Joël argued that the Orthodox had no reason to secede from the official Jewish community since Orthodoxy and Reform accepted the same basic religious principles. However, while the *Oberat* had left these common principles vague, Joël sought to give a more precise definition of what he called the "dogma" of Judaism. He noted that religious dogmas were not specific to Judaism, since "every positive religion has specific dogmas [*gewisse Dogmen*] which it regards as fixed and the denial of which causes one to cease being a member [of the religion]."[64] Like Mendelssohn, Joël rejected the binding authority of Maimonides's Thirteen Principles of Judaism, observing that important later Jewish thinkers had disputed them.[65] But rather than seeing all of Jewish law as of binding authority or describing general categories of Jewish beliefs as had Mendelssohn, Joël identified two specific theological "beliefs" (*Glaubensvorstellungen*) as Jewish dogma.[66]

We have seen that the question of how to conduct the synagogue service was central to the emergence of German Jewish denominationalism. Joël's approach to dogma clearly reflected his concern with debates over the synagogue, as indicated by the title of his essay ("On the Question of Religious *Worship*"—emphasis mine). For this reason, rather than presenting dogmas as conceptual principles upon which Judaism was constructed, Joël interpreted dogmas as creedal declarations expressed in the synagogue liturgy,[67] and sought out the theological principles that formed the foundation of this liturgy.

Joël concluded that Judaism rested on two dogmas: a monotheistic representation of God and the existence of prophecy and revelation.[68] He contended that both the Orthodox and the Reform liturgy affirmed these dogmas and that they therefore belonged to the same confession. Disputes between the Orthodox and Reformers over whether the liturgy should express hope for a personal messiah who would re-

---

62 See *Protokolle und Aktenstücke der zweiten Rabbiner-Versammlung* (Frankfurt: E. Ullmann, 1845), 20.

63 Manuel Joël, *Zur Orientirung in der Cultusfrage* (Breslau: Schletter'sche Buchhandlung, 1869).

64 Joël, 10.

65 Joël.

66 Joël, 11. Joël notes the importance of history in Judaism, but does not consider belief in specific historical narratives to be Jewish dogma. See Joël, 7–8.

67 On creed vs. dogma, see above, note 1.

68 Joël, 11–12.

store the Jews to Palestine and rebuild the Temple in Jerusalem or over the question whether or not sacrifices had atoning power were not disputes over fundamental Jewish dogmas. Rather, they were minor disagreements, since these ideas had what Joël called a "contested history" within Judaism.[69] As such, arguments over these ideas did not provide legitimate grounds for tearing the community apart.

Joël acknowledged that Reformers and the Orthodox had radically different interpretations of the two dogmas of Judaism he had enumerated. This was most evident with the dogma of prophecy/revelation which Orthodoxy understood to involve God's revelation of the Written and Oral Torah to Moses, an idea that Reformers largely rejected. However, for Joël, this was not a problem, for while monotheism and prophecy/revelation were Jewish dogmas, they were never given a "final, fixed form," and thus there was much latitude in how to interpret them. Joël claimed that this vagueness was "salutary for Judaism" as it allowed different points of view to be included in the tradition.[70] Yet he made clear that there were limits to how these dogmas could be interpreted. The Bible and what Joël called "later Jewish religious texts" constituted a definite boundary to Judaism. As such, he asserted that "ideas that clearly and unquestionably [*fraglos*] contradict the letter and spirit of Jewish religious texts" lay outside Judaism.[71]

Like Mendelssohn, Joël rejected dogmas in the sense of irrational revealed truths commanded by God whose affirmation was a precondition for salvation. However, unlike Mendelssohn he focused his account of the principles of Judaism on particular—albeit vaguely defined—creedal declarations of belief as expressed in the liturgy, which Joël was happy to call "dogma." The reason for this shift was that Joël's primary concern was to provide a basis for Jewish communal unity in the face of the threat of Orthodox secession, where the synagogue service was a central point of contention.

Abraham Geiger's Reform: *Judenheit* as Basis of Communal Unity

Shortly after the appearance of Joël's pamphlet, Abraham Geiger published a long, devastating attack on it titled "Etwas über Glauben und Beten: Zu Schutz und Trutz" ("On Belief and Prayer: Protection and Defence").[72] Twelve years earlier, Gei-

69 Joël, 11.

70 Joël, 10.

71 Joël, 10. Joël uses the term *Schriftwidriges*, but in light of the fact that he considers not only the Bible, but also later Jewish religious texts as constituting a clear boundary of Judaism, I think he means Jewish sacred texts more generally, especially rabbinic texts.

72 Abraham Geiger, "Etwas über Glauben und Beten: Zu Schutz und Trutz," *Jüdische Zeitschrift für Wissenschaft und Leben* 7 (1869): 1–59. Geiger's review clearly had a personal dimension. At the time, he and Joël were competing for a rabbinical position in Berlin, which Geiger was ultimately offered.

ger had published his groundbreaking *Urschrift und Übersetzungen der Bibel in ihrer Abhängigkeit von der innern Entwicklung des Judenthums* (*Original Text and Translation of the Bible in Its Dependence on the Inner Development of Judaism*). The key argument of the *Urschrift* was that different schools of Judaism had rewritten the Bible to fit their needs.[73] This reflected a radical expansion of Geiger's understanding of the evolution of Judaism that he had expressed two decades earlier in a critique of Hirsch's *Neunzehn Briefe*. There, Geiger had emphasised the development of Judaism by contending that the talmudic rabbis used fanciful exegesis to modify the biblical letter to make it speak "for their time."[74]

Geiger began "Etwas über Glauben und Beten" by accusing Joël of incoherence. How could Joël claim that Judaism had dogma without fixed forms, but then claim that this lack of fixed forms had been beneficial for Judaism? As Geiger put it, "dogma without fixed forms, concepts without conceptual determination, signify nothing."[75] However, Geiger's deeper criticism was animated by his critical understanding of history. He noted that Joël sought to set boundaries on the interpretation of Jewish dogma by asserting that interpretations which "clearly and unquestionably" violated the spirit or the letter of the Bible must be rejected. But Geiger observed that what clearly violated the spirit and letter of the Bible for one person might not for another. Who got to decide which interpretations of dogma were legitimate and which were not?[76]

Geiger noted that Joël had claimed that not just the Bible, but also "later Jewish religious texts" constituted a "definite boundary" of legitimate Jewish ideas. But Joël did not specify which later Jewish texts he was referring to. Geiger observed that the Orthodox could easily claim that these texts included not just the Talmud, but even the *Šulḥan ʿArukh*, concluding that any religious ruling that contravened the *Šulḥan ʿArukh* lay outside Judaism.[77]

See Max Wiener's introduction in Abraham Geiger, *Abraham Geiger and Liberal Judaism: The Challenge of the Nineteenth Century*, rev. ed., trans. Ernst J. Schlochauer, ed. Max Wiener (Cincinnati: Hebrew Union College Press, 1981), 65–66.

73 For discussion of Geiger's *Urschrift*, see Susannah Heschel, *Abraham Geiger and the Jewish Jesus* (Chicago: University of Chicago Press, 1998), 76–83, 121–22; Ken Koltun-Fromm, *Abraham Geiger's Liberal Judaism: Personal Meaning and Religious Authority* (Bloomington: Indiana University Press, 2006), 44–50; Nahum M. Sarna, *Studies in Biblical Interpretation* (Philadephia: Jewish Publication Society, 2000), 161–72.

74 See Abraham Geiger, "Recension der *Briefe über Judenthum* (3)," *Wissenschaftliche Zeitschrift für jüdische Theologie* 3, no. 1 (1837): 90. For discussion of Geiger's complex early attitudes towards rabbinic exegesis, see Jay M. Harris, *How Do We Know This? Midrash and the Fragmentation of Modern Judaism* (Albany, NY: State University of New York Press, 1995), 157–65; Heschel, *Abraham Geiger and the Jewish Jesus*, 25–26; Koltun-Fromm, *Abraham Geiger's Liberal Judaism*, 40–63.

75 Geiger, "Etwas über Glauben und Beten," 7.

76 Geiger, 14.

77 Geiger.

Geiger then turned to Joël's specific account of Jewish dogma. He noted that while Joël claimed that Judaism had universally acknowledged dogmas that received their imprimatur in the liturgy, he enumerated only two such dogmas: monotheism and prophecy/revelation. However, Geiger observed that the traditional Jewish liturgy included a plethora of other ideas, including the resurrection of the dead, belief in angels, and the existence of miracles. "Why did Joël not include these as dogmas of Judaism?" he asked.[78] Geiger further argued that while Joël asserted that the expectation of a personal messiah and the atoning power of sacrifices had a disputed history, this did not reflect the historical record. Until the modern period, almost no Jews seriously doubted the coming of a messiah who would lead the Jews to Palestine, rebuild the Temple, and restore the sacrificial cult and these ideas were clearly affirmed in the traditional Jewish liturgy.[79]

Geiger then criticised Joël's claim that the dogmas of monotheism and prophecy/revelation had never been disputed. Maimonides claimed that monotheism required the belief that God is incorporeal and that a person who believed that God has a body was no longer a Jew. But his older contemporary Rabbi Abraham ibn Daud responded that there were men better than Maimonides who believed that God had a body and Jewish corporealists were never excluded from the synagogue. Moreover, while Joël made monotheism a principle, presumably to exclude the Christian belief in the Trinity and the Incarnation, Kabbalah posited a God composed of ten persons.[80]

Geiger concluded that Judaism has no dogma, a position he ascribed to Mendelssohn. But interpreting Mendelssohn as contending that revealed ritual law is the defining feature of Judaism, Geiger called Mendelssohn's approach an "unhappy thought" that "turned [Judaism] into a lifeless skeleton," since for Geiger, the ritual law had lost its relevance.[81]

In rejecting dogma, Geiger did not deny the importance of doctrines in Judaism. Indeed, he regarded doctrines as Judaism's defining characteristic. What Geiger was contesting was the notion that any *specific*, *fixed* ideas were constitutive of Judaism. Instead, he adopted a teleological view of Judaism, seeing its religious ideas as continually developing and undergoing purification. For Geiger, Judaism's "spiritual power" (*Geistesmacht*) derived from its capacity to tolerate and nourish a diversity of ideas, even contradictory ones, and to continually evolve new, more potent concepts.[82]

But, Geiger asked, how did one determine which ideas were Jewish and which were not? The standard could not be conformity to some arbitrary definition of "Judaism" (*Judentum*). Rather, what made an idea Jewish was that the person introduc-

78 Geiger, 7.
79 Geiger.
80 Geiger, 9.
81 Geiger, 7.
82 Geiger, 7–8.

ing it "feels connected to *Judenheit* and [...] knows himself to be in accordance with the spirit [*Geist*] prevailing in *Judenheit* [...] even if he departs from conventional approaches in essential questions."[83]

83 Geiger, 9. Geiger's position in this text seems to contradict views he presents in other texts. In 1858, he published a fictional correspondence titled *Über den Austritt aus dem Judenthume: ein aufgefundener Briefwechsel* (*On Secession from Judaism: A Newly Discovered Correspondence*). This correspondence was between a Jewish merchant who had recently converted to Christianity and a Reform rabbi who questions the decision. The rabbi asks the convert how he could abandon Judaism with its "basic principle" (*Grundgedanke*) of belief in a "unique, holy, living God" in favour of Christianity with its belief in irrational doctrines such as the Trinity, the Incarnation, Original Sin, and Vicarious Atonement through faith in Jesus. See Abraham Geiger, *Abraham Geiger's nachgelassene Schriften*, ed. Ludwig Geiger (Berlin: Louis Gerschel Verlagsbuchhandlung, 1875), 1:234–35. Shortly after Geiger's essay appeared, a Jewish convert to Christianity named Martin Maass attacked it. Noting that the rabbi in Geiger's dialogue did not consider Jewish ceremonial law to be binding, Maass contended that absent ceremonial law, Geiger's Reform Judaism was nothing but "abstract Deism" or "vestigial Voltairianism" (*zurückgebliebenen Voltairianismus*) that lacked anything distinctly Jewish. Geiger responded to Maass with a 35-page "open letter" in which he asserted that "the actual kernel of Judaism is pure faith in God [*reine Gottesglaube*] and whoever firmly cleaves to this basic kernel is a Jew." See Abraham Geiger, *Über den Austritt aus dem Judenthume: Offenes Sendschreiben an Herrn. M. Maaß* (*On Secession from Judaism: Open Letter to Mr. M. Maass*) (Breslau: J. U. Kern, 1858), 10. Geiger then detailed a series of doctrines that he deemed foundational in Judaism, including belief in a unique, holy, living God who is the creator and preserver of the universe; God as the source of laws governing nature and human life; God being revealed in all stirrings of human good; God as the source of the spiritual development of humanity throughout all of history; that the divine spirit exists in all human strivings in art and science; that humans are the only beings created in the image of God; that human beings have two divine gifts, free will and reason, by which they strive for perfection; that human beings alone can commune with God; that God's fundamental commandment is to love one's neighbour as oneself; that the messianic task is to perfect humanity so that all humanity will revere the unique God in peace and love, etc. Geiger called these doctrines the "faith content [*Glaubensbehalt*] of Judaism" and made it clear that these beliefs were never systematised. Nonetheless, his emphasis on certain doctrines being essential in Judaism seems to differ from his more open-ended, amorphous account of Jewish doctrines laid out in "Etwas über Glauben und Beten." Similarly, in an 1846 work on Jewish religious instruction, Geiger presents three principles of Judaism, namely belief in: (1) a unique holy God; (2) human beings having superior dignity on account of their being created in God's image; (3) human beings being called to work towards the universal brotherhood of mankind united in love. See Geiger, "Abhandlung aus den Programmen der jüdischen Religionsunterrichtsanstalt," in Geiger, *Abraham Geiger's nachgelassene Schriften*, 1:323, cited in Brämer, "Abraham Geiger- skeptischer Pionier einer Glaubenslehre des Reformjudentums?," in *Yearbook of the Maimonides Centre for Advanced Studies 2018*, ed. Bill Rebiger (Berlin: De Gruyter, 2018), 228–29. The differences between "Etwas über Glauben und Beten," and these other texts may be due to Geiger's different aims in each text which leads him to employ different methodological approaches. In *Über den Austritt aus dem Judenthume: Offenes Sendschreiben*, Geiger seeks to give Judaism positive content in order to distinguish it from Deism and in "Abhandlungen aus den Programmen der jüdischen Religionsunterrichtsanstalt" he seeks to specify principles of Judaism for the instruction of Jewish youth. As such he presents his own normative, theological conception of Judaism. By contrast, in "Etwas über Glauben und Beten" he uses an academic, historical approach to refute Joël's assertion that Judaism can be defined by specific dogmas.

Geiger's term *Judenheit* is parallel to the term *Christenheit*, which is generally translated as "Christendom." Just as *Christenheit* signifies all the members of the Christian religion, so Geiger's *Judenheit* signifies all the members of the Jewish religion. For Geiger, one could never be certain from the outset what constituted an authentic Jewish idea, and this was subject to change. For example, he observed that at one time, Kabbalah had enjoyed wide acceptance within Judaism, but that at present, its fantastic excesses had "been fought and overcome."[84] Instead, Geiger asserted that history was decisive and that Jews should "trust Judaism's power to spiritually weed out the foreign and smooth out perversions." The greatest danger was trying to draw clear boundaries to Judaism, as this would end up suffocating it. Granting Jewish thinkers the full freedom to develop their ideas would be what would preserve Judaism's life and vitality for the future. For Geiger, identification with Jewish religion in general rather than affirming specific ideas or practices constituted the basis of Jewish communal unity.

Geiger concluded by addressing the problem of schism. He acknowledged that his approach would not deter Orthodox secessionists, since they defined Judaism by adherence to all the laws in the *Šulḥan ʿArukh* and considered Jews who denied this to be outside of Judaism. However, he asserted that Joël's attempt to avoid schism by arbitrarily selecting vaguely defined dogmas which he declared to be binding on all Jews was no solution, since the Orthodox would simply reject this.[85] Geiger made clear that while he was not in favour of schism, he did not fear it.[86] Protestantism had brought schism to Christianity, but in doing so, it had freed itself from Catholicism's stifling hierarchy and laid the ground for spiritual progress and *Bildung*. What was most important for the future of Judaism was freedom. As he put it: "In religious life, freedom certainly precedes unity [*geht sicher die Freiheit vor der Einheit*]!"[87]

84 Geiger, "Etwas über Glauben und Beten," 9.

85 Geiger, 15–16. For his part, Joël was likely not concerned with convincing Orthodox Jews of his views, but rather with convincing the governmental authorities so that they would deny Orthodox Jews the right to secede from the official Jewish community.

86 Geiger, 22, 25. Geiger's attitude towards religious schism was highly ambivalent. Liberles notes that in 1831, when Reform was a small minority, he had favoured Reformers forming separate communities. Between 1839 and 1841, Geiger vacillated on the question. By 1846, as Reform had established a dominant position among German Jews, he opposed secession. However, as we can see from his 1869 essay, Geiger did not oppose secession at all costs. See Liberles, *Religious Conflict*, 165–67. See also Michael Meyer, "Universalism and Jewish Unity in the Thought of Abraham Geiger," in *The Role of Religion in Modern Jewish History*, ed. Jacob Katz (Cambridge, MA: Association for Jewish Studies, 1975), 91–107.

87 Geiger, "Etwas über Glauben und Beten," 25–26.

Conclusion

Writing in the eighteenth century, Mendelssohn addressed the question of dogma in Judaism from an interreligious perspective as a means of distinguishing Judaism from Christianity. For Mendelssohn, Judaism's lack of dogma involved its rejecting the idea that affirming irrational revealed doctrines commanded by God and confirmed by an authoritative religious body was a precondition for salvation. According to Mendelssohn, Judaism taught that all human beings could attain salvation by practising natural religion which could be known universally through reason or common sense. Mendelssohn also denied that Judaism was defined by an authoritative list of articles of faith which, if denied, made one a heretic. Mendelssohn's own account of the three components of Judaism referred, for the most part, to general classes of principles and expressed his analytical understanding of Judaism rather than being a normative means of excluding heretics. In defining Judaism as he did, Mendelssohn sought to show that Judaism was more consistent with the enlightened principles of rationality and tolerance than was Christianity and hence that Jews were eligible for civil inclusion. His approach reflected his maskilic view that Judaism was an enlightened religion of reason (*Vernunftreligion*).[88]

In the nineteenth century, the debate over dogma in Judaism came to be shaped by intrareligious concerns; namely, the problem of Jewish communal unity. Several writers, including the members of the Reform-controlled *Oberat* in Karlsruhe, the Positive-Historical rabbi Manuel Joël, and the Neo-Orthodox rabbi Samson Raphael Hirsch all affirmed that Judaism had dogma in the sense of a set of fixed doctrines and/or practices. This claim served as a justification either for pan-Jewish communal unity or for Orthodox Jewish sectarianism. Joël's view reflected the Positive-Historical commitment to Jewish communal unity, while Hirsch's view reflected his Orthodox commitment to the authority of the timeless, unchanging, revealed laws of the Torah.

Geiger's approach to the question of dogma in Judaism was the most radical. He rejected dogma not only in the Christian sense, but also in the sense in which Joël and Hirsch had defined it; namely, the idea that specific doctrines were constitutive of Judaism. While not denying the centrality of doctrines in Judaism, Geiger saw the basis of Jewish communal unity in what he called *Judenheit*, which was grounded in a subjective sense of belonging to a religious community rather than espousing certain beliefs or engaging in specific practices. Geiger's emphasis on the centrality of intellectual freedom lead him to a much weaker commitment to Jewish communal unity than that espoused either by the Reform-controlled *Oberat* in Karlsruhe or

---

88 Mendelssohn acknowledged that some Christians, such as Unitarians, adhered to a religion which, like Judaism, was rational, and he declared: "We must almost regard Unitarians as fellow believers." For Mendelssohn, the difference between Orthodox Christian teachings and Unitarianism was much greater than the difference between Judaism and Unitarianism. See Mendelssohn, *JubA*, 7:102–6; *Moses Mendelssohn: Writings*, 26–30.

by Joël. This nineteenth-century debate over whether or not Judaism has dogma is thus a pivotal debate intimately tied to the problem of Jewish communal unity.

A fascinating sequel to the nineteenth-century debate occurred in the first third of the twentieth century and involved many of the leading German Jewish scholars and rabbis of the time, including Leo Baeck, Julius Guttmann, Max Wiener, and Alexander Altmann. Here the question of dogma in Judaism again returned to being an interreligious one. Exploring this debate will have to await another occasion.[89]

Bibliography

Altmann, Alexander. "Moses Mendelssohn's Concept of Judaism Reexamined." In *Von der mittelalterlichen zur modernen Aufklärung: Studien zur jüdischen Geistesgeschichte*, 234–48. Tübingen: Mohr, 1987.

Baeck, Leo. "Does Traditional Judaism Possess Dogmas?" In *Studies in Jewish Thought: An Anthology of German Jewish Scholarship*, edited by Alfred Jospe, 41–53. Detroit: Wayne State University Press, 1981.

Baron, Salo. "Freedom and Constraint in the Jewish Community: An Historic Episode." In *Essays and Studies in Memory of Linda R. Miller*, edited by Israel Davidson, 9–24. New York: Jewish Theological Seminary, 1938.

Bleich, Judith. "Jacob Ettlinger, His Life and Works: The Emergence of Modern Orthodoxy in Germany." PhD diss., New York University, 1974.

Blutinger, Jeffrey. "'So-Called Orthodoxy': The History of an Unwanted Label." *Modern Judaism* 27, no. 3 (2007): 310–28.

Brämer, Andreas. "Abraham Geiger- skeptischer Pionier einer Glaubenslehre des Reformjudentums?" in *Yearbook of the Maimonides Centre for Advanced Studies 2018*, Edited by Bill Rebiger, 207–30. Berlin: De Gruyter, 2018.

Brämer, Andreas. *Rabbiner Zacharias Frankel: Wissenschaft des Judentums und konservative Reform im 19. Jahrhundert.* Hildesheim: Olms, 2000.

Carlebach, Elisheva. "La communauté juive et ses institutions au début de l'époque moderne." In *Aux Origines du Judaïsme*, edited by Julien Darmon and Jean Baumgarten, 358–89. Paris: Actes Sud, 2012.

Darlap, Adolf, and Karl Rahner. "Dogma." In *Encyclopedia of Religion*, 2nd ed., edited by Lindsay Jones, 4:2387–90. Detroit: Macmillan, 2004.

Frankel, Zacharias. "Über Reformen im Judenthume." *Zeitschrift für die religiösen Interessen des Judenthums* 1, no. 1 (1844): 3–27.

Freudenthal, Gideon. *No Religion without Idolatry: Mendelssohn's Jewish Enlightenment.* Notre Dame, IN: University of Notre Dame Press, 2012.

Friedländer, David. *Der Prediger.* Berlin: Friedrich Maurer, 1788.

Geiger, Abraham. *Abraham Geiger and Liberal Judaism: The Challenge of the Nineteenth Century.* Rev. ed. Translated by Ernst J. Schlochauer. Edited by Max Wiener. Cincinnati: Hebrew Union College Press, 1981.

89 On this debate, see von der Krone, "Jüdische Wissenschaft und modernes Judentum: eine Dogmendebatte," in *Die "Wissenschaft des Judentums": Eine Bestandsaufnahme*, ed. Thomas Meyer and Andreas Kilcher (Paderborn: Wilhelm Fink Verlag, 2015), 125–38.

Geiger, Abraham. *Abraham Geiger's nachgelassene Schriften.* Edited by Ludwig Geiger. 5 vols. Berlin: Louis Gerschel Verlagsbuchhandlung, 1875.
Geiger, Abraham. "Etwas über Glauben und Beten: Zu Schutz und Trutz." *Jüdische Zeitschrift für Wissenschaft und Leben* 7 (1869): 1–59.
Geiger, Abraham. "Recension der Briefe über Judenthum (3)." *Wissenschaftliche Zeitschrift für jüdische Theologie* 3, no. 1 (1837): 74–91.
Geiger, Abraham. *Über den Austritt aus dem Judenthume: Offenes Sendschreiben an Herrn. M. Maaß.* Breslau: J. U. Kern, 1858.
Gottlieb, Michah. *Faith and Freedom: Moses Mendelssohn's Theological-Political Thought.* New York: Oxford University Press, 2011.
Gottlieb, Michah. *Faith, Reason, Politics: Essays on the History of Jewish Thought.* Brighton, MA: Academic Studies Press, 2013.
Gottlieb, Michah. "Oral Letter and Written Trace: Samson Raphael Hirsch's Defense of the Bible and Talmud." *Jewish Quarterly Review* 106, no. 3 (2016): 316–51.
Harris, Jay M. *How Do We Know This? Midrash and the Fragmentation of Modern Judaism.* Albany, NY: State University of New York Press, 1995.
Harvey, Warren Zev. "Hasdai Crescas and Moses Mendelssohn on Beliefs and Commandments." In *Moses Mendelssohn: Enlightenment, Religion, Politics, Nationalism*, edited by Michah Gottlieb and Charles H. Manekin, 79–89. Bethesda: University Press of Maryland, 2015.
Heschel, Susannah. *Abraham Geiger and the Jewish Jesus.* Chicago: University of Chicago Press, 1998.
Hirsch, Samson Raphael. *Collected Writings.* Edited by Marc Breuer, Jacob Breuer, Meta Bechhofer, and Elliot Bondi. 9 vols. New York: Feldheim, 1984–2012.
Hirsch, Samson Raphael. *Gesammelte Schriften.* Edited by Mendel Hirsch. 6 vols. Frankfurt am Main: J. Kauffmann, 1902–12.
Hirsch, Samson Raphael. *Neunzehn Briefe über Judentum.* 2nd ed. Frankfurt: J. Kauffmann, 1889.
Hirsch, Samson Raphael. *Sefer Šemeš Marpeʾ: šeʾelot u-tešuvot, ḥiddušei ha-Šas, iggerot u-mikhtavim.* Edited by Eliyahu Meir Klugman. New York: Mesorah, 1992.
Hirsch, Samson Raphael. *The Nineteen Letters about Judaism.* Translated by Karin Paritzky. Edited by Joseph Elias. New York: Feldheim, 1995.
Jacobs, Louis. *Principles of the Jewish Faith: An Analytical Study.* London: Vallentine Mitchell, 1964.
Joël, Manuel. *Zur Orientirung in der Cultusfrage.* Breslau: Schletter'sche Buchhandlung, 1869.
Katz, Jacob. *A House Divided: Orthodoxy and Schism in Nineteenth-Century Central European Jewry.* Translated by Ziporah Brody. Hanover, NH: Brandeis University Press, 1998.
Katz, Jacob. *Divine Law in Human Hands: Case Studies in Halakhic Flexibility.* Jerusalem: Magnes Press, 1998.
Katz, Jacob. *From Prejudice to Destruction: Anti-Semitism, 1700–1933.* Cambridge, MA: Harvard University Press, 1980.
Katz, Jacob. "Orthodoxy in Historical Perspective." In *Studies in Contemporary Jewry, Volume 2*, edited by Peter Medding, 3–17. Bloomington: Indiana University Press, 1986.
Katz, Jacob. *Tradition and Crisis: Jewish Society at the End of the Middle Ages.* Translated by Bernard Dov Cooperman. Syracuse: Syracuse University Press, 2000.
Kayserling, Meyer. *Moses Mendelssohn: sein Leben und seine Werke.* Leipzig: H. Mendelssohn, 1862.
Kellner, Menachem. *Dogma in Medieval Jewish Thought: From Maimonides to Abravanel.* Oxford: Oxford University Press, 1986.
Kellner, Menachem. *Must a Jew Believe Anything?* 2nd ed. Oxford and Portland, OR: Littman Library of Jewish Civilization, 2006.

Koltun-Fromm, Ken. *Abraham Geiger's Liberal Judaism: Personal Meaning and Religious Authority.* Bloomington: Indiana University Press, 2006.
Krone, Kerstin von der. "Jüdische Wissenschaft und modernes Judentum: eine Dogmendebatte." In *Die "Wissenschaft des Judentums": Eine Bestandsaufnahme*, edited by Thomas Meyer and Andreas Kilcher, 115–38. Paderborn: Wilhelm Fink Verlag, 2015.
Liberles, Robert. *Religious Conflict in Social Context: The Resurgence of Orthodox Judaism in Frankfurt Am Main, 1838–1877.* Westport, CT: Greenwood Press, 1985.
Mendelssohn, Moses. *Gesammelte Schriften: Jubiläumsausgabe.* Edited by Alexander Altmann et al. 25 vols. Stuttgart and Bad Cannstatt: Frommann-Holzboog, 1971–.
Mendelssohn, Moses. *Jerusalem, or On Religious Power and Judaism.* Translated by Allan Arkush. Edited by Alexander Altmann. Hanover, NH: University Press of New England, 1983.
Mendelssohn, Moses. *Moses Mendelssohn: Writings on Judaism, Christianity, and the Bible.* Translated by Curtis Bowman, Elias Sacks, and Allan Arkush. Edited by Michah Gottlieb. Hanover, NH: University Press of New England, 2011.
Mendelssohn, Moses. *Moses Mendelssohn's Hebrew Writings.* Translated by Edward Breuer. Edited by Edward Breuer and David Sorkin. New Haven: Yale University Press, 2018.
Mendes-Flohr, Paul R., and Jehuda Reinharz, eds. *The Jew in the Modern World: A Documentary History.* 3rd ed. New York: Oxford University Press, 2011.
Meyer, Michael. *Response to Modernity: A History of the Reform Movement in Judaism.* New York: Oxford University Press, 1988.
Meyer, Michael. "Universalism and Jewish Unity in the Thought of Abraham Geiger." In *The Role of Religion in Modern Jewish History*, edited by Jacob Katz, 91–107. Cambridge, MA: Association for Jewish Studies, 1975.
Philipson, David. *The Reform Movement in Judaism.* New York: Macmillan, 1907.
Plaut, W. Gunther. *The Rise of Reform Judaism: A Sourcebook of Its European Origins.* New York: World Union for Progressive Judaism, 1963.
Protokolle und Aktenstücke der Zweiten Rabbiner-Versammlung. Frankfurt, E. Ullmann, 1845.
Sacks, Elias. *Moses Mendelssohn's Living Script: Philosophy, Practice, History, Judaism.* Bloomington: Indiana University Press, 2017.
Sarna, Nahum M. *Studies in Biblical Interpretation.* Philadelphia: Jewish Publication Society, 2000.
Schorsch, Ismar. *From Text to Context: The Turn to History in Modern Judaism.* Hanover, NH: University Press of New England, 1994.
Schwarzschild, Steven. *The Pursuit of the Ideal: Jewish Writings of Steven Schwarzschild.* Edited by Menachem Kellner. Albany, NY: State University of New York Press, 1990.
Shapiro, Marc B. *The Limits of Orthodox Theology: Maimonides' Thirteen Principles Reappraised.* Oxford and Portland, OR: Littman Library of Jewish Civilization, 2004.
Silber, Michael. "Orthodoxy." In *The Yivo Encyclopedia of Jews in Eastern Europe*, edited by Gershon Hundert, 2:1292–97. New Haven: Yale University Press, 2008.
Speight, R. Marston. "Creeds: An Overview." In *Encyclopedia of Religion*, 2nd ed., edited by Lindsay Jones, 3:2052–54. Detroit: Macmillan 2004.
Steinheim, Salomon. *Moses Mendelssohn und seine Schule.* Hamburg: Hoffmann und Campe, 1840.
Williams, M. E. "Dogma." In *New Catholic Encyclopedia*, 2nd ed., edited by Thomas Carson, 1:810–11. Detroit: Thomson/Gale, 2002.

Anne Fiebig

From Scepticism to Tolerance of "the Other": The Example of Yeshayahu Leibowitz

Introduction

The European-born Israeli thinker Yeshayahu Leibowitz was as provocative as he was popular among the Israeli public, and he influenced more than one generation of society both in Israel and abroad. Born in Riga in 1903, he studied chemistry in Berlin, medicine in Cologne and Heidelberg, and later took his MD in Basel. After participating in the religious-Zionist *Mizrachi* movement in Germany, he later migrated to Palestine, where he lived from 1934 until his death in Jerusalem in 1994. Leibowitz was widely interpreted as provocative yet highly topical due to his harsh criticism of the Israeli government—particularly after the Six-Day War (1967)—and his interpretation of the Jewish religion, which was "a highly resilient form of Jewish religiosity, capable of enduring the most vigorous philosophical and ethical criticism from the Enlightenment to our time."[1] To this day, there is discussion as to whether Leibowitz can be defined as a philosopher and whether he was part of the tradition of serious Israeli philosophy, as his roots were in natural science. Perhaps this is one reason why his thought has regrettably never gained much popularity outside of Israel. Yet Leibowitz is worth engaging with. At the very least, he was an important thinker in a generation and time that was occupied with asking: "What would be the character of the new Jewish society?"[2] He is worth engaging with because of his vast spectrum and originality of thought about the Israeli state and the role of its religious Jewry, and because of his interesting personal development from a young ambitious religious-Zionist thinker into a near-existentialist.[3] Finally, he is worth engaging with because his interpretation of Jewish religion allows space for a modern form of Jewish religiosity within a mostly secular Jewish democratic state, as he himself said that his approach was not historical, but contemporary, for the "here and now."[4]

This article was intended as part of an ongoing dialogue with Roi Benbassat on how to interpret scepticism in Leibowitz.

1 Roi Benbassat, "Yeshayahu Leibowitz: Jewish Existentialism," *Religious Studies* 51, no. 2 (2015): 141.
2 Ernst Akiva Simon, "Are We Israelis Still Jews? The Search for Judaism in the New Society," in *Arguments and Doctrines: A Reader of Jewish Thinking in the Aftermath of the Holocaust*, ed. Arthur A. Cohen (New York: Harper & Row, 1970), 392. Leibowitz was looking for the role of religious Jewry and of faith in general within an autonomous Jewish statehood.
3 See Benbassat, "Jewish Existentialism."
4 Yeshayahu Leibowitz (1953), *Judaism, Human Values, and the Jewish State*, ed. Eliezer Goldman, trans. Eliezer Goldman, Yoram Navon, Zvi Jacobson, Gershon Levi and Raphael Levy (Cambridge,

https://doi.org/10.1515/9783110618839-013

The highly topical nature of Leibowitz's thought can inspire a generation of young religious Israelis who aspire to be both followers of the *Halakhah* and modern citizens of a democratic state.

In 2017, Roi Benbassat published an article that identified scepticism in the thought of the religious thinker.[5] The essential question of this paper will be: If Leibowitz can provide an example of Jewish faith and scepticism, does he also provide an example of tolerance that stems from his sceptical approach? This question is based on the premise that scepticism is a route to tolerance. Inquiries into the roots of toleration[6] have shown that an attitude that doubts or questions the possibility of knowledge opens the possibility of tolerating, for example, differences in religion or culture.[7] The topic raises particular difficulties because Leibowitz, unlike other religious thinkers, never analysed the great phenomenon of tolerance himself. As in the analysis of scepticism, we have to look for implicit evidence in the literary heritage of this great thinker.

I will first give an introduction detailing Leibowitz's definition of the Jewish religion and present his assumption of the co-existence of different value systems which is a crucial element of the analysis of tolerance in his thought. After this, I will present some arguments which have been made in the analysis of scepticism in Leibowitz's thought. As scepticism has been widely discussed as a route to tolerance,[8] I will then look for tolerant approaches in Leibowitz's philosophy. I will then show evidence of errors in Benbassat's approach to scepticism in Leibowitz's thought and will further present a new methodological approach to analysing his sceptical

MA and London: Harvard University Press, 1995), 1. I will always write the essay's date of publication, as there is a wide variety of dates and Leibowitz's early thought differs from his later thought.

5 See Roi Benbassat, "Jewish Faith and Scepticism—The Example of Yeshayahu Leibowitz," in *Yearbook of the Maimonides Centre for Advanced Studies 2017*, ed. Bill Rebiger (Berlin: De Gruyter, 2017), 161–81.

6 I am using the words "tolerance" and "toleration" as synonyms.

7 See Giuseppe Veltri, "The Limits of Scepticism and Tolerance," *Bollettino della società filosofica italiano* 2 (2017): 49. Referring to, for example, Jean Bodin, John Locke's *Letter Concerning Toleration* (1689), and the *Essais* of Michel de Montaigne (1580–88), Veltri states: "[The] attitude to doubt (starting from the recognition of the human incapacity to discern the truth and based on the admission that humans can err in choosing the right way) or questioning the possibility of knowledge opens up the possibility of recognizing differences in religion, culture, and economy." And further: "Despite differences among them, these texts contributed to the development of the question of tolerance [...]. At their base, they concern liberty of conscience, *libertas philosophandi*, and the possibility of erring from the correct way."

8 See, for example, Sam Black, "Locke and the Skeptical Argument for Toleration," *History of Philosophy Quarterly* 24, no. 4 (2007): 355–75; Bican Sahin, "A Defense of Toleration on the Basis of Scepticism: The Case of Michel de Montaigne," *Hacettepe University Journal of Economics and Administrative Sciences* 23, no. 2 (2005): 283–304.

attitude. Finally, I will present further important questions regarding the practice and principle of tolerance within Leibowitz's thought.[9]

1 Leibowitz's Definition of Jewish Religion

One essential element which enables us to integrate scepticism and tolerance in Leibowitz's thought was his conception of Jewish faith.[10] This was original when compared with previous Jewish thinkers and quite radical for an Orthodox thinker like Leibowitz. In contrast to the most popular post-*Haskalah* Jewish thinkers, "Leibowitz did not attempt to reconcile Judaism within universal moral values, basically positing it instead as conflicting with all aspects of humanism."[11] He launched a so-called "Leibowitzian" or "Copernican" revolution[12] in Jewish religious thought that was constituted in an absolute removal of theological and metaphysical elements from

---

9 Regarding the current state of research, allow me to make some general remarks. In Israel, there are numerous discussions about Leibowitz's definition of faith and statehood, comparisons to other philosophers—such as, for example, Kant, Kierkegaard, or Levinas—and analyses of his relationship to individualism, subjectivity, and even feminism. Outside the Israeli state, not much has been published about Leibowitz's thought, and hardly anything in Germany. From his literary legacy, only one large anthology of his essays has been translated into English (Leibowitz, *Judaism, Human Values, and the Jewish State*). Also translated were an anthology of his commentaries on the weekly Torah portion (Yeshayahu Leibowitz, *Accepting the Yoke of Heaven: Commentary on the Weekly Torah Portion*, trans. Shmuel Himelstein [Jerusalem: Urim Publications, 2002]) and his monograph about Maimonides's faith (Yeshayahu Leibowitz, *The Faith of Maimonides*, trans. John Glucker [Tel-Aviv: MOD Books, 1989]). There are several translations of his works into French and Spanish, but only two translations into German from his entire literary corpus. See Yeshayahu Leibowitz, *Vorträge über die Sprüche der Väter: Auf den Spuren des Maimonides*, trans. Grete Leibowitz (Obertshausen: Context-Verlag, 1984); Yeshayahu Leibowitz and Michael Shashar, *Gespräche über Gott und die Welt*, 3rd edition (Frankfurt am Main: Insel Verlag, 1994). I do not know of any other German text about this Israeli thinker, with the following exceptions: one text about axiology in Leibowitz's thought (Ottfried Fraisse, "Ent-Geisterung der Rheologie. Zur Axiologie von Yeshayahu Leibowitz' Denken," In *Alles Wirkliche Leben ist Begegnung. Festschrift zum vierzigjährigen Bestehen von Studium Israel e.V.*, ed. Johannes Ehmann, Joachim J. Krause, and Bernd Schröder [Leipzig: Evangelische Verlagsanstalt, 2018], 305–17), one lexical entry (Matthias Morgenstern, "Jeschajahu Leibowitz," in *Metzler Lexikon jüdischer Philosophen: Philosophisches Denken des Judentums von der Antike bis zur Gegenwart*, ed. Andreas B. Kilcher, Ottfried Fraisse, and Yossef Schwartz [Stuttgart: J.B. Metzler, 2003], 403–7), and a contribution to an anthology about religion and democracy (Ines Jaqueline Werkner et al., eds., *Religionen und Demokratie. Beiträge zu Genese, Geltungund Wirkung eines aktuellenpolitischen Spannungsfeldes* [Wiesbaden: VS Verlag, 2009]).

10 See Benbassat, "Jewish Faith and Scepticism," 163.

11 Benbassat, "Jewish Existentialism," 142.

12 See Avi Sagi, *Jewish Religion after Theology*, trans. Batya Stein (Boston: Academic Studies Press, 2009), 46; Ronny Miron, "Yeshayahu Leibowitz on the Possibility of Religious Subjectivity: Critique and Defense," in *Watchtower Religionswissenschaft. Standortbestimmung im wissenschaftlichen Feld*, ed. Anne Koch (Marburg: Diagonal-Verlag, 2007), 95.

Jewish religion.[13] For professor of philosophy Avi Sagi, this revolution took place in two steps: "First, in the denial of theology, and second, in the denial of 'religious facts.'"[14] Jewish religion, for Leibowitz, was reduced to nothing but the execution of the *Halakhah:* the service of God. This "revolution" also contains the Jewish religion's shift from a metaphysical and theological realm to a strictly normative realm. That means that Jewish religion was one value system among others, unjustified by any truth claims. It was hence absolutely non-cognitive. However, this radical standpoint was not always the key concept of Leibowitz's thought.

Like so many philosophers, Leibowitz underwent a transformation from an "early" to a "later" thought. Often ignored by people concerned with his thinking, this development is a crucial point within the understanding of his system of thought, because it includes Leibowitz's definition of faith, which is the foundation of reading scepticism and tolerance in his thought. In his earlier thinking, one finds a distinction between faith and belief, in which belief is a cognitive determination of our minds, while faith is a conative determination.[15] In 1953, Leibowitz was still presenting religious praxis as a superstructure rising above the "cognitive and emotional components, or the spiritual and mental worlds of the Jewish religion,"[16] arguing that the categories of religious cognition and sensibility are interpretations of the system of the Torah and its *mitzvot*. In his later writings, Leibowitz makes no such distinction and argues strongly that there are *no* cognitive or emotional elements in the Jewish religion at all: "For Judaism, faith is nothing but its system of *mitzvot*, which was the embodiment of Judaism. [...] Faith, in Judaism, is the religion of *mitzvot*, and apart from this religion Jewish faith does not exist."[17] While in his earlier writings, Leibowitz still makes a vague distinction between faith and *mitzvot*—at least philologically—in his later writings, there is no faith beside, above, or under the *mitzvot*. He accuses people "who contend that one can separate faith from religion" and who think that "adherence to Torah and *mitzvot* is a kind of superstructure built upon faith"[18] of naiveté. This "radicalisation" in his definition of faith is proof of a development in Leibowitz's thought from the "early Leibowitz" to the more radical "later Leibowitz." This transformation is most distinct in Leibowitz's view of the composition of a Jewish state, but it can also be found in his use of terms from or interpretations of the fundaments of Jewish religion. Only within this development can Yeshayahu Leibowitz's thought be understood, as it radically changed between his early and his later expectations, while the resilient turn around happened sometime

13 See Avi Sagi, "Yeshayahu Leibowitz—A Breakthrough in Jewish Philosophy: Religion without Metaphysics," *Religious Studies* 33, no. 2 (1997): 205 f.; Eliezer Goldman, "Introduction," in Leibowitz, *Judaism*, vii–xxxiv, xix.

14 Sagi, "Yeshayahu Leibowitz—A Breakthrough," 206.

15 See Benbassat, "Jewish Faith and Scepticism," 161 f.

16 Leibowitz (1953), *Judaism*, 4 f.

17 Leibowitz (1981), *Judaism*, 38.

18 Leibowitz (1981), *Judaism*, 44.

in the 1950s and 1960s.[19] And it is only within this *clear* definition that scepticism and tolerance can be detected in Leibowitz's thought. Because of this transformation, his texts must be read with sincere attention. Terms like "*Halakhah*," "faith," and "religion" and their meanings change over the course of time, as Hannah Kasher illustrated in 2000.[20] For this analysis, I will concentrate on Leibowitz's later and more popular position.

One of the most important distinctions for Leibowitz was the distinction between the holy and the profane: "There is no holiness outside the sphere of divinity."[21] He stated that nothing is holy in itself, but there is only that which is holy to God, which is the practice of the *Halakhah*, because "God has nothing in his world beyond the four ells of *Halakhah*."[22] The abrogation of the distinctive category of holiness and the imputation of holiness to certain things, persons, locations, institutions, events, military loyalty, or the nation itself is, in Leibowitz's eyes, idolatrous: "By distinguishing the sacred from the profane the *Halakhah* functions as a bulwark against

---

19 See Moshe Hellinger, "A Clearly Democratic Religious-Zionist Philosophy: The Early Thought of Yeshayahu Leibowitz," *Journal of Jewish Thought and Philosophy* 16, no. 2 (2008), 270: "As a matter of fact, the late Leibowitz is a radical liberal thinker, whereas the early Leibowitz is a radical democratic thinker. In his early teaching Leibowitz stresses collective principles, such as popular sovereignty and the decision of the majority, and mobilizes them for the purpose of not only shaping the political character of the desirable Hebraic state but also guiding the specific community of the faithful observant. In contrast, his later teaching emphasizes liberal, individualistic elements, such as freedom of thought and speech, human rights (of Israelis and Palestinians alike), and so forth."

20 Kasher shows that in his early thought, Leibowitz used the word *emunah* to mean a "mental quality" and "the practical training that may bring the mind closer to *emunah* is the study of Torah and observance of the practical commandments" (Yeshayahu Leibowitz, *Torah and Commandments in Our Times* [Hebrew] [Tel Aviv: Schocken, 1954], 55; this article was first given as a lecture in a seminary, see Hannah Kasher, "On Yeshayahu Leibowitz's Use of Religious Terminology," *Journal of Jewish Thought and Philosophy* 10 [2000], 30 f). In his later writings, Leibowitz consistently uses the term *emunah* to mean "practice of the Halakhah," because—as he states—"faith [*emunah*] is identical with its system of commandments, in which it was embodied" (Yeshayahu Leibowitz, *Faith, History, and Values* [Hebrew] [Jerusalem: Academon, 1982], 11–12, in Kasher, 34). Kasher described the later meaning of the term *emunah* as a non-lexical "stipulative" meaning, which is probably closer to its biblical meaning of "steadfastness" or "dependability." See Kasher, 59: "As we shall see, in his early works, Leibowitz normally treated such words as *dati* (religious), *emunah* (faith), Torah, 'God,' 'fear and love of God' as concepts which a priori 'well-defined things,' using these words in accordance with their lexical definitions. In time, he came to view them as mere terms and, having constructed his special, personal words of values, endowed them with stipulative meaning in keeping with that world."

21 Leibowitz (1953), *Judaism*, 24.

22 Babylonian Talmud, Tractate *Berakhot* 8a: "Once I heard what Rabbi Ḥiyya bar Ami said in the name of Ulla: Since the day the Temple was destroyed, the Holy One, Blessed be He, has only one place in His world, only the four cubits of *halakha* alone, from which I understood the significance of the four cubits of *halakha*, and I pray only where I study" (William Davidson Talmud).

idolatry [...]. For idolatry is simply the representation of things profane as sacred."[23] As an idolatrous act, the prevalent imputation of holiness on unholy categories in the education of young people and the dissemination of public information laid the foundations for dangerous misbehaviour.[24]

Beneath the fact that faith was absolutely non-cognitive, for Leibowitz was the fact that God was absolutely transcendent. In this transcendence, he was in agreement with Maimonides, whom he had frequently discussed.[25] But how does the religious Jew approach a transcendent God? Leibowitz faces this problem by making a distinction between faith *li-šmah* and faith *lo᾽ li-šmah* and a theory of distinct value systems. This distinction between two "grades" of religiosity did already exist in rabbinic literature.[26] The first grade is the sanctification of God *li-šmah* (for its own sake), which means honouring God without expecting personal benefit, or, vice versa, punishment (in the case of non-tribute). The sanctification of God *lo᾽ li-šmah* (not for its own sake) means honouring God because one expects a personal benefit.[27] This includes spiritual needs such as "'happiness,' 'perfection,' 'morality,'"[28] or material needs such as "'life, children, nutriment,' or the redemption of Israel."[29] This assumption of faith *li-šmah* and faith *lo᾽ li-šmah* is imbedded in Leibowitz's definition of the plurality of value systems. Hence, Judaism is one normative system of values which co-exists among a plurality of other value systems. This theory of parallelism is crucial, because it is only within this system of plurality that faith *li-šmah* is possible. Only in a world that provides different systems that satisfy

---

23 Leibowitz (1953), *Judaism*, 25. This also applies to mysticism, which he calls incompatible with *halakhic* Judaism; Leibowitz (1953), *Judaism*, 24 ff. See also Leibowitz (1981), *Judaism*, 46.

24 One notable example was the reprisal attack on Kibiyeh on October 13, 1953, where Israeli forces attacked Kibiyeh in retaliation for the murder of a Jewish mother and two of her children. In this kickback, more than fifty inhabitants of the village were killed. Leibowitz strongly criticised this attack, calling it unjustified, not because a certain kind of traditional "Jewish morality" forbade Jews to retaliate, but because retaliation was forbidden *per se* (see Leibowitz [1953–54], *Judaism*, 189). For Leibowitz, this action, executed by young Israeli soldiers, was only possible through the imputation of holiness on profane categories such as the nation and its security, because for the sake of that which is holy, "man is capable of acting without any restraint," see Leibowitz (1953–54), *Judaism*, 190.

25 "For Maimonides, God is 'true Being' by Himself and for Himself, and not for the sake of the world. [...] Maimonides posits a pure concept of a transcendental God whose link with the world is not naturalistic causation." Yeshayahu Leibowitz, "Maimonides: The Abrahamic Man," *Judaism* 6, no. 2 (1957): 151 f.

26 See, for example, Babylonian Talmud, tractate *Ta῾anit* 7a: "It is taught in a baraita that Rabbi Bena'a would say: Anyone who engages in Torah for its own sake, his Torah study will be an elixir of life for him, as it is stated: 'It is a tree of life to them who lay hold upon it' (Prov. 3:18) [...]. And anyone who engages in Torah not for its own sake, e. g., for self-aggrandizement, his Torah will be an elixir of death for him, as it is stated: 'My doctrine shall drop as the rain' (William Davidson Talmud). See Goldman, "Introduction," xviii.

27 See Goldman, xviii.

28 Leibowitz (1977), *Judaism*, 63.

29 Leibowitz (1972), *Judaism*, 55.

human needs does the Jewish religion stay unaffected for itself, and only when religion stays unaffected and it is not idolatrously used to satisfy human needs is the "pure" faith, faith *li-šmah*, for its own sake, possible. Let me introduce you to Leibowitz's example of the religious father which describes life within different value systems:

> Throughout his life, he [the religious Jew] rises early in the morning to observe the Mitzvah of prayer with the congregation even when he feels no need to "pour out his complaint before God," and perhaps has never felt this need in his life. [...] He prays, it may be, contrary to his perception that there is no relation between his prayer and the events which befall him or occur in the history of Israel. If one day he or one of his children falls ill, he consults the physician and resorts to the science of medicine as would any atheist [...]. To provide for his livelihood he makes use of all the means employed to this end by the atheist [...]. The Jew who regularly prays his daily prayer does so in recognition of the obligation to pray, as an aspect of his obligation to serve God.[30]

In different realms of life, the religious practitioner can choose between different value systems and because he believes in the doctor's knowledge in a moment of illness and despair, his faith stays unaffected (he does not, for example, do service to God or pray in the expectation of good health = *lo' li-šmah*). Faith *li-šmah* and *lo' li-šmah* exist side by side due to religious opportunism. Because not all humans can reach the faith *li-šmah*, only those who follow *Halakhah* can reach this level. As real faith is nothing but the practice of the *Halakhah*, Leibowitz assigns the human interpretation of the Torah and the tradition to the area of *lo' li-šmah*. Also, the messianic vision of salvation is *lo' li-šmah* and is merely a construct to fulfil a psychological need.

Leibowitz's definition of faith as purely the service of God, the distinction between faith *li-šmah* and faith *lo' li-šmah*, and the assumption of co-existing value systems laid the foundations for introducing scepticism and tolerance into his thought. Within Leibowitz's definition of value systems, his axiology will play an important role later in this paper.

2 Jewish Faith and Scepticism

Benbassat presented Leibowitz as a representative of a religious approach "that not only accepts scepticism as an essential feature of religious development, but also regards a lack of scepticism as corruptive for a genuinely faithful position."[31] The comparison to scepticism already creates a problem, as scepticism has been defined differently in different periods and from different groups of people: "If, on the one hand, scepticism makes a forceful case for our knowledge being limited, on the

30 Leibowitz (1977), *Judaism*, 64 f.

31 Benbassat, "Jewish Faith and Scepticism," 161.

other, it refuses every delimitation of 'reality' and implication of authority."[32] In his analysis, Benbassat considered scepticism to be the ancient form of *epochē*. In the ancient interpretation of Sextus Empiricus—to whom he refers—*epochē* is "the state of mind in which one neither denies nor affirms any belief or knowledge—as leading to the most perfect tranquillity of the mind":[33] *ataraxia*. In the context of Yeshayahu Leibowitz, Benbassat reinterprets *epochē* not as the traditional means of achieving *ataraxia*,[34] but as a means "of achieving purified, genuine religious faith,"[35] which he translates, in Leibowitz's case, to the achievement of the highest form of Jewish faith, faith for its own sake (*li-šmah*). A sceptical form of *epochē* hence leads to faith *li-šmah*. The essential element which enables the integration of scepticism in Leibowitz's thought, Benbassat claims, is his religious conception of the Jewish faith as a pure conative determination.[36]

For this analysis, Benbassat uses the *later* definition of Jewish faith in Leibowitz's thought. In his later thought, as described above, Leibowitz does not draw any distinction between faith as a "superstructure rising above the *mitzvot*"[37] and the practice of the *mitzvot* itself. He states that faith is this very engagement, or, as Benbassat notes, "one's commitment to live under the constraints of religious practice."[38] This definition, Benbassat claims, does not even depend on prior acknowledgement of God's existence, because religious practice is "prior to any cognitive and emotional aspect of Jewish religiosity."[39] That means that faith in Judaism "does not rely on the cognitive determination (belief or knowledge) of God's existence."[40] Benbassat concludes with the assumption "that a religious practitioner is allowed (in Judaism) to be sceptical with regard to God's existence."[41] He continues that "in Leibowitz's view, it is the *Halakhah* that posits God as its source and not vice versa."[42] In this interpretation, a Jewish practitioner is allowed to avoid affirming or denying the existence of God and other assumptions with regard to God's nature and deeds. He is also allowed to avoid affirming or denying God's creation of the

32 Giuseppe Veltri, *Alienated Wisdom: Enquiry into Jewish Philosophy and Scepticism* (Berlin and Boston: De Gruyter, 2018), 286 f.
33 Benbassat, "Jewish Faith and Scepticism," 161. He is referring to Sextus Empiricus, *The Skeptic Way: Sextus Empiricus's Outlines of Pyrrhonism*, trans. Benson Mates (New York: Oxford University Press, 1996), 3.
34 See Sextus Empiricus, *Outlines of Pyrrhonism*, 3.
35 Benbassat, "Jewish Faith and Scepticism," 161.
36 See Benbassat, 163.
37 Leibowitz (1953), *Judaism*, 6.
38 Benbassat, "Jewish Faith and Scepticism," 164.
39 Leibowitz (1953), *Judaism*, 4, in Benbassat, "Jewish Faith and Scepticism," 164.
40 Benbassat, 165.
41 Benbassat.
42 Benbassat.

world as written in the Bible.[43] For *epochē*—as the chosen sceptical approach—means "neither to affirm nor to deny it—as long as it is a matter of mere cognition."[44] Because this form of *epochē* is a means "of achieving purified, genuine religious faith,"[45] this sceptical approach, Benbassat states, might not only be valid, but may even be necessary for reaching faith *li-šmah.*[46] Ultimately, even if a person trusts the words written in the Bible, this belief would not have any religious significance within Jewish faith, because the only *essential* feature of the Jewish faith is the adherence to its practice, which does not rely on any cognition at all. Thus, "faith remains indifferent to any cognitive data"[47] and therefore, the Jewish religion tolerates several viewpoints.[48] The same attitude counts in regard to the origin of the Scripture or "religious" events such as the Israelites' reception of the law from God on Mount Sinai or divine providence, because "the world and all it contains are insignificant before God."[49] Within this interpretation, a man of faith can—and must—be sceptical.

In his analysis, Benbassat stresses Leibowitz's assumption of truth. The truth value of the proclamation of the laws is not affirmed as a sheer fact, but as a matter of individual will: "That the *law* is rendered divine by one's willingness to accept it as such."[50] And for Leibowitz, this feature generally applies to the realm of any moral imperatives or duties: the validity is established only by one's subjective choice. "Truth" in the realm of values is a matter of personal intention. The truth of science is not a value, and "[v]alues are not anchored in reality. They are what man aspires to impose upon reality."[51] Benbassat states:

> A man of faith can be sceptical with regard to the assumption of the Halakhah as a divine command, considered as a matter of cognition, but he can still affirm this assumption within the realm of values, in which assumptions do not determine whether something is objectively true, or whether something is "real," but rather reflect one's subjective determination to impose his or her values upon reality. In the realm of values, the truth value of our assumptions does not refer to the state of affairs in reality, but to one's will and intention.[52]

As the Jewish faith is a consequence of a subjective irrational value determination and is independent of any knowledge or truth claim, scepticism can be seen as a natural companion of faith and therefore fights fundamentalism regarding the belief

---

43 Also, he claims that we are in any case unable to understand the real meaning of the first section of the Bible. Therefore, as a matter of cognition, there is nothing to affirm or to deny.
44 Benbassat, "Jewish Faith and Scepticism," 165.
45 Benbassat, 161.
46 See Benbassat, 165.
47 Benbassat, 169.
48 See Benbassat.
49 Leibowitz (1988), *Judaism*, 45.
50 Benbassat, "Jewish Faith and Scepticism," 174.
51 Leibowitz (1976), *Judaism*, 139; see Benbassat, "Jewish Faith and Scepticism," 174.
52 Benbassat, 175.

in God's creation of the world or the perception of the Torah as a divine command. By accepting fundamental truth claims as true, one renders religion a service to oneself, and in Leibowitz's definition, any end besides the worship of God is idolatry. Therefore, scepticism towards general beliefs in Judaism "is not only possible, but also crucial for the maintenance of a genuine religious position,"[53] which means a "pure" Jewish faith for the sake of itself (*li-šmah*).

3 From Scepticism to Tolerance

Just as "religious faith is generally perceived as being in contrast with sceptical approaches," so it is the same with tolerance. The question is therefore how to determine tolerance for someone who has absolute faith and for whom the possibility of being mistaken in this faith does not arise. Gabriel Marcel states that for someone who has absolute faith, it is insufficient "to determine the principle on which we can base a tolerance which is really a counter-intolerance, but which is not, at the same time, the expression or mark of a complete skepticism, but rather the living incarnation of a faith."[54] The root of tolerance in a believer's stance, he states, is "an expression of love's mediation of the 'blinded consciousness' of unbelief and God's salvific end."[55] For a believer, serving the divine will means "to act as a mediator between it and the other consciousness whom I assume is blinded."[56] With this definition, Gabriel charts a course into the understanding of tolerance as a theological virtue.[57] Alan Udoff introduces another theological concept of tolerance as an active "waiting for the other": "In this reception, the other is received as (potentially) en route to the point that has already been attained by the one who receives."[58] As fullness and power lie in faith, the faithful are waiting for the enrouted, who belongs to the future, and therefore tolerance is placed into the realm of hope.[59] These may be two valid and plausible definitions of religious tolerance. However, as we already stated, the theological and metaphysical elements of the Jewish religion have been entirely removed from Yeshayahu Leibowitz's thought. Hence, for Leibowitz, theological concepts of tolerance will not work and tolerance cannot be considered a theological virtue; therefore, it seems to be a logical consequence to consider that Leibowitz's

---

53 Benbassat, 180.
54 Gabriel Marcel, "The Phenomenology and Dialectic of Tolerance," in *Creative Fidelity*, trans. Robert Rosthal (New York: Farrar Straus Giroux, 1970), 216.
55 Alan Udoff, "Tolerance," in *20th Century Jewish Religious Thought: Original Essays on Critical Concepts, Movements, and Beliefs*, ed. Arthur A. Cohen and Paul Mendes-Flohr (Philadelphia: First JPS edition, 2009; originally published as *Contemporary Jewish Religious Thought: Original Essays on Critical Concepts, Movements, and Beliefs* [New York: Scribner, 1987]), 992.
56 Marcel, "Phenomenology and Dialectic of Tolerance," 216.
57 See Udoff, "Tolerance," 993.
58 Udoff, 993.
59 Udoff.

concept of tolerance originated in the sceptical elements of his thought. This is especially reasonable because some religious thinkers have already been discussed within the framework of scepticism and tolerance, such as, for example, Montaigne.[60]

Now, introducing the term "tolerance" is another difficulty, as tolerance—like scepticism—is *not* clearly defined. Orienting myself using the distinction between "concept" and "conceptions of justice" by John Rawls,[61] I distinguish between the *concept* of tolerance in general and different *conceptions* in particular. The concept of tolerance therefore constitutes an umbrella term rising above a vast spectrum of conceptions, which existed in different places and times and which can exist in the same place and time all at once. The scale of conceptions of toleration[62] reaches from a very pragmatic or negative standpoint (bearing, suffering) at one end of the scale to an almost utopian or positive standpoint (welcoming and support) on the other.[63] While this paper gives no room for a deeper analysis of the specific conception (or conceptions) of tolerance in Leibowitz's thought, I will examine general indications of tolerance. Researchers sometimes recognise the distinction between a

---

60 See Alan Levine, "Skepticism, Self and Toleration in Montaigne's Political Thought," in *Early Modern Skepticism and the Origins of Toleration*, ed. Alan Levine (Lanham, MD: Lexington Books, 1999), 51–76.

61 See John Rawls, *A Theory of Justice* (Cambridge, MA: Belknap Press, 1971).

62 Some people differentiate between "tolerance" as theory and "toleration" as the distinct act of tolerance, but I see no benefit in this.

63 This paper has no room to explain all the historical definitions of the term and the meaning of tolerance, and there is no history or "encyclopaedia of tolerance" to which I can refer, but I will give a very short insight into its current trends in this footnote. The spectrum of the modern definitions of tolerance stretches from sheer "bearing" (see John Horton, "Why the Traditional Conception of Toleration Still Matters," *Critical Review of International Social and Political Philosophy* 14, no. 3 [2011]: 289–305), through a "respect concept" of tolerance (see Rainer Forst, ed. *Toleranz: Philosophische Grundlagen und gesellschaftliche Praxis einer umstrittenen Tugend* [Frankfurt and New York: Campus Verlag, 2000]; Rainer Forst, *Toleranz im Konflikt: Geschichte, Gehalt und Gegenwart eines umstrittenen Begriffs* [Berlin: Suhrkamp Verlag, 2003]; Matthias Kaufmann, "Einleitung," in *Integration oder Toleranz? Minderheiten als philosophisches Problem*, ed. Matthias Kaufmann [Freiburg and Munich: Karl Alber, 2001]: 11–30), to the point of "recognition" (see Anna Elisabetta Galeotti, *Toleration as Recognition* [Cambridge: Cambridge University Press, 2002]; David Heyd, "What Toleration Is Not," paper presented at the "Tolerance and Intolerance as Challenge in Past and Present" conference, Universität Hamburg, March 27, 2019, https://lecture2go.uni-hamburg.de/l2go/-/get/v/24408) and even "welcoming and support" (see Monika Kirloskar-Steinbach, *Toleranz im interkulturellen Kontext* [Nordhausen: Traugott Bautz, 2005]; Susan Mendus, "The Changing Face of Toleration," *Reset DOC*, June 26, 2014, https://www.resetdoc.org/story/the-changing-face-of-toleration/). As an all-encompassing concept, tolerance is an object of concern in very different fields of research including sociology, politics, theology, philosophy, philology, and religious or migration studies and therefore the definitions differ radically. Cary J. Nederman points out that "the myth of liberal 'uniqueness'" should be broken and complains about the "historical and geographical rigidity of the main stream of toleration scholarship." For this reason, he gives some "alternative theoretical approaches to the rights-laden, modern, Western discourse of liberalism": Cary J. Nederman, "Toleration in a New Key: Historical and Global Perspectives," *Critical Review of International Social and Political Philosophy* 14, no. 3 (2011): 351.

"public" or "political" and an "individual" agency of tolerance, although they often ignore it. Although Leibowitz gives enough material to support an examination of the degree of public or political tolerance in his descriptions of a modern democratic Jewish state, in this analysis, I will look at the individual degree of tolerance of "the other" within the context of a liberal democratic state. Before I begin, I state that tolerance is not a value determination itself, but a strategy of agency, and the degree of tolerance is *dependent* on other value determinations, such as, for example, faith, ethics, or any other value determination a person is willing to follow.[64] As mentioned before, there exists a long tradition of connecting between scepticism and tolerance, defining tolerance as the "child of doubt."[65] However, the history of this connection will not be the topic of this paper.[66]

The Creative Halakhah

As Leibowitz himself never examined the issue of tolerance himself (unlike, for example, Montaigne), we have to look for "implicit" indications of tolerance in his thought. In my earlier research, I discovered some interesting features of Leibowitz's philosophy that leave room for a tolerant attitude. One of these indications is his stance concerning the flexibility of the *Halakhah*. In his opinion—and against an orthodox point of view—the *Halakhah* has to be "creative" in order to adjust to contemporary society.[67] Because the "'Torah state' is not a reality—neither in the historical past, nor in the present or in any foreseeable future,"[68] the principles of the *Halakhah* required restatement. But Leibowitz was sceptical about whether the religious elite was capable of prescribing ways of life for the new politically independent community. He criticised them for being narrow-minded and strongly called on them to seek new rules for Jewish religion. He denied the ideas of the Reform movement and called on the *halakhic* community to invent new answers to upcoming questions when the religious elites failed to do so. The reason for this "creative" *halakhic* approach might come from Leibowitz's unbroken desire for the Jewish religion to survive in the modern world. One example of the "creative *Halakhah*" is his attitude regarding the rights of women:

64 See, for example, Hamid Reza Yousefi, "Tolerance und Individuum," in *Toleranz im Weltkontext: Geschichten—Erscheinungsforme—Neue Entwicklungen*, ed. Hamid Reza Yousefi and Harald Seubert (Wiesbaden: Springer, 2013), 221; Nederman, "Toleration in a New Key."

65 Susan Mendus, "Introduction," in *Justifying Toleration: Conceptual and Historical Perspectives*, ed. Susan Mendus (Cambridge: Cambridge University Press, 1988), 6.

66 For the evident connection between scepticism and tolerance, see, for example, Levine, ed., *Early Modern Skepticism and the Origins of Toleration.*

67 See Goldman, "Introduction," xxiv.

68 Yeshayahu Leibowitz, "People, Religion and the State" [Hebrew], 154, in Hellinger, "A Clearly Democratic Religious-Zionist Philosophy," 278.

> Nowadays even our religious society is part of a world in which all political and public issues are shared by men and women. Thus it is "the way of women" to participate in public affairs. Jewish religious society will not be able to survive if, for pseudo-religious reasons, we continue to deprive women of their due rights. This is the point at which we—those of us resolved to practice Torah—cannot perpetuate the halakhic decisions of our fathers dating from a social reality which differed radically from our own.[69]

Leibowitz also follows this sceptical stance concerning the study of the Torah and the Talmud. He states that the Jewish religious community cannot refuse to allow women to study Torah and Talmud, although women's studies are of no religious significance:

> The perpetuation of this attitude within Judaism and the Jewish religion is intolerable in the Jewish world of today. The religious Jewish public to whom these remarks are addressed belongs to a society whose culture is common to both men and women. Such is our mode of Jewish existence as well. Therefore, barring women today from Talmud Torah segregates Judaism from the spiritual reality shared by Jews of both sexes. [...] What is really important is that Talmud Torah should be available to the entire religious community, the duty and privilege of both men and women.[70]

To summarise, Leibowitz is sceptical concerning the fathers' *halakhic* decisions, which date from a social reality that is radically different from his own, and calls for a more tolerant up-to-date approach and the restatement of the *Halakhah*. Leibowitz's tolerant approach concerning homosexuality was also well-known. He declared that "homosexuals were members of the Jewish community like all other Jews, and should strive to follow the commandments, even if there were some that they could not abide by."[71]

Co-existence of Valid Value Systems

Rejecting the roots of a sceptical approach, Avi Sagi nevertheless found evidence for a so-called pluralism in Leibowitz's thought. He raises the question of whether the Jewish religion can develop tolerance for non-religious Jews without sacrificing fun-

---

69 Leibowitz (1980), *Judaism*, 130 f.

70 Leibowitz (1980), *Judaism*, 129–30; see also Leibowitz and Michael Shashar, *Gespräche über Gott und die Welt*, 3rd ed. (Frankfurt am Main: Insel Verlag, 1994), 163 f. It is necessary to note Tamar Ross's criticism of the distinctions and limits that Leibowitz makes in this argumentation: see Tamar Ross, "The Status of Woman in Judaism: Several Reflections on Leibowitz's Conception of the Mechanism by Which *Halakha* Adjusts to Reality" [Hebrew], in *Yeshayahu Leibowitz: His World and Philosophy*, ed. Avi Sager (Jerusalem: Keter, 1995), 148–62.

71 Yaakov Ariel, "Gay, Orthodox, and Trembling: The Rise of Jewish Orthodox Gay Consciousness, 1970s–2000s," *Journal of Homosexuality* 52, no. 3/4 (2006/7): 96–97.

damental traditions.[72] In Sagi's interpretation, tolerance is *not* a product of relativism or scepticism, since man can only tolerate what he rejects.[73] Instead, he recognises two other reasons for tolerance: a *utilitarian* reason and a *value-based* reason. According to Sagi, Leibowitz offers the answer of tolerance through his separation between a normative religious value system and co-existent non-religious value systems, as mentioned before. Leibowitz's acceptance of a pluralistic Israeli society containing religious practitioners and non-religious Jews hence arises from two assumptions. First, he assumes that cultures are always subject to dynamic processes. Thus, Jewish society is always developing and a culture is not only defined by its religious members. Secondly, Leibowitz develops a form of tolerance that is not based on excluding others as "alien," but rather exclusively transfers the "foreignness" to the religious realm, while all other realms are shared:

> Our commitment to our values is not measured by the extent of our rejection of the other's worlds but by the willingness to endorse these values consistently in our life and by our unwavering commitment to them. Loyalty to our values is related to our disposition toward them and not necessarily to their cognitive superiority.[74]

Sagi can take this stance because Leibowitz formulates a co-existence of parallel value systems that have no general hierarchy and are only ordered through individual preference. The choice of one value system is not the negation of another, for the other value system may indeed be void for the selecting subject, but important for another individual who, because of another tradition or cultural identity, would opt for it in this very situation. This approach is underpinned through Leibowitz's axiology. Leibowitz conveys a non-cognitive approach to the adoption of values. He claims that values cannot be rationally anchored; they are instead the result of subjective irrational decisions.[75] Because the choice of one value system is absolutely non-cognitive, all value systems have to be held as valid. Yonatan Brafman calls this parallelism of value systems the "relativistic 'polytheism' of competing values."[76] Through this theory of genuine value conflict and the subjectivity of axiological decisions, Leibowitz leaves room for tolerance. Let us look at an example where Leibowitz adapts the theory of genuine value conflict. In an interview with the *Journal of Palestine Studies* in 1989, he stated: "We have the fact of two peoples, each of them believing in their heart that this country is theirs. Therefore, it's either partition or

---

72 See Sagi, *Jewish Religion after Theology*, 4.

73 See Sagi, 8.

74 Sagi, 63.

75 See Yonatan Brafman, "Yeshayahu Leibowitz's Axiology. A 'Polytheism of Values' and the 'Most Valuable Value,'" *Journal of Religious Ethics* 43, no. 1 (2015): 146–68.

76 Brafman, 165. However, Brafman reveals a paradox in Leibowitz's axiology: while affirming a relativistic polytheism of competing values, Leibowitz nonetheless implies that the religious value is the "most valuable value."

war to the finish."[77] This stance comes from the assumption that values cannot be rationally adjudicated and this leads to a tolerance of others' value decisions. Further: "The existence of the 'other' cannot be denied"[78] and when "Palestine consider themselves to be a people, [...] that is the decisive point."[79] This tendency shows more tolerance for "the other" embodied in the Muslim society of Israel and Palestine than is found in most Jewish Orthodox thinkers.[80]

Avi Sagi recognises Leibowitz as not only a tolerant thinker, but perhaps even a pluralistic one.[81] In contrast to Sagi's distinction between scepticism and tolerance, there are clear correlations between the sceptical approach and the "value-based reason" that Avi Sagi wants to separate from scepticism. Nonetheless, it is important to stress here that his assumption of a plurality of value systems allows for tolerance among different value orientations, but does not allow for pluralism *within* value orientations. Though Leibowitz strongly suggests changes within the *halakhic* system, he does however condemn non-Orthodox *halakhic* change as inauthentic.[82] As the religious elites failed to answer the questions of statehood, it was the *halakhic* community's task to provide these answers. Leibowitz was against the changes of Reform Judaism, which "regards the Halakhah as a husk hiding the essential core of religion."[83] Though Leibowitz is an axiological pluralist, an individual can only have one absolute value.

Concerning my definitions of negative and positive conceptions of tolerance, I place Leibowitz within the tradition of tolerance as a positive and even supportive conception. The negative or pragmatic conceptions of tolerance stem from the premise that one can only tolerate what one rejects and that toleration is "an only grudging acceptance of [...] difference," which means an "acceptance of it only as a practical expedient."[84] This pragmatic conception of tolerance was often associated with the politically instituted toleration during the French religious wars of the sixteenth century.[85] Leibowitz, in contrast, even supported "the difference" in different religious and social paths. Because of his sceptical approach, Leibowitz was open "to

77 Yeshayahu Leibowitz and John P. Egan, "Yeshayahu Leibowitz: Liberating Israel from the Occupied Territories," *Journal of Palestine Studies* 15, no. 2 (1986): 105.

78 Leibowitz and Egan, 107.

79 Leibowitz and Egan, 105.

80 This tendency towards tolerance should be further examined because it seems to me that this tolerance stems from a more pragmatic conception of tolerating others in order not to be punished for these events later. See Goldman, "Introduction," xxxf.

81 See Sagi, *Jewish Religion After Theology*, 52f.

82 I thank Yonathan Brafman for this remark.

83 Goldman, "Introduction," xvii.

84 Alan Levine, "Introduction: The Prehistory of Toleration and Varieties of Skepticism," in *Early Modern Skepticism and the Origins of Toleration*, 6.

85 See Levine.

the possibility that the thing in question, whether it be an idea or practice, may represent some truth or good."[86] With this stance, he actually defends a liberal attitude.

4 Another Approach to Scepticism in Leibowitz's Thought

In this analysis so far, we have asserted that scholars have already identified Leibowitz as a sceptical thinker. We also asserted that scepticism is a route to tolerance and that this approach is suitable for finding arguments in support of tolerance in Leibowitz's thought. Afterwards, we looked at examples of tolerance of "the other" in Leibowitz's thought. However, we have not revealed which sceptical approach leads to which conception of tolerance, as "toleration cannot be founded on pure scepticism, [and] all the paths which flow from scepticism do not necessarily lead to toleration."[87] In his monograph *Alienated Wisdom*, Giuseppe Veltri states: "There are two possible political consequences of a sceptical attitude: either it aims to avoid every constriction of society and to adopt a completely *ataraxic* attitude, or it accepts every possible and plausible solution without precluding anything."[88] Clearly, Veltri is here referring to the distinction between the different schools of scepticism made by Sextus Empiricus in book 1 of *Outlines of Pyrrhonism*.[89] Veltri further states: "I would call the first radical Pyrrhonism which does not support any kind of social life; the second is the basis of tolerance as the limit of behaviour in a multifarious cultural society."[90] With this assumption in mind, I want to revoke Benbassat's conclusion that Leibowitz was a Pyrrhonian sceptic. I want to show that the claim of Pyrrhonian scepticism, compared to the ancient attitude of *epochē* and the "suspension of judgement," should be replaced by the more plausible attitude that Leibowitz was someone who "accepts every possible and plausible solution without precluding anything."[91] I do not think that Leibowitz's objective was a "state of mind in which one neither denies nor affirms any belief or knowledge," aiming at faith *li-šmah*. I think that Leibowitz indeed supported people *believing* or *not believing* in things and individual truth and that he did not advocate "suspension of judgement." As Benbassat states: "Whether one believes in the cosmological meaning of the opening passage of the Bible or prefers scientific theories

86 Levine, 7. Levine associated a positive sceptical approach with "openness" and contrasts "openness" to "toleration in a strict sense." I define an open approach as the positive conception of toleration/tolerance.
87 Levine, 2.
88 Veltri, *Alienated Wisdom*, 287; see also Veltri, "The Limits of Scepticism and Tolerance," 41.
89 See, for example, Sextus Empiricus, *Outlines of Pyrrhonism*, especially chapters 1–12 and 29–34.
90 Veltri, "The Limits of Scepticism and Tolerance," 41.
91 Veltri, 41.

in this respect, the Jewish religion, in principle, tolerates both viewpoints."[92] And further: "This tolerant characteristic with regard to cognitive assumptions is rendered possible by acknowledging that the only essential feature of the Jewish religion is the adherence to its practice, which, as stated, does not rely on any cognition."[93] Faith, in Leibowitz's thought, is hence *not dependent* on the "suspension of judgement," but fully independent of any individual judgement, even individual or collective truth claims, for the Jewish faith is nothing but the practice of the *Halakhah* and has no cognitive meaning at all. One may "conduct [oneself] in accord with the plausible [and] employ it in everyday life" and one does not have to "live without belief, following the laws, customs, and [...] natural pathé."[94] Hence, neither scepticism nor one's judgments and beliefs "jeopardise faith,"[95] as faith is isolated from any cognition.

Leibowitz is not a Pyrrhonist. He does indeed note the multiplicity of views on the subject under consideration, as the Pyrrhonists do, and he likewise seeks to keep an open mind. But unlike the Pyrrhonists, who do not know whether they can know anything or not and ultimately suspend judgement, Leibowitz had true concerns and beliefs about human problems such as the responsibility of religious Jewry, the duty of education, or the ideal politics of an Israeli state. In Leibowitz's thought, *we*, the people, know: "But the little we know, we know."[96] And even in the realm of Jewish faith, which is independent of every cognition and meaning in the world, the religious practitioner at least *knows* that God is "the giver of the Torah."[97] As a scientist and a philosopher, Leibowitz himself had true beliefs outside Jewish faith and could not ignore or suspend prevailing problems such as what the character of the new Jewish society would be. Leibowitz even supported individuals having their own beliefs and values in different realms of life and accepting "every possible plausible solution without precluding anything."

For this assertion, I will repeat the example of the religious father:

> If one day he or one of his children falls ill, he consults the physician and resorts to the science of medicine as would any atheist [...]. To provide for his livelihood he makes use of all the means employed to this end by the atheist [...]. The Jew who regularly prays his daily prayer does so in recognition of the obligation to pray, as an aspect of his obligation to serve God.[98]

To provide for his livelihood and that of his children, the religious practitioner uses all the means that the atheist employs to this end. He believes in the doctor's abilities, and because life gives him the choice between several value systems, he

---

92 Benbassat, "Jewish Faith and Scepticism," 169.
93 Benbassat, 169.
94 Sextus Empiricus, *Outlines of Pyrrhonism*, 33.
95 Benbassat, "Jewish Faith and Scepticism," 180.
96 Leibowitz, *Judaism* (1976), 137.
97 See Leibowitz, *Judaism* (1953), 5. This is a known paradox in Leibowitz's philosophy.
98 Leibowitz, *Judaism* (1977), 64 f.

does not fall into idolatry by sanctifying the religious realm. Within this very division between the religious realm and all other realms of life, only the "purified, genuine religious faith,"[99] aiming at faith *li-šmah*, is possible. Leibowitz hence believed in scientific knowledge and individual truth claims, but for him, religion regards this knowledge with indifference.[100]

For the sake of God as an absolute value, "anything in the world might be renounced, for anything else is only relatively, conditionally valuable"[101]—for which Abraham and the *ʿAqedah* serve as the best example. This attitude does not negate evaluability (nihilism), but strengthens an individualistic perspective of truth. Besides the "instrumental truth," the truth of science, which is "a datum within science,"[102] "truth" in the realm of values—as stated before—is a matter of personal intention. A value determination is hence an individual decision. "A person's understanding and feeling derive from his own subjectivity, which differs from anyone else's,"[103] and this counts for "the set of beliefs and the religious experience"[104] as much as for "consciousness, sensibility, and valuation [as] radically private."[105] Leibowitz hence does not claim, as the Pyrrhonists do, that we do not know that we know, but rather that every person holds his "truth" for himself. Leibowitz does not negate truth, nor does he ascribe truth to faith. As Goldman states, for Leibowitz, it is "absurd to regard revelation as a surrogate or supplement for natural knowledge."[106]

Conclusion

By shifting the Jewish religion from a metaphysical theological realm to a strictly normative realm, defining the Jewish faith as a closed normative realm co-existing with other closed normative realms, and his definition of individual truth, Leibowitz laid the foundation of a plausible form of scepticism and tolerance. The key concept is the premise that there is no universal hierarchy between value systems, but that

99 Benbassat, "Jewish Faith and Scepticism," 161.
100 See Judaism (1976), 141. Here, Leibowitz is referring to Maimonides's writings, particularly the *Guide of the Perplexed* and the *Codex Maimoni.*
101 Benbassat, "Jewish Faith and Scepticism," 170.
102 Leibowitz, *Judaism* (1976), 139.
103 Leibowitz, *Judaism* (1953), 9.
104 Leibowitz, *Judaism* (1953), 10; see also Goldman, "Introduction," xv.
105 Leibowitz, *Judaism* (1953), 9.
106 Goldman, "Introduction," xiv. Religious thinkers have often been discussed as sceptic fideists. Fideists believe that the only truth available to us can be revealed by foregoing rational inquiry and relying solely on faith. Leibowitz is no fideist either. See Levine, "Scepticism, Self and Toleration," 54; Richard Amesbury "Fideism," in *The Stanford Encyclopedia of Philosophy* (Fall 2017 Edition), ed. Edward N. Zalta. https://plato.stanford.edu/archives/fall2017/entries/fideism/.

the human prioritisation of the value system is time-, place-, and need-bound.[107] As the value determinations cannot be rationally adjudicated, all of them are *valid*. The choice of a value system in one particular situation does not negate the other value systems, but allows them to exist in parallel, for "truth" in the realm of values is a matter of personal intention. There is a constant and genuine conflict of values, which the individual faces in every new decision in his life. Leibowitz's pluralistic definition is understandable, for he found himself in the difficult situation of filling his *own* value system with inherent value and of giving legitimacy to this value system within a modern society and politics.[108] Thereby, he not only followed his aim of strengthening Jewish faith within a modern Israeli state, but he also left room for a positive and supportive form of tolerance as a form of "accepting the other."

This paper has shown that scepticism may be found in Leibowitz's thought and that this sceptical attitude can be considered as a route that leads to tolerance. But while Benbassat concentrated on the so-called Pyrrhonian tradition of scepticism (*epochē* and "suspension of judgement"), I assume that, rather, the clearly sceptical methodology of Yeshayahu Leibowitz's thought requires further analysis. We should examine the object of "methodical doubt at the beginning of every philosophical debate and treatise concerning all aspects of human life or, alternatively, an epistemological attitude which doubts the possibility of knowledge."[109] In the analysis of tolerance in Leibowitz's thought, there are further questions to be dealt with. Where there has been a development from the "earlier Leibowitz" to the "later Leibowitz," there might also be a development within his sceptical and tolerant attitude. At the very least, the "target audience" of scepticism and tolerance could have shifted from one group to another. In the history of tolerance, its blind spots have already been emphasised by Christian Laursen in his paper "Blind Spots in the Toleration Literature," in which he posits John Milton, Benedict Spinoza, Denis Veiras, John Locke, and Pierre Bayle as examples of the limitations of tolerant thinking.[110] Tolerance, in his interpretation, always goes hand in hand with intolerance. Clearly, one will find blind spots (for example, Christianity externally, Reform Judaism internally) in Leibowitz's thinking too, which should further be examined. Ultimately, Yeshayahu Leibowitz was popular for his harsh criticism of the Israeli government. Regarding the room for tolerance and individual truth in his thought, one may examine the political consequences that Leibowitz suggested for the state of Israel and investigate his ideas of an ideal Israeli state.

In "The Limits of Scepticism and Tolerance," Giuseppe Veltri refers to a BBC broadcast given by Bertrand Russell in which he defines scepticism as a powerful means of combating fanaticism, stating that the alternative to fanaticism is a rational

---

107 See Sagi, *Jewish Religion after Theology*, 49 f.

108 See Sagi, 46.

109 Veltri, *Alienated Wisdom*, 287.

110 John Christian Laursen, "Blind Spots in the Toleration Literature," *Critical Review of International Social and Political Philosophy* 14, no. 3 (2011): 307–22.

scepticism that is open to scientific discovery: “Fanatics are unwilling to accept scientific discoveries made by their enemies, and therefore soon fall behind those whose outlook is more cosmopolitan.”[111] Scepticism is therefore considered as a strategy which leads to a form of success. In this methodology, Leibowitz’s definition of faith leads to a sceptical attitude that can be seen as a strategy for developing a tolerant attitude towards “the other.” The statement that “fanatics are unwilling to accept scientific discoveries [...] and therefore soon fall behind” could also have been made by Yeshayahu Leibowitz, whose aim was to examine a new role of religious Jewry in the modern Israeli state in order to avoid “falling behind.” He defines this new role as a flexibility, which—for example—shows itself in the modification of a “creative” *Halakhah* to strengthen Jewish faith in a modern secular society.[112]

Bibliography

Amesbury, Richard. “Fideism.” In *The Stanford Encyclopedia of Philosophy* (Fall 2017 Edition), edited by Edward N. Zalta. https://plato.stanford.edu/archives/fall2017/entries/fideism/.

Ariel, Yaakov. “Gay, Orthodox, and Trembling: The Rise of Jewish Orthodox Gay Consciousness, 1970s – 2000s.” *Journal of Homosexuality* 52, no. 3/4 (2006/7): 91–109.

Benbassat, Roi. “Jewish Faith and Scepticism—The Example of Yeshayahu Leibowitz.” In *Yearbook of the Maimonides Centre for Advanced Studies* 2017, edited by Bill Rebiger, 161–81. Berlin/Boston: De Gruyter, 2017.

Benbassat, Roi. “Yeshayahu Leibowitz: Jewish Existentialism.” *Religious Studies* 51, no. 2 (2015): 141–63.

Black, Sam. “Locke and the Skeptical Argument for Toleration.” *History of Philosophy Quarterly* 24, no. 4 (2007): 355–75.

Brafman, Yonatan. “Yeshayahu Leibowitz’s Axiology. A ‘Polytheism of Values’ and the ‘Most Valuable Value.’” *Journal of Religious Ethics* 43, no. 1 (2015): 146–68.

Empiricus, Sextus. *The Skeptic Way: Sextus Empiricus’s Outlines of Pyrrhonism.* Translated by Benson Mates. New York: Oxford University Press, 1996.

Forst, Rainer. *Toleranz im Konflikt: Geschichte, Gehalt und Gegenwart eines umstrittenen Begriffs.* Frankfurt am Main: Suhrkamp Verlag, 2003.

Forst, Rainer, ed. *Toleranz: Philosophische Grundlagen und gesellschaftliche Praxis einer umstrittenen Tugend.* Frankfurt and New York: Campus Verlag, 2000.

Fraisse, Ottfried. “Ent-Geisterung der Rheologie. Zur Axiologie von Yeshayahu Leibowitz‘ Denken.” In *Alles Wirkliche Leben ist Begegnung. Festschrift zum vierzigjährigen Bestehen von Studium Israel e.V.*, edited by Johannes Ehmann, Joachim J. Krause, and Bernd Schröder, 305–17. Leipzig: Evangelische Verlagsanstalt, 2018.

Galeotti, Anna Elisabetta. *Toleration as Recognition.* Cambridge: Cambridge University Press, 2002.

Goldman, Eliezer. “Introduction.” In Yeshayahu Leibowitz, *Judaism, Human Values and the Jewish State*, edited by Eliezer Goldman, translated by Eliezer Goldman, Yoram Navon, Zvi Jacobson,

111 Bertrand Arthur William Russell, “Why Fanaticism Brings Defeat,” *The Listener* 40 (September 23, 1948): 452–53, in Veltri, “The Limits of Scepticism and Tolerance,” 39. This text is a transcript from a BBC broadcast.

112 I would like to thank the anonymous reviewers as well as Yonatan Brafman and Yoav Meyrav for their helpful suggestions in refining this essay.

Gershon Levi, and Raphael Levy, vii–xxxiv. Cambridge, MA and London: Harvard University Press, 1992.

Hellinger, Moshe. "A Clearly Democratic Religious-Zionist Philosophy: The Early Thought of Yeshayahu Leibowitz." *Journal of Jewish Thought and Philosophy* 16, no.2 (2008): 253–82.

Heyd, David. "What Toleration Is Not." Paper presented at the "Tolerance and Intolerance as Challenge in Past and Present" conference, Universität Hamburg, March 27, 2019. https://lecture2go.uni-hamburg.de/l2go/-/get/v/24408.

Horton, John. "Why the Traditional Conception of Toleration Still Matters." *Critical Review of International Social and Political Philosophy* 14, no. 3 (2011): 289–305.

Kasher, Hannah. "On Yeshayahu Leibowitz's Use of Religious Terminology." *Journal of Jewish Thought and Philosophy* 10 (2000): 27–55.

Kaufmann, Matthias. "Einleitung." In *Integration oder Toleranz? Minderheiten als philosophisches Problem*, edited by Matthias Kaufmann, 11–30. Freiburg and Munich: Karl Alber, 2001.

Kirloskar-Steinbach, Monika. *Toleranz im interkulturellen Kontext.* Nordhausen: Traugott Bautz, 2005.

Laursen, John Christian. "Blind Spots in the Toleration Literature." *Critical Review of International Social and Political Philosophy* 14, no. 3 (2011): 307–22.

Leibowitz, Yeshayahu. *Accepting the Yoke of Heaven: Commentary on the Weekly Torah Portion.* Translated by Shmuel Himelstein. Jerusalem: Urim Publications, 2002.

Leibowitz, Yeshayahu. *Faith, History, and Values.* [Hebrew.] Jerusalem: Academon, 1982.

Leibowitz, Yeshayahu. *Judaism, Human Values, and the Jewish State.* Edited by Eliezer Goldman. Translated by Eliezer Goldman, Yoram Navon, Zvi Jacobson, Gershon Levi, and Raphael Levy. Cambridge, MA and London: Harvard University Press, 1995.

Leibowitz, Yeshayahu. "Maimonides: The Abrahamic Man." *Judaism* 6, no. 2 (1957): 148–54.

Leibowitz, Yeshayahu. *The Faith of Maimonides.* Translated by John Glucker. Tel-Aviv: MOD Books, 1989.

Leibowitz, Yeshayahu. *Torah and Commandments in Our Times.* [Hebrew.] Tel Aviv: Schocken, 1954.

Leibowitz, Yeshayahu. *Vorträge über die Sprüche der Väter: Auf den Spuren des Maimonides.* Translated by Grete Leibowitz. Obertshausen: Context-Verlag, 1984.

Leibowitz, Yeshayahu, and John P. Egan. "Yeshayahu Leibowitz: Liberating Israel from the Occupied Territories." *Journal of Palestine Studies* 15, no. 2 (1986): 102–8.

Leibowitz, Yeshayahu, and Michael Shashar. *Gespräche über Gott und die Welt.* 3rd edition. Frankfurt am Main: Insel Verlag, 1994.

Levine, Alan. "Introduction: The Prehistory of Toleration and Varieties of Skepticicm." In *Early Modern Skepticism and the Origins of Toleration*, edited by Alan Levine, 1–20. Lanham, MD: Lexington Books, 1999.

Levine, Alan. "Skepticism, Self and Toleration in Montaigne's Political Thought." In *Early Modern Skepticism and the Origins of Toleration*, edited by Alan Levine, 51–76. Lanham, MD: Lexington Books, 1999.

Marcel, Gabriel. "The Phenomenology and Dialectic of Tolerance." In *Creative Fidelity.* Translated by Robert Rosthal, 210–21. New York: Farrar Straus Giroux, 1970.

Mendus, Susan. "Introduction." In *Justifying Toleration: Conceptual and Historical Perspectives*, edited by Susan Mendus, 1–19. Cambridge: Cambridge University Press, 1988.

Mendus, Susan. "The Changing Face of Toleration." *Reset DOC*, June 26, 2014. https://www.resetdoc.org/story/the-changing-face-of-toleration/.

Miron, Ronny. "Yeshayahu Leibowitz on the Possibility of Religious Subjectivity: Critique and Defense." In *Watchtower Religionswissenschaft. Standortbestimmung im wissenschaftlichen Feld*, edited by Anne Koch, 95–118. Marburg: Diagonal-Verlag, 2007.

Morgenstern, Matthias. "Jeschajahu Leibowitz." In *Metzler Lexikon jüdischer Philosophen: Philosophisches Denken des Judentums von der Antike bis zur Gegenwart*, edited by Andreas B. Kilcher, Ottfried Fraisse, and Yossef Schwartz, 403–7. Stuttgart: J.B. Metzler, 2003.
Nederman, Cary J. "Toleration in a New Key: Historical and Global Perspectives." *Critical Review of International Social and Political Philosophy* 14, no. 3 (2011): 349–61.
Rawls, John. *A Theory of Justice*. Cambridge, MA: Belknap Press, 1971.
Ross, Tamar. "The Status of Woman in Judaism: Several Reflections on Leibowitz's Conception of the Mechanism by Which *Halakha* Adjusts to Reality." [Hebrew.] In *Yeshayahu Leibowitz: His World and Philosophy*, edited by Avi Sagi, 148–62. Jerusalem: Keter, 1995.
Russell, Bertrand Arthur William. "Why Fanaticism Brings Defeat." *The Listener* 40, September 23, 1948: 452–53. Transcript of a BBC Broadcast.
Sagi, Avi. *Jewish Religion after Theology*. Translated by Batya Stein. Boston: Academic Studies Press, 2009.
Sagi, Avi. "Yeshayahu Leibowitz—A Breakthrough in Jewish Philosophy: Religion without Metaphysics." *Religious Studies* 33, no. 2 (1997): 203–16.
Sagi, Avi, ed. *Yeshayahu Leibowitz: His World and Philosophy*. [Hebrew.] Jerusalem: Keter, 1995.
Sahin, Bican. "A Defense of Toleration on the Basis of Scepticism: The Case of Michel de Montaigne." *Hacettepe University Journal of Economics and Administrative Sciences* 23, no. 2 (2005): 283–304.
Simon, Ernst Akiva. "Are We Israelis Still Jews? The Search for Judaism in the New Society." In *Arguments and Doctrines: A Reader of Jewish Thinking in the Aftermath of the Holocaust*, edited by Arthur A. Cohen, 388–401. New York: Harper & Row, 1970.
Udoff, Alan. "Tolerance." In *20th Century Jewish Religious Thought: Original Essays on Critical Concepts, Movements, and Beliefs*, edited by Arthur A. Cohen and Paul Mendes-Flohr, 987–1006. Philadelphia: First JPS edition, 2009. Originally published as *Contemporary Jewish Religious Thought: Original Essays on Critical Concepts, Movements, and Beliefs*. New York: Scribner, 1987.
Veltri, Giuseppe. *Alienated Wisdom: Enquiry into Jewish Philosophy and Scepticism*. Berlin and Boston: De Gruyter, 2018.
Veltri, Giuseppe. "The Limits of Scepticism and Tolerance." *Bollettino della società filosofica italiana* 2 (2017): 37–56.
Werkner, Ines Jaqueline et al. *Religionen und Demokratie. Beiträge zu Genese, Geltungund Wirkung eines aktuellenpolitischen Spannungsfeldes*. Wiesbaden: VS Verlag, 2009.
Yousefi, Hamid Reza. "Tolerance und Individuum." In *Toleranz im Weltkontext: Geschichten—Erscheinungsforme—Neue Entwicklungen*, edited by Hamid Reza Yousefi and Harald Seubert, 221–29. Wiesbaden: Springer, 2013.

Giuseppe Veltri

Jesuit Strategic Scepticism toward Religious Self-Consciousness and Commonalities: Chinese Muslims, Jews, and Christians in the Eyes of Matteo Ricci (1552–1610) and His "Translator" Nicolas Trigault (1557–1628)

In 1616, a book about an expedition to China was published by the Walloon Jesuit Nicolas Trigault (Trigau(l)tius in Latin), whose Chinese name was Jin Nígé, 金尼閣.[1] This book was a report of the Chinese "expedition," to use his own title, made by the famous Jesuit Matteo Ricci.[2] Ricci's "expedition" is relatively important because 1) it testifies to the Muslim and Christian presence in China in the early modern period and that of the Jews in Kaifeng in the province of Henan in the sixteenth and seventeenth centuries and 2) it gives us a portrait of three "European" or occidental religions (which are referred to as "sects") and their development in the eyes of an intellectual Christian who was a missionary traveller. If the reports could qualify as more or less historically plausible, they could help us to understand religious commonalities between social groups in their past and present and their consciousness of belonging to a religion. Yet Ricci and Trigault pursued different goals, revealed by their different styles and strategies, which are the object of the present paper.

The interpretative categories used here will be narratively illustrated later in the context in which they are adopted. Yet it should first be noted that sceptical strategies are dialectical tools, mostly tacitly implied, which aim to demolish or at least weaken an argumentation, a belief, or the claimed verity of a fact. Ancient scep-

1 I would like to thank my friend and colleague Michael Friedrich, professor of Sinology at Universität Hamburg, for inviting me to the Hamburg Shanghai Forum (Universität Hamburg and Fudan University) from April 10 to 13, 2019. His invitation and his accurate and fruitful advice were the initial motivation for my interest in the story of the Jesuit mission in China in the early modern period. My reading of the text is exclusively devoted to its literary strategy and is concerned with little, if any, historical research on the circumstances of the origins and propagation of both writings. I have not addressed the problem of their plausibility, nor will I deal with Chinese primary sources to contrast or confirm the observations and claims of both Jesuits. My approach is from the perhaps narrower angle of the history of ideas and strategies. I refer the reader to the vast literature on the topic. For the history of the Jesuits, see now the recent volume by Markus Friedrich, *Die Jesuiten. Aufstieg, Niedergang, Neubeginn* (Munich: Piper, 2016) and the literature mentioned there.

2 Nicholas Trigault, *De Christiana expeditione apud Sinas suscepta ab Societate Jesu. Ex P. Matthaei Riccij eiusdem societatis commentarijs, libri V: Ad S.D.N. Paulum V. In quibus sinensis regni mores, leges, atque instituta, & novae illius Ecclesiae difficillima primordia accurate & summa fide describuntur* (Leiden, 1616). See further details below.

https://doi.org/10.1515/9783110618839-014

tics did not aim to reconstruct the theory, idea, arguments, or philosophy they were defeating. In the following analysis, I will introduce the concept of *strategic scepticism* as a tactic for disavowing authorities, facts, ideas, or groups with the intention of suggesting alternative authorities and interpretations as premises for new dogmatisms. As is well known, the philosophical term "sceptic" philologically means "researcher," without the methodological implication of the double aspect of the Scholastic *pars destruens* as a premise of the *pars construens*.[3] In the present case, I suggest understanding *strategic scepticism* in the sense of the current strategic procedure used in spreading "fake news"; namely, insinuating doubts and uncertainty, but supporting other concurrent systems of power.[4]

In what follows, I will present and discuss the report on the three religious groups in the eleventh paragraph of the first chapter of Trigault's book and the corresponding parts of Ricci's original Italian, mostly scattered throughout his diary. I will also raise questions about the internal logic, additions, points of contrast, and possible plausibility of both the original Italian version and Trigault's Latin translation.

Ricci's Text and Trigault's Translation

To begin with, I must answer the implicit question of why we should consider a translation together with the original, or *Vorlage*. The answer is simple: the Latin translation of the book was highly influential, as the history of its printings and translations testifies: it was printed four times (in 1615, 1616, 1617, and 1619); there were three French versions (published in 1616, 1617, and 1618); it was published in German (1617), Spanish (1621), and Italian (1621); excerpts from it appeared in English in *Hakluytus Posthumus* or *Purchas his Pilgrimes, contayning a History of the World in Sea Voyages and Lande Travells, by Englishmen and others* (4 vols., 1625);[5] and finally, a complete English version was published in 1953. In contrast, the first Italian edition of Matteo Ricci's work was published in Macerata, Ricci's birthplace, as late as 1911.[6] The world of scholarship had to wait three more decades for his com-

3 See Giuseppe Veltri, *Alienated Wisdom: Enquiry into Jewish Philosophy and Scepticism* (Berlin and Boston: De Gruyter, 2018), 146–56.

4 See below.

5 Full title: *Hakluytus Posthumus in fiue bookes. The first, contayning the voyages and peregrinations made by ancient kings, patriarkes, apostles, philosophers, and others, to and thorow the remoter parts of the knowne world: enquiries also of languages and religions, especially of the moderne diuersified professions of Christianitie. The second, a description of all the circum-nauigations of the globe. The third, nauigations and voyages of English-men, alongst the coasts of Africa [...] The fourth, English voyages beyond the East Indies, to the ilands of Iapan, China, Cauchinchina, the Philippinæ with others [...] The fifth, nauigations, voyages, traffiques, discoueries, of the English nation in the easterne parts of the world [...] The first part Hakluytus posthumus or Purchas his pilgrims* (London, 1625).

6 *Opere storiche del padre Matteo Ricci S.I*, ed. Pietro Tacchi (Macerata: F. Giorgetti, 1911–13).

plete works: in 1942, the first volume of *Fonti Ricciane*, edited by Pasquale d'Elia, appeared, and the second and third volumes were published in 1949.[7]

In the very first lines of the book, Trigault writes:[8]

> Ricci's commentaries were intended for Europe, and for fear that the record of the labor of so great a man might be lost amid the perils of such a long journey, and the dangers of overland travel on pack animals, it was decided to translate them *into Portuguese*, and to retain a copy of what he had written in Italian.[9]

Despite Trigault's statement about the danger of the text being lost, the reason for publishing the Latin translation before the Italian edition is not clear. We must remember that the Christian Portuguese presence in China (primarily in Macau) played a major role in this period, and we should also remember that Ricci, together with Michele Ruggieri, composed the first *Dicionário Português–Chinês (Pu-Han cidian)* in 1594,[10] although it remained unpublished until 2001. From the point of view of market economy, we can guess that one important reason was that the translation into Portuguese had the potential to conquer the market. However, I do not think so. To explain his translation, Trigault made recourse to his personal commitment to his mission in China; he claimed that he himself was the sole authority regarding the transfer of knowledge between China and the occident. He openly writes:

> The status of the Christian Mission to China at that time demanded that a Procurator be sent to Europe *to promote the mission interests*, and being selected for this office, my first thought was to read the manuscript of father Matthew Ricci's commentaries and translate them into Latin. My reasons for thinking so were, first, because I realized that this task could not be done by anyone *who was not well acquainted with the affairs of the Mission*, or with the various parts of the country that are mentioned, and secondly, as we have already said, because it was necessary to *fill in various parts that were left unfinished*, also to add certain items and to amplify others, which our good father, in his modesty, had either omitted entirely, or only touched upon in passing. And so, although the sea voyage was long, the weather clear, and the sea calm, the work of transla-

7 *Fonti Ricciane: documenti originali concernenti Matteo Ricci e la storia delle prime relazioni tra l'Europa e la Cina (1579–1615)*, ed. Pasquale M. d'Elia, 3 vols (Rome: La Libreria dello Stato, 1942–49). There are of course new translations of the original into modern languages as well as into Chinese, which I will not take into consideration for obvious reasons.

8 Here, I am using the English translation offered in Matthew Ricci, *China in the Sixteenth Century: The Journals of Matthew Ricci, 1583–1610*, trans. Louis J. Gallagher (New York: Random House, 1953), downloaded from http://www.questia.com/read/11620863/china-in-the-sixteenth-century-the-journals-of-matthew.

9 Ricci, *China in the Sixteenth Century*, xiii. All emphases are mine.

10 Michele Ruggieri and Matteo Ricci, *Dicionário Português–Chinês: 葡漢辭典 (Pu-Han cidian): Portuguese–Chinese Dictionary*, ed. John W. Witek (Lisbon and San Francisco: Biblioteca Nacional Portugal, Instituto Português do Oriente and Ricci Institute for Chinese–Western Cultural History, University of San Francisco, 2001). It is beyond my intent and capacity to answer the question of whether Ricci was a co-author of this dictionary or only its front man.

> tion was no small undertaking, and I realized that I had endeavored to accomplish something that demanded more leisure and more quiet than was customary among a crew of noisy sailors.[11]

We will return to Trigault's "real" intention at the end of this essay. An intention that is, in my opinion, secondary—or perhaps entirely absent—is to acquaint his audience with the richness and multifariousness of the Chinese world. One should also exclude the possibility that Trigault intended to offer an "ethnography" of Chinese customs and beliefs, similar to modern researchers' claims concerning Christian reports about the Jews in the seventeenth century.[12] Yet, ethnography was not a seventeenth-century science; the central aspect of collecting customs and beliefs was not a concern with expressions of human diversity as represented by ethnic divisions and cultural and social cults and beliefs, but almost always a polemic and apologetic literature. The report about Chinese philosophy and customs shows some interesting parallels in their polemical and apologetic aspects to the Christian European interest in the so-called *ceremonialia* of the Jews and Turks. We will analyse them in the conclusion. Here, it is enough to note that it must have been evident to the reader that the intention of this report is exclusively concentrated in the phrases "to promote the mission interests" and "fill in various parts [of Ricci's works] that were left unfinished."[13] It is clear that Trigault wanted to establish a permanent Jesuit mission to disseminate Christian Roman Catholic tradition in China, given his claim that he is presenting an accurate report on the mission. Trigault goes beyond Ricci's account: he claims that his own additions are also based on evidence from eyewitnesses:

> As for myself, I can assure you that what I have added, I have seen with my own eyes, or have obtained it from the true report of other fathers, who either witnessed it themselves, or approved of it for the annals of the Mission. I have not only visited China but have traversed six of its principal provinces, seen all the Mission Centers and, as I believe, garnered a thorough knowledge of the affairs of the Mission in general. We thought it best to inform you of all this, even at length, lest you be led into doubt by the contrary opinions appearing in the various writings thus far edited concerning the Kingdom of China.[14]

The historical accuracy of the report—namely, how precise it is in its historical details or, on the contrary, how in some ways it was ingeniously fabricated as though by an inventive storyteller—remains unclear. It is not my task to answer this question, but rather to present some thoughts and considerations that occurred to me following my reading of both versions, leaving to other scholars, who are assuredly much more acquainted with Chinese history in the early modern period, the heavy

11 Ricci, *China in the Sixteenth Century*, xiii.

12 See, e.g., Yaacov Deutsch, *Judaism in Christian Eyes: Ethnographic Descriptions of Jews and Judaism in Early Modern Europe* (Oxford: Oxford University Press, 2012) and my criticism in Veltri, *Alienated Wisdom*, 69–89.

13 Both texts are quoted from Ricci, *China in the Sixteenth Century.*

14 Ricci, xiv.

task of looking into both books and the possible sources and materials used by each author.

The Report on Chinese Religion, Muslims, Jews, and Christians

At the end of the first chapter, Trigault reports on Ricci's meetings with Muslims, Jews, and Christians in China ("11. Signs of the Doctrines of the Saracens, the Jews, and of Christianity among the Chinese"). In the *Fonti Ricciane*, the corresponding text is very short. Ricci begins with a preface on the religions which he called sects "which are around the religion," meaning the ancient Chinese religion. His admiration cannot be underestimated, as he compares the ancient Chinese religion and philosophy with those of the Greco-Roman world by emphasising the superiority of the Chinese. He believes the ancient Chinese concept of divinity to be superior because it is not blemished by sexual obscenities like the Greco-Roman Olympus. Chinese adepts, on the contrary, followed reason. As he was of the school of a typical Scholastic interpretation of religious consciousness, Ricci expressed his hope that the ancient Chinese believers had reached salvation through natural law. Peter Duigman asserts that "the Jesuits [...] had a more flexible system of moral philosophy and a broader concept of natural law than did other Catholic missionaries."[15] I am somewhat perplexed by this general statement—a Jesuit commitment of sorts to relativism *à la* Montaigne—when confronted with the evident goal of Ricci's note about ancient philosophy and religion. He states that the antique Chinese (and only they) could reach this natural salvation through their good works, "the love of their homeland, of the public good, and the benefit of the people."[16]

Ricci has high esteem for ancient Chinese philosophy, which in his opinion cannot be considered to be inferior to European philosophy (*senza niente cedere ai più famosi dei nostri philosophi antichi*). Nevertheless, the conclusion of the paragraph is clear: the corrupt nature of human beings without divine grace can only produce idolatry and atheism. The door to true Christianity is hence open. That is not relativism, but a sceptical strategy for indirectly attacking Greek philosophy by creating an opponent, i.e., Chinese philosophy, that demolishes the uniqueness of Greek reason on behalf of a philosophy that is even better due to its superior morality. The *pars construens* is of course Christianity, the only religion in which faith, morals, and reason are perfectly suited to the goal of salvation.

15 Peter Duigman, "Early Jesuit Missionaries: A Suggestion for Further Study," *American Anthropologist* 60 (1958): 729.

16 "Le buone opere che fecero quei primi Cinesi per amore della patria, del bene pubblico et utilità del popolo" (Nr. 170).

As for Trigault, he intriguingly refers to salvation according to the Scholastic doctrine of *conscientia:*

> One can confidently hope that in the mercy of God, many of the ancient Chinese found salvation in the natural law, assisted as they must have been by that special help which, as the theologians teach, is denied to no one who does what he can toward salvation, according to the *light of his conscience*. That they endeavored to do this is readily determined from their history of more than four thousand years, which really is a record of good deeds done on behalf of their country and for the common good.[17]

This self-consciousness is an important element of Thomas Aquinas's theology of other religions, which states that a person can be saved by the fact of belonging to the true religion, a motif which goes back to the Islamic philosopher Abū Ḥāmid Muḥammad ibn Muḥammad al-Ghazālī (1058–1111). There is no space to go into this concept in more detail here, which at first glance attempts to be a solution to the problem of the diversity of monotheisms and revealed religions. I said "at first glance," because the hidden reason for introducing it is not relativism, but the strategic way of introducing the Christian truth: the non-Christian can only reach salvation via natural law and "according to the light of his conscience." How can the conscience be illuminated if not by the light of the Gospel?[18]

Trigault emphatically mentions the ancient Chinese religion and philosophy, although his tone is less enthusiastic than Ricci's. In this context, he reports on the three Chinese "sects." Listing the Chinese cults, i.e. the practical observances, he reduces them to "the Literati, the Sciequia, and the Laucu,"[19] whose common element is the Chinese script used by the Japanese, Koreans, and Formosans who belong to these three sects. Here, for the first time, he speaks of the "traces of Saracen, Judaic, and Christian worship evident in China."[20] At first glance, it is not evident why Ricci and Trigault list these three religions in a non-chronological order.

17 Ricci, *China in the Sixteenth Century*, 93.
18 On the principle of *conscientia*, see Giuseppe Veltri, "The Principle of *Coscientia:* The Averroist Joshua ha-Lorqi" (in preparation).
19 That is, the Confucians, Buddhists (from Shijia), and Daoists (from Laozi). I am grateful to Michael Friedrich for this information.
20 Ricci, *China in the Sixteenth Century*, 94: "In this chapter, we shall treat only of the triple cult of the Chinese as distinguished from all other pagan sects. [...] Chinese books enumerate only three cults or systems of religious observance for the whole world and this people knows of no others. These are, the Literati, the Sciequia, and the Laucu. All Chinese and all people of the surrounding nations who make use of Chinese writing – the Japanese, the Koreans, the Leuquici or Formosans, and the Cochin Chinese – belong to one or other of these three sects."

Muslims

The mention of the ancient Chinese religion is also the context in which Ricci introduces the three groups or sects present in China. The Muslims are presented as making a decisive contribution to the decadence initiated by the Chinese, coming from Persia and pouring into China in increasing numbers throughout the generations, leading to the formation of many thousands of families (*e si moltiplicarono tanto per generatione ne' suo figlioli e nipoti, che già hanno impita la Cina con molte migliaia di famiglie*).[21] Ricci adds that they are not missionaries and are obedient to Chinese law. They are ignorant and therefore irrelevant for the Chinese authorities, who allowed them to study and become graduates and magistrates of the Empire, and accordingly, they abandoned their religion, with exception of eating pork, to which they were not accustomed (*non gli restando altrio di essa che il non mangiare care di porco per non essere avezzi ad essa*).[22]

The corresponding paragraph in Trigault is also devoted to Muslims. For the most part, he translates Ricci's main ideas with some more particular narrative elements on the introduction of Islam into China: their character as outsiders on the one side, and their full integration into society on the other.[23] The reader understands that integration is tantamount to conversion, although the commonality between converted and non-converted Muslims is the custom of avoiding the consumption of pork. Are they outsiders or insiders? This is a paradox I cannot understand.

Jews

Ricci states that very few Jewish families have remained in China and that they are only present in Henan (Kaifeng) and the province of Hanceufu.[24] These families possess a parchment scroll, which they consider very precious, bearing the Pentateuch of Moses written in Hebrew letters without vocalization. They still practise the rite of circumcision and do not eat pork or meat with nerves, according to their old tradition.

---

21 Nr. 171.

22 Nr. 171.

23 It is not my intention here to give a historical analysis of what Ricci and Trigault wrote, but only to follow their argumentation.

24 On the history of Chinese Jews, see Hyman Kublin, *Studies of the Chinese Jews: Selected Articles from Journals East and West* (New York: Paragon, 1971), with contributions from Laurence Kramer, Donald Leslie, and Adolf Neubauer, among others; Anson H. Laytner and Jordan Paper, eds., *The Chinese Jews of Kaifeng: A Millennium of Adaptation and Endurance* (Lanham, MD: Lexington Books, 2017).

Trigault's report substantially differs from Ricci's in the context of the three religions or sects: here, he introduces narratives of the Jews' first introduction into China by retelling Ricci's meeting with a Jew of Kaifeng:

> Some few years after our Society had taken a fixed residence in Pekin, a man who was Jewish by race and by profession, learning of the reputation of Father Ricci, decided to pay him a visit. He had read about Ricci and his companions in a book about the Europeans, written by a Chinese scholar. The man's name was Ngai and he was born in Chaifamfu, the metropolitan city of the Province of Honan. At that time he was on his way to Pekin to take the examinations for the doctorate. He had already attained to the order of the licentiate. Being a Jew and having read in the book mentioned that our Fathers were not Saracens and that they believed in only one God of heaven and earth, he concluded that we must be believers in, and followers of, the Mosaic Law. On entering our home he seemed to be quite excited over the fact, as he expressed it, that he professed the same faith that we did. His whole external appearance, nose, eyes, and all his facial lineaments, were anything but Chinese. Father Ricci took him into the church and showed him a picture above the high altar, a painting of the Blessed Virgin and the child Jesus, with John the Precursor, praying on his knees before them. Being a Jew and believing that we were of the same religious belief, he thought the picture represented Rebecca and her two children, Jacob and Esau, and so made a humble curtsy before it. He could not refrain, as he remarked, from doing honor to the parents of his race, though it was not his custom to venerate images. This happened on the Feast of St. John The Baptist.[25]

It is not important to ascertain whether or not this story is true. The goal is clear: to underline the Jew's ignorance. An important part of the story is that the Jew Ngai cannot read Hebrew:[26]

> When he brought the visitor back to the house and began to question him as to his identity, it gradually dawned upon him that he was talking with a believer in the ancient Jewish law. The man admitted that he was an Israelite, but he knew of no such word as Jew. It would seem from this that the dispersion of the ten tribes penetrated to the extreme confines of the East. Later on Ngai saw a royal edition of the Bible, printed by Plantin, and though he recognized the Hebrew characters he could not read the book.[27]

It is obviously possible that the Jew of Kaifeng did not understand Hebrew, but not certain. In 1895, Adolf Neubauer published an article on the Jews of China[28] which predominantly focused on Hebrew documents brought from China. These manuscripts are, in his opinion, very important because of their Persian influence. Neubauer was also of the opinion that Kaifeng's Jews had not mastered Hebrew. His argument relies upon a nineteenth-century report and also, of course, upon Ricci's report. I am not entirely convinced, because Ricci had only met one Jew; namely, Ngai, who

---

25 Ricci, *China in the Sixteenth Century*, 107–8.

26 On Ricci and Trigault's lack of accuracy, see Paul Pelliot, "Le juif Ngai, informateur du P. Mathieu Ricci," *T'oung Pao* 20, no. 1 (1920–1921): 32–39.

27 Ricci, *China in the Sixteenth Century*, 107–8.

28 Adolf Neubauer, "Jews in China," *Jewish Quarterly Review* 8 (1895): 123–39.

does not really seem acquainted with the religion of his fathers. Moreover, why would they conserve Hebrew texts if nobody could read them?

Another cliché is the typical Christian topic of the Messiah:

> Ricci then sent the same brother on a second visit, this time to the ruler of the synagogue, with a letter, informing him that he had all the books of the Old Testament at his house in Pekin, together with a copy of the New Testament, containing the life and works of the Messiah, who he asserted had already appeared. The Archisynagogus took exception to this last remark and replied that the Messiah would not come for another ten thousand years.

I think that the synagogue speaker committed an error in speaking of ten thousand years before the coming of the Messiah. Perhaps this is an error on Ricci and Trigault's part, because according to Jewish belief, the age of the world amounts to 6,000 years.

This is similar to the fantastic story that Ricci was conferred the dignity of high priest (which cannot be attained!):

> He added, however, that since they had heard so much of Ricci's reputation and learning, they would confer upon him the dignity of high priest of the Synagogue, if he would join their faith and abstain from eating pork.

The whole story ended—of course, I would say—with the Jews' conversion to Christianity:

> Three other Jews came from that same city to Pekin later on, and as they were detained here for several days on business and were ready to receive the Christian faith, they decided to be baptized. One of these was a nephew of our first visitor, on the father's side. Our Fathers received them with hospitality and taught them a great deal about the Scriptures that was unknown to their Rabbis. When they learned that the Messiah had really come, they knelt down and prayed before his statue, as Christians do, and they were delighted with a compendium volume of the Christian faith, and with the other books on Christian doctrine, written in Chinese, which they brought back to their people.[29]

The reference to the Jewish scholars is a clear topos of Christian apologetics: the Jews do not know the most important parts of the Scriptures because they are hermeneutically and exegetically in the Christians' possession.

Ricci's strategy is to illustrate the "ignorance" of a few Jews in order to imply that of the Chinese Jews in general. This Latin argumentation of *ab uno disce omnes* (Virgil, *Aeneid* 2:65–66)—which means that the whole may be inferred from one instance—is a sophism based on the pseudo-logical and pseudo-scientific inference according to which an attribute of a single object/substance may be ascribed to the whole. I do not have to explain how this well-known "principle" could be dangerous to human society.

---

29 Ricci, *China in the Sixteenth Century*, 108–9.

Christians

For Ricci, Christianity also does not merit a long presentation, although there were many Christians families in the Northern provinces,[30] who flourished in terms of number, education, and military strength (*e fiorivano tanto in numero di famiglie, lettere et armi*). Yet it is for this very reason that according to Ricci, they were suspected of rebellion, "perhaps accused by the Muslims who are in every part our enemies." This provoked mass conversions to Islam (whose converts are here called Turks), Judaism, or even atheism (whose converts are here called Gentiles), and their churches become temples for idolatry. Ricci states that their heirs have preserved the tradition of making the sign of the cross over food and drink, but they are also afraid of being considered their descendants and are ignorant of what this cross means. In contrast, the corresponding passage in Trigault is much longer, but he does not offer any novel information, speculating more on the word "cross" and its meaning for the Chinese.[31]

Names of the "Sects"

Near the end of the report, an important element is recorded by both Ricci and Trigault: the names of the "sects" according to Chinese custom. Ricci wrote:

> All these foreigners [i. e., Muslims, Jews, and Christian] are called *hoeihoei* [回回] by the Chinese, a name whose origin we cannot ascertain.[32] They call the *Muslims* ["macomettani"] *hoeihoei of the three laws*, they call the Jews *hoeihoei who cut the nerves of the meat which they eat*, the *Chris-*

---

30 See Hui Li, "Jesuit Missionaries and the Transmission of Christianity and European Knowledge in China," *Emory Endeavors in World History* 4 (2012): 48–63, downloaded from http://history.emory.edu/home/undergraduate/endeavors-journal/volume-4.html.

31 "Now to offer some evidence of the traces of Christianity, which we do very willingly, believing it should be of special interest to our friends in Europe. [...] There was no special word in their language to express the idea, and so our Fathers had to give them a Chinese word for it. In doing so they chose the Chinese character expressing the number ten, which is written in the form of a cross, thus †. It probably was not without the guidance of divine providence that a Christian in our day, selecting a name for the cross for the Chinese, should hit upon the very word that their ancestors had invented because of a similar paucity of symbolic expressions. In each instance the word selected was Scie-cu, meaning, the tenth. In so doing they really were not far afield from the practice observed in sacred literature, in which the cross is expressed by the letter T, which is even a more exact form than the one in question. When the conversation came around to designating the cross by this particular symbol, our friend the Israelite remembered that in Chaifamfu, the capital of his district, and in Lincin, a trading post in the Province of Sciantum, there were certain strangers whose ancestors came from abroad and who observed the religious custom of venerating a cross. They were accustomed, he said, to make a gesture with the hand in the form of a cross over their food and drink. He had no realization of the significance of such a rite and assured us that neither had the people in question who were accustomed to use it."

32 See Victor Escroignard, "Les Sarrasins des trois normes ou San Jiiao huihui," *Acta Orientalia Academiae Scientiarum Hungaricae* 45 (1991): 253–70.

> *tians* are called *hoeihoei* of the number ten because the letter of ten in China is a perfect cross and here there is no name that either uses for the word "cross."[33] Beyond the name *isai*, which means *iesuini*, the Muslims ["*maomettani*"] call those Christians *terzai*. I heard from an Armenian that in Persia they call Armenians by this name. It may be that the Christians originally came from Armenia.

Trigault's "translation" is accordingly a correction and an addition of further details:

> The three sects first discussed, called barbarian sects by the Chinese, are all designated by one common term, and their followers are all known as *Hoei-hoei*. What the origin of that name may be is still a puzzle to us. The more learned Chinese generally distinguish them as follows:
>
> The Saracens are called Hoei because they abstain from pork.
>
> The Jews are Hoei also because they do not eat the portion of meat containing the thigh nerve. This custom was introduced by the Jews because Jacob was stricken in that nerve.
>
> Those who venerate the cross are called Hoei because they do not eat of the flesh of animals with rounded hoofs.
>
> Although the Chinese and the Saracens and the Jews all partake of the flesh of horses, asses, and mules, this last class probably refrain from such meats in accordance with a custom peculiar to their race. The Chinese have other names also for this third class.
>
> In general, they use the word *Hoei* with reference to those who follow the cult of the cross, while both Chinese and Jews say *Hoei*, referring to the Saracens. By this they mean that the Saracens belong to all three sects since they have adopted something of their belief from gentiles, Jews, and Christians. Besides the ordinary term Isai or Jesuini which the Saracens apply to all Christians, here in China the followers of the ancient cult of the cross are known to them as Terzai. The origin of this term is also uncertain, except that we have learned from a certain Armenian that in Persia the Armenians call the Christians by that same name. It might be concluded from this, with some probability, that the devotees of the cross had their origin in Armenia, and at various times found their way into China from the West. This could have happened when the vast armies of the Tartars poured over China, about the time, as it would seem, when the Venetian Marco Polo found his way hither.[34]

My surprise in reading this rendition of Trigault is his complete rewording of the text, which resolves some problems and complicates others. There are no "barbarian sects" (*tres Barbarorum sectas, ut Sinae appellant*) in Ricci, and the Muslims are called "Mohammedans" and not "Saracens." "Saracen" was a medieval term which had been used since the twelfth century, while "Mohammedan" was more recent, a product of the sixteenth century.[35] Ricci defines the Muslims according to the three laws, which, as can be seen from another text (Nr. 725), were reported by a Jew called Nghaettien in 1605: "law" means "religion" and refers to Judaism, Nestorian Christianity, and atheism, i. e., Buddhism. Trigault defines Muslims by their food customs: they do not eat pork.

---

33 *Fonti Ricciane*, Nr. 174.

34 Ricci, *China in the Sixteenth Century*, 112

35 See David Thomas and John A. Chesworth, eds., *Christian–Muslim Relations. A Bibliographical History. Volume 11: South and East Asia, Africa and the Americas (1600–1700)* (Leiden: Brill, 2016), 434.

Ricci defines the Jews as those who "cut the nerves of the meat which they eat." This is obviously erroneous, because the Jewish *kashrut* did not and does not address or include *every* nerve, but, according to the biblical book of Genesis, pertains only to the sciatic nerve: "Therefore the Children of Israel are not to eat the *sciatic nerve* to this day, because he struck *Jacob's* leg on the *sciatic nerve*" (32:33). Perhaps Ricci's aim was to find a difference between Muslims and Jews in the eyes of the Chinese because neither of them eat pork.

The complication is much more intriguing because those who venerate the cross, i.e., the Christians, are also defined by Trigault in reference to food customs: "Those who venerate the cross are called *Hoei* because they do not eat of the flesh of animals with rounded hoofs." This is a blatant error, because the *Jews* did not eat animals with rounded hoofs "according to their customs," which are prohibited as food in Jewish law (Leviticus 11), and they decisively do not venerate the cross.

The names of the sects testify only that the commonalities between the sects are based on food customs, as well as the differences. In other words: nobody knows what the difference is.

Conclusion: Consciousness of Commonalities

In 1940, the *Archivum Historicum Societatis Iesu*[36] published a somewhat trenchant article by Edmond Lamalle attacking Nicolas Trigault. The author clearly defines Trigault's work as "propaganda" for his agency in favour of the mission in China. As I suggested at the beginning of this paper, Trigault did indeed produce his reworked translation of Ricci's work as pure propaganda for his missionary project. The book is not a report to narrate what Matteo Ricci did, but a reworked propagandic account intended to obtain money for the mission. Ricci's notes and experiences, hence, were instrumentalised to this goal. I do not intend to go beyond my task as a historian of ideas, but it is strange to describe the presence of Muslims, Jews, and "other" Christians based only on their differences in food customs.

I cannot understand why Ricci and Trigault devoted so few lines to the Muslims, a majority among the minorities, while Trigault mainly focused on presenting the Jews in an unequivocally negative light. There is an obvious negative vein in Ricci's original Italian work on the Muslims, which he defended with an indirect historical reference to the Turkish attacks on Christians, hence the sentence "the Muslims are in every part our enemies."[37] Yet Ricci offered a very short report on all of them and did not negatively privilege the Jews in the context of the three "sects."

---

36 Edmond Lamalle, "La propagande du P. Nicolas Trigault en faveur des missions de Chine (1618)," *Archivium Historicum Societatis Iesu* 9 (1940): 49–120.
37 Nr. 171.

It is important to underline that the use of strategic scepticism is always intended to demolish facts and logical discourse in order to establish new dogmatism and imperial power. Ricci's real aim is, of course, to keep his account about his meetings with religious groups in China, if any, very brief, because they are not the core of the book and are not representative of the Chinese *Weltanschauung* in this period. Trigault's real aim is to discredit all three religions or sects, the remnants of which—particularly in the case of the Jews—are incapable of understanding their own religion and writings. This gives a basis for asking European supporters to give money for a further a Jesuit mission to China. The Jesuit missionary show must go on.

Bibliography

d'Elia, Pasquale M., ed. *Fonti Ricciane: documenti originali concernenti Matteo Ricci e la storia delle prime relazioni tra l'Europa e la Cina (1579–1615)*. 3 vols. Rome: La Libreria dello Stato, 1942–1949.

Deutsch, Yaacov. *Judaism in Christian Eyes: Ethnographic Descriptions of Jews and Judaism in Early Modern Europe*. Oxford: Oxford University Press, 2012.

Duigman, Peter. "Early Jesuit Missionaries: A Suggestion for Further Study." *American Anthropologist* 60 (1958): 725–32.

Escroignard, Victor. "Les Sarrasins des trois normes ou San Jiao Huihui." *Acta Orientalia Academiae Scientiarum Hungaricae* 45 (1991): 253–70.

Friedrich, Markus. *Die Jesuiten. Aufstieg, Niedergang, Neubeginn*. Munich: Piper, 2016.

Kublin, Hyman, ed. *Studies of the Chinese Jews: Selected Articles from Journals East and West*. New York: Paragon, 1971.

Lamalle, Edmond. "La propagande du P. Nicolas Trigault en faveur des missions de Chine (1616)." *Archivium Historicum Societatis Iesu* 9 (1940): 49–120.

Laytner, Anson H., and Jordan Paper, eds. *The Chinese Jews of Kaifeng: A Millennium of Adaptation and Endurance*. Lanham, MD: Lexington Books, 2017.

Li, Hui. "Jesuit Missionaries and the Transmission of Christianity and European Knowledge in China." *Emory Endeavors in World History* 4 (2012): 48–63.

Neubauer, Adolf. "Jews in China." *Jewish Quarterly Review* 8 (1895): 123–39.

Pelliot, Paul. "Le juif Ngai, informateur du P. Mathieu Ricci." *T'oung Pao* 20, no. 1 (1920–1921): 32–39.

Purchas, Samuel. *Hakluytus Posthumus in fiue books [...]* London, 1625.

Ricci, Matteo [Matthew]. *China in the Sixteenth Century: The Journals of Matthew Ricci, 1583–1610*. Translated by Louis J. Gallagher. New York: Random House, 1953.

Ricci, Matteo. *Opere storiche del padre Matteo Ricci*. Edited by Pietro Tacchi. 2 vols. Macerata: F. Giorgetti, 1911–13.

Ruggieri, Michele, and Matteo Ricci. *Dicionário Português–Chinês:* 葡漢辭典 *(Pu-Han cidian): Portuguese–Chinese Dictionary*. Edited by John W. Witek. Lisbon and San Francisco: Biblioteca Nacional Portugal, Instituto Português do Oriente and Ricci Institute for Chinese–Western Cultural History, University of San Francisco, 2001.

Thomas, David, and John A. Chesworth, eds. *Christian–Muslim Relations. A Bibliographical History. Volume 11: South and East Asia, Africa and the Americas (1600–1700)*. Leiden: Brill, 2016.

Trigault, Nicholas. *De Christiana expeditione apud Sinas suscepta ab Societate Jesu. Ex P. Matthaei Riccij eiusdem societatis commentarijs, libri V: Ad S.D.N. Paulum V. In quibus sinensis regni mores, leges, atque instituta, & novae illius Ecclesiae difficillima primordia accurate & summa fide describuntur.* Leiden, 1616.

Veltri, Giuseppe. *Alienated Wisdom: Enquiry into Jewish Philosophy and Scepticism.* Berlin and Boston: De Gruyter, 2018.

Veltri, Giuseppe. "The Principle of *Coscientia:* The Averroist Joshua ha-Lorki." Forthcoming.

Part II: **Reports**

Activities and Events

compiled by Maria Wazinski

Annual Lecture
Opening of the Fourth Academic Year
30 October 2018

Julie R. Klein, Villanova University/USA
Scepticism in Spinoza and the Project of Critique

Spinoza famously opposes political rebellion, or what we would call "revolution," on the grounds that it reproduces, and may even worsen, just the conditions it was intended to replace. Yet Spinoza's theoretical and practical philosophy is marked by philosophical rebellion and conceptual revolution: he invites, even demands, that we criticise and move beyond our usual horizons, lest we reproduce old problems. In this talk, Julie Klein considered Spinoza's relation to scepticism and his practice of critique as a philosophical strategy. She focused particularly on his account of the freedom to philosophise in order to clarify his view of what makes philosophising possible.

Regular Events

In order to explore the different manifestations of scepticism in different contexts, the Maimonides Centre for Advanced Studies organises internal events in which junior and senior fellows as well as members of the Centre discuss their research. These events are not open to the public.

Dialectical Evenings

The Dialectical Evening is an informal meeting every four weeks (in fortnightly rotation with the Reading Evening) for discussions and readings which is designed to promote dialectical culture and sceptical thought within the research unit. Members of the Maimonides Centre and occasional guests convene to challenge, doubt, and explore theses in various subject areas.

 https://doi.org/10.1515/9783110618839-016

[DE 24] 4 September 2018
Israel Netanel Rubin and Bill Rebiger
Why Is A Sceptic Worse Than An Atheist?

In contrast to the religious philosophers who exploited scepticism to serve faith, such as Pascal in his famous gamble, Saʿadya Gaon expresses views in *The Book of Beliefs and Opinions* which indicate a firm position against scepticism. In Saʿadya's opinion, the sceptic who claims that "nothing possesses any reality whatever" and therefore, for example, that the problem of the eternity of the world cannot be resolved is even worse than a heretic who claims that the world is definitely eternal or an Epicurean who teaches that the world was created by accident. It is enough to refute such heretics philosophically, but sceptics must "be struck so painful a blow that they would weep and cry out. When they finally acknowledge the experience of... the smart of a blow, they will have admitted the reality of sensation." Why does the sceptic upset Saʿadya so much? During the Dialectical Evening, the convenors discussed Saʿadya's perception of scepticism as cynicism and the position of Jewish rationalism that the belief in reason and its ability to reach certainty precedes even the belief in God, for it is God who wants man to know Him through reason.

[DE 25] 20 November 2018
Máté Veres and Yoav Meyrav
"The Second Speaker Must not be Heard"

> In response to him who said, "Do not pass judgement until you have heard both sides," Zeno stated the contrary thesis, with this sort of argument: "The second speaker must not be heard, whether the first speaker proved his case (for the inquiry is then finished) or did not prove it (for that is just like his not having complied when summoned, or his having complied by talking nonsense). But either he proved his case or he did not prove it. Therefore the second speaker must not be heard" (Plutarch of Chaeronea, *On Stoic Self-contradictions*, 1034E).

We tend to assume that, other things being equal, one stands to benefit from hearing both sides on any given question. Yet it appears that, if Plutarch is to be trusted—and that is a strong "if"—the founder of the Stoic school condemned the practice of offering counterarguments. In this Dialectical Evening, the convenors discussed the possible motivations behind Zeno's striking claim and its background in his epistemological theory; they then left history behind and reflected on whether we can conceive of situations in which it is indeed rational not to attend to opposing considerations.

[DE 26] 22 January 2019
Daniel Davies and Florian Lützen
Al-Ġazālī (d. 1111) and Maimonides (d. 1204) on the Simile of the Water Clock—Knowledge and Negative Theology

- Al-Ġazālī says in *The Highest Goal:* "The cause that causes the movement of the spheres, the stars, the sun, and the moon according to a known calculation is like that aperture, which necessitates the flow of water according to a known measure. That the movement of the sun, the moon, and the stars lead to temporal results on earth is similar to the fact that the movement of the water leads to those movements that result in the ball's falling, indicating that the hour has come."
- Maimonides in the *Guide for the Perplexed* says: "Whenever the onlooker sees a movement, he obtains new knowledge. And by not ceasing to look on and gradually to obtain increase of knowledge, he acquires in this way a knowledge of the whole of the instrument. However, if you suppose that the motions of this instrument are infinite, the onlooker could never contain them in his knowledge."

Both al-Ġazālī and Maimonides used the simile of the water clock. Al-Ġazālī, in two of his later works, used it to explain how humans can relate to the divine attributes, whereas Maimonides employed it in the *Guide* to illustrate how one might think about the relation between God's knowledge and creation and to indicate why it is impossible for humans to grasp such a relationship.

In this Dialectical Evening, the convenors first looked at each of the two cases, the methodology underlying them, and how the two scholars presented their arguments. Subsequently, they discussed ensuing questions about the comparability or incommensurability of the cases presented. Can we look at the world as a mechanical device, or is the simile rather an allegory alluding to a higher purpose concerning the limitations of knowledge?

[DE 27] 26 February 2019
Haim Kreisel and Daniel Davies
Reading Maimonides: Esotericism vs. Exotericism. The Case of God's Knowledge of Particulars

In the first part of the presentation Haim Kreisel gave a brief introduction to the esoteric reading of Maimonides's *Guide of the Perplexed*: the justification for this approach and its history. He then discussed Maimonides's theory of divine providence and tried to show why Maimonides's presentation strongly suggests that he holds an esoteric position on the subject—namely, his approach is a completely naturalistic one. Given the fact that Maimonides presents his view of God's knowledge of particulars within the context of his discussion of divine providence, he further argued that Maimonides hints to an esoteric position on this topic as well. Daniel Davies responded by suggesting that this conclusion is not warranted, and that one should read Maimonides's view of God's knowledge of particulars exoterically, irrespective of his position on divine providence.

[DE 28] 19 March 2019
Carlos Lévy and Giuseppe Veltri
Philo of Alexandria: The First Jewish Sceptic?

The studies on scepticism are currently flourishing, but the benefit from them for the analysis of Philo is quite small. In studies on and companions to scepticism, the focus is mainly on Sextus Empiricus, a thinker who lived two centuries after Aenesidemus (ca. middle of first century BCE), who refounded scepticism almost three centuries after Pyrrho. Philo ought to be considered a privileged witness in the history of scepticism, but he is not, though he is the closest to Aenesidemus from both a historical and a geographical point of view.

The first question of this Dialectical Evening was: why this quasi omission nowadays, while a century ago Hans von Arnim devoted an entire book to the analysis of sceptical tropes in Philo?

Second: are there sceptical strategies in Philo, or something else? In the first case, are they comparable to those of Greek pagan texts? In the second, can we escape overly general concepts such as eclecticism and utilitarianism?

Third: what are the possible meanings of "Jewish scepticism" and can Philo be considered the first "Jewish sceptic?" What are the elements of a Jewish sceptical tradition?

Fourth: fideism is a concept of great importance in the history of Western culture, especially during the Renaissance. Can it be affirmed that Philo laid the foundations of this concept?

[DE 29] 4 June 2019
José María Sánchez de León Serrano and Jason Yonover
Genealogical Scepticism

Each of us was born in a particular place at a particular moment, and may boast of a unique trajectory from there to here. Common sense would indicate that the special trajectory specific to someone has a decisive influence on their beliefs. Yet what does this mean for the truth of said convictions? This sort of question is in the background of genealogical thinking. Genealogical thinking questions the validity of a given belief, concept, or idea by tracing its emergence from contingent circumstances. Thus, we might criticise a philosophical conception by showing how it originated in a historical and cultural context that no longer corresponds to ours. Can genealogical critiques ever be valid? Do they commit the so-called "genetic fallacy," the allegedly misled assessment of a claim based on its origin alone? And is the scepticism genealogical thinking may lead to—where we begin to broadly question our convictions given their apparently contingent origins—justified?

[DE 30] 2 July 2019

Aryeh Botwinick and Daniel Boyarin

A Rabbinic Genealogy: Rabbi Akiva as a Precursor of Maimonides's Negative Theology

Two overarching themes concerned the convenors during this Dialectical Evening. The first was to enquire if there are indigenous Rabbinic sources that could be said to serve as precedents for Maimonides's negative theology. The second theme (which is closely wrapped up with the first one) was whether there could be said to be a genre or sub-genre of Talmudic discourse that could be categorised as a philosophical *maḥloqet*—a philosophical division of opinion over a second-order philosophical question. Aside from the issue of whether one can deduce the possible from the impossible which was dealt with at the Dialectical Evening, five immediate examples that come to mind are *brera* (Does the arrow of time move in two directions simultaneously—from future to past as well as from past to future?); *zeh we-zeh gorem* (What is the most elegant way to theorise overdetermination—patterns of multiple causation?); Is nominalism true or false? and, related to this, how does one demarcate between identity and difference? and, *Kul milsa' d'amar raḥmana' lo' s'evid, ei 'avid mehani o lo' mehani*, which forms the subject matter of a *maḥloqet* between Abaye and Rava in Temurah 4b, but which is echoed elsewhere in the Babylonian Talmud. This latter topic raises the spectre of how to theorise the relationship between the normative and the descriptive. Unearthing a Rabbinic pedigree for Maimonides's negative theology is important not only academically, but also existentially. Following the sterling example of Joseph Levenson in his classic work, *Confucian China and its Modern Fate*, we can assume that unless Maimonides had established for himself a homegrown Rabbinic pedigree for negative theology (that in an important sense there was an *equivalence* between the two), he would not have done it. He could only do it because *Ḥazal* had been there first. The anxiety of influence can also work to curb the desire for innovation.

[DE 31] 6 August 2019

Reimund Leicht and Ariel Malachi

Is Logic a Tool for Producing Certainty and/or Conviction in Judah ha-Levi's *Sefer ha-Kuzari?*

Judah ha-Levi's *Sefer ha-Kuzari* is a work about controversy that arouses controversy. If its author wanted to use this literary dialogue between the king of the Khazars and a philosopher, a Christian, a Muslim, and finally a Jew to show that philosophers (and historians of philosophy, for that matter) are not able to reach a real consensus about the most fundamental questions, the interpretation history of this book can be taken as a proof in itself: What is its basic intention? What is the appropriate way to describe its literary genre? What were its intellectual sources? Does the book form a literary and intellectual unity, or is it piecemeal? Is this a book of philosophy, and if so, to which philosophical strand does it belong? Or is it rather a religious critique of philosophy? Is it anti-philosophy in a philosophical garb? Is it a piece of medieval rationalism, anti-rationalism, or both at the same—or different—times?

In spite of all these uncertainties, there can be little doubt that the *Sefer ha-Kuzari* is intended to convince. Moreover, the question of how to produce conviction and certainty looms large in a book telling the story of a pagan king who converts to Judaism after he finds it to be more convincing than all the competing alternatives (philosophy, Christianity, and Islam). But how does Judah ha-Levi argue for his ideas and how is conviction and perhaps even certainty achieved? What is the role of rational argumentation based upon and drawn from the tools provided by philosophical logic and epistemology in a dialogue that opposes Jewish religion to other worldviews? The apparent ambivalence between the religious motives of the book and the necessity of using rational, logical, and philosophical tools of persuasion leads to the question of whether and how logic can serve as a tool to produce certainty and/or conviction in Judah ha-Levi's *Sefer ha-Kuzari*.

[DE 32] 20 August 2019
Stephan Schmid and Yoav Meyrav
Metaphysics and Truth?

For a long and venerable tradition in philosophy, metaphysics—the study or "science" of being *qua* being and its ultimate causes—has been considered to be not only the most fundamental, but also the most noble form of intellectual activity. At the same time, various philosophers have doubted the very possibility of this science: even if there were any, the purported objects of metaphysical investigation, they maintained, are too remote to be grasped by us through our limited cognitive capacities. This kind of suspicion towards metaphysics is well-known, but relies on the question what metaphysics is supposed to be. In this evening, the convenors took up this issue by exploring a logically prior question regarding the allegedly essential connection between metaphysics and truth.

Reading Evening

The Reading Evening is an informal meeting every four weeks (in fortnightly rotation with the Dialectical Evening). Fellows and researchers read and discuss primary texts that are specifically relevant to their respective projects. Each meeting, one fellow or research team member selects and presents a text of particular importance for her research. In reading together, the group benefits from the expertise of the individual researcher.

[RE 27] 21 August 2018
Ronny Vollandt
Divine Determination and Wisdom

The Reading Evening offered a critical re-evaluation of the very beginning of commentary on the Book of Genesis by Saʿadya Gaon. In contrast to Saʿadya's translation, the commentary of which the former was initially part survived in a very sketchy form. In 1984, Moshe Zucker furnished an edition of the first half of the book of Genesis, in which he assembled hundreds of Genizah fragments and brought the commentary back to life for modern scholarship. Since then, numerous additional texts have appeared that can improve our understanding of Saʿadya's exegesis and its context. The passages that were read attempt to prove that the biblical account of creation is a universally acceptable theory, contrary to the belief of some sceptics, which can be defended also on the grounds of scientific literature.

[RE 28] 25 September 2018
Rachel Aumiller
The Body of Belief: Augustine's *Confessions* and Derrida's *Circumfession*

In this Reading Evening, several passages from Augustine's *Confessions* were compared to passages in Jacques Derrida's *Circumfession*: an experimental homage to and parody of the *Confessions*.

While Augustine describes the decisive event in which he turns toward the God of his mother, Derrida reflects on the moment when he turns away from the God of his mother (his "cut with Kippur"). And yet, both (de)conversion stories are also confessions of the failed attempt to leave behind a former way of life, the memories of which reverberate on the narrators' bodies.

Augustine and Derrida question how belief first enters us in our infancy. Belief is articulated on our skin through both a gentle and violent touch: the caress of the caregiver, the switch of the schoolmaster. Because belief exists on the level of bodily memory, it is difficult to shed when we later sceptically engage our inherited values. Our skin continues to cling to beliefs that we denounce.

The Reading Evening explored (1) Derrida's sceptical framing of his relationship to Judaism, (2) dogma in the form of bodily affect/memory, (3) the disjunction between what we say we believe/disbelieve and bodily affect.

[RE 29] 9 October 2018
Yoav Meyrav
Is God an Intellect? Themistius between Aristotle and the Platonists

Themistius (c. 317–388 CE), a philosopher and political figure in the imperial court in Constantinople, composed paraphrases of several of Aristotle's works. These paraphrases were pedagogical tools intended to elucidate Aristotle's arguments, but also functioned as vehicles for Themistius's own philosophical innovations. Since his paraphrases adopt Aristotle's voice, it is difficult to discern at which point in the text Aristotle ends and Themistius begins. What is more, since Themistius's philosophical training was independent from the established schools of his time, his doctrinal affiliation has been the subject of much scholarly debate, especially concerning the question whether he was a Peripatetic or a Platonist. Whereas formulating the question in this manner is somewhat over-reductive, it might nevertheless be valuable to explore how Themistius engages with those of Aristotle's ideas that are known to have been criticised in the Platonic circles with which Themistius was familiar.

In this Reading Evening, selected passages were read from a forthcoming English translation of Themistius's paraphrase of Aristotle's *Metaphysics* 12 alongside passages from Aristotle himself and a few passages from Plotinus (204/5–270 CE), whose *Enneads* shaped the Platonic school in Late Antiquity. Yoav Meyrav explored the possibility that in his paraphrase, Themistius implicitly responds to some Platonic critiques of Aristotle, especially concerning Aristotle's identification between God and Intellect and its various ramifications. He thus, perhaps, attempts to soften the notion of divine otherness and resist the idea of divine transcendence. Consequently, his paraphrase might have had a moderating effect on the sceptical expressions of these elements in medieval Arabic and Hebrew thinkers.

[RE 30] 11 December 2018
Behnam Zolghadr
Denying the Undeniable: Contra Aristotle's Arguments for the Law of Non-Contradiction

Aristotle devoted book Gamma of his *Metaphysics* to the defence of what we now know as the law of non-contradiction. His prominent defence of this law was so influential that it became the dominant view among western philosophers, albeit with very few arguable exceptions, such as Hegel and some of his followers. It is difficult to find a defence of the law after Aristotle's which is worth mentioning. However, with the contemporary developments in logic and metaphysics, some philosophers advocate views holding some contradictions to be true, and thus they challenge the law of non-contradiction. Hence, we may want to look back to the origins of the law of non-contradiction and question whether this so-called undeniable law is deniable or not. In this Reading Evening, the most important parts of Aristotle's arguments for the law of non-contradiction were explored. These arguments are not as crucial as they are usually thought to be. Then, the first arguments for explosion—i. e. that from an arbitrary contradiction, anything may follow—were discussed. These belong to Avicenna and the twelfth-century French logician William of Soissons. It is worth mentioning that Aristotle never argued for explosion. Thus, some passages of Avicenna's *Metaphysics* of *The Healing* were read as well.

[RE 31] 15 January 2019
Avraham Rot
Affective Scepticism: Between Descartes and Spinoza

Descartes excludes the passions from both the exposition and the domain of applicability of his method of doubt. Questioning the reality of an external world, he arguably presupposes the reality of internal experience. As with respect to other Cartesian convictions that have come to dominate modern thought, Spinoza's philosophy indicates an alternative way of thinking. Though Spinoza is commonly regarded as an anti-sceptical thinker, he presents us in effect with a more radical and truthful form of scepticism than Descartes. While on Descartes's account, the affects cannot be doubted and doubt is not significantly affective, on Spinoza's account the affects themselves are expressions of doubt while it is in fact clear and distinct ideas that cannot be doubted with earnestness. Cartesian scepticism is transitory and narrowly cognitive, and hence less radical than Spinozist scepticism, which is all encompassing, ongoing and, at once, cognitive, embodied, and affective. To assess these tentative claims, which potentially challenge received interpretations of Spinoza's philosophy, as well as to further explore the differences between the forms of scepticism respectively at work in the writings of Descartes and Spinoza, relevant excerpts from Descartes's *The Passions of the Soul* and Spinoza's *Treatise on the Emendation of the Intellect* and *Ethics* were read. Special attention was given to the emotional logic governing the relations between the affects of hope, fear, confidence, and despair as it pertains to doubt, the lack thereof, and certainty.

[RE 32] 5 February 2019
Steven Harvey
The Path of the *Falāsifah* to Knowledge and Ultimate Human Happiness… or Not

How did the medieval Aristotelians seek certain knowledge and pursue ultimate human happiness? How did they see their task as philosophers and scientists? Could intellectual perfection be sought and found in the confines of one's own abode, with little more than parchment, ink, and a reed brush? As one respected contemporary historian of medieval philosophy put it: [Jewish Aristotelians of the thirteenth and fourteenth centuries,] "following their two philosophical authorities, Maimonides and Averroes, sought truth and human perfection through the orderly study of Aristotle's books on logic, natural science, and metaphysics, as they were paraphrased and explained by Averroes." And, one might ask: "If so, and if such bookish learning could be carried out alone in one's home, was the goal to learn what others have written and transmitted? Or, perhaps, was it to correct this knowledge and/or to progress to new knowledge?"

Some of the selected texts strengthen the conventional view that all the knowledge needed for attaining truth may be found in the Aristotelian corpus as interpreted by the great Greek, Islamic, and Jewish commentators; others suggest that these texts should be studied carefully and corrected when necessary; and still others hold that all such study is, in fact, in vain, for the desired knowledge is illusory and cannot be attained by humans. But some of the texts allude to another way of searching for and attaining knowledge, one of observation and experience. To what extent were observation, experience, and experimentation a part of the medieval quest for certain knowledge?

[RE 33] 16 April 2019
Nancy Abigail Nuñez Hernández
"Elusive Knowledge" by David Lewis

Epistemic contextualism is a position developed by contemporary epistemologists to deal with the problem of philosophical scepticism, which in turn is understood as a problem that results from imposing unattainable standards for knowledge. In response to such excessive demands, epistemic contextualism proposes that knowledge attribution sentences of the form "*S* knows that *p*" are context-sensitive, so their truth conditions can vary from one context to another. By claiming that the same sentence can have different truth-conditions in different contexts, epistemic contextualism aims to show that there are contexts (both ordinary and scientific) in which we are safe from the sceptical threat because in those contexts, we do not have to meet unattainable standards for knowledge. This position has had a deep influence not only in epistemology, but also in philosophy of language and other areas of contemporary philosophy.

In this Reading Evening, one of the papers that laid the foundations of epistemic contextualism, David Lewis's "Elusive Knowledge" (1996), was studied.

[RE 34] 30 April 2019
Jason Yonover
Spinoza and Nietzsche on the Erroneous Belief in Free Will (Selections from *Ethics*, Letter 58, and *Twilight of the Idols*)

Despite a highly suggestive letter in which Nietzsche labels Spinoza his only "precursor," there is surprisingly little commentary on the relationship between the two thinkers. In this Reading Evening, the possibility was explored that the most important connection between Nietzsche and Spinoza has been missed and is methodological in nature. It was done on the basis of their rejections of free will in which, notably, they both not only (1) deny that we may boast of such a faculty, but furthermore (2) clarify the origins of our erroneous belief to the contrary (and indeed in a closely related fashion). Throughout questions pertaining to scepticism that are of epistemological interest were particularly explored.

[RE 35] 21 May 2019
Bakinaz Abdalla
Prophecy, Philosophy, and Truth in the Thought of Isaac Albalag

Was the world created in time or does it exist from eternity? This question provoked heated disputations in medieval philosophical and theological writings. Answering this question in favour of either creation or the eternity doctrine had epistemological and theological implications respectively. Perhaps a less problematic answer would be that the universe is both created and eternal. A philosopher who is faithful to both science (philosophy) and revelation may find it legitimate to deem the teachings of Scripture and philosophy, though contradictory, to be true. Odd though this suggestion may appear, it is not unprecedented. The Jewish philosopher Isaac Albalag (thirteenth century) admits that prophetic knowledge and philosophy are sometimes in conflict. In cases where reason and revelation prove to be contradictory, as with the issue of eternity versus creation, one admits their veracity; neither philosophy nor religion overrides the other.

To support this claim, Albalag advances an unusual theory of prophecy that views prophets as extraordinary human beings capable of apprehending a body of knowledge that surpasses human knowledge in both quantity and quality. In some cases, this knowledge contradicts the conclusions of demonstration. Albalag's theory of prophecy, together with his view of the relationship between religion and philosophy, led many scholars to read his thought in terms of the double truth doctrine which was advocated by medieval Latin Averroists.

In this Reading Evening, the double truth hypothesis was investigated. The goal was to examine whether Albalag's theory of prophecy is compatible with the fundamental metaphysical premises and strong sceptical elements of his sole treatise, *Sefer Tiqqun ha-Deʿot*.

[RE 36] 25 June 2019
Uta Lohmann
Scepticism in the Educational Philosophy of Joel Löwe as a Representative of the Berlin Haskalah

Joel Bril Löwe (1762–1802) was one of the most important proponents of the Berlin Jewish Enlightenment. He was a member of a small group of maskilim who spent a few years in close proximity to Moses Mendelssohn and who were regarded as his "true disciples." The most outstanding characteristics of this group of young maskilim include their activities as translators of some of the Holy Scriptures and their engagement as private or public teachers.

The Reading Evening focused on the influence of Mendelssohn's philosophy on Löwe. How did the Haskalah's fundamental thoughts about humankind being destined for perfection (*Bestimmung des Menschen zur Vervollkommnung*) and Mendelssohn's sceptical posture towards writtenness and language shape Löwe's pedagogical concepts, linguistic approaches, and translation method? What relevance did scepticism have to the educational philosophy (*Bildungsphilosophie*) of the Haskalah in general? What did scepticism mean for the perception of knowledge and for the pursuit of genuine truth in a pedagogical setting in particular? Last but not least: How can scepticism be systematically described not as a philosophical idea, but as a cultural form of application-oriented strategies and concepts of everyday (educational) practice?

Occasional Events

Conferences

12–14 November 2018
Scepticism and Anti-Scepticism in Jewish Averroism
Convenors: Racheli Haliva (Universität Hamburg/Germany), Yoav Meyrav (Universität Hamburg/Germany), Daniel Davies (Universität Hamburg/Germany)

"Jewish Averroism" is a concept that usually refers to Jewish philosophers from the thirteenth century onwards who philosophised within the context of Averroes's interpretation of Aristotelian philosophy. At the same time, many of them were committed to Maimonides's legacy of reconciling philosophical investigation with the Law of Moses. In order to settle this apparent tension, they interpreted Judaism in light of Averroes's Aristotelianism on the assumption that Judaism and true philosophy must always coincide.

Although Averroes's philosophy and commentary were attractive to some circles, others found it unsatisfactory or simply threatening to the traditional way of life. These thinkers responded with various critiques of Averroes, his followers, and the Averroistic approach to philosophy and its relationship to revealed religion.

The purpose of the conference was to crystallise the understanding of Jewish Averroism as a philosophical and cultural phenomenon. Special emphasis was put upon the Jewish Averroists's engagement with Maimonides's apparent sceptical approach, mainly concerning metaphysical knowledge about which Averroes is patently dogmatic. Finally, lectures about various aspects of anti-Averroism served as a necessary counterbalance.

Most of the conference's lectures can be viewed online at Universität Hamburg's lecture2go website: https://lecture2go.uni-hamburg.de/en_US/l2go/-/get/l/5089.

Programme

Keynote Lecture
Steven Harvey (Bar-Ilan University, Ramat-Gan/Israel)
Was Al-Ġazālī an Avicennist? Some Provocative Reflections on Jewish Averroism

Panel: What is Jewish Averroism?
Daniel J. Lasker (Ben-Gurion University of the Negev, Beer-Sheva/Israel)
Averroism and Interreligious Polemics in Late Medieval Iberia

Giovanni Licata (Scuola Normale Superiore di Pisa/Italy)
Similarities and Dissimilarities between Jewish and Latin Averroism

Reimund Leicht (Hebrew University of Jerusalem/Israel)
Who and What is a Jewish Averroist? Remarks on Averroes's First Jewish Readers

Panel: Reason, Scriptures, and Jewish Law
David Lemler (Université de Strasbourg/France)
The Interpretation of the Garden of Eden among Averroistic Maimonideans and the Scope of Human Knowledge

Mercedes Rubio (POLIS Jerusalem/Israel)
Is Maimonides's Biblical Exegesis Averroistic?

Shalom Sadik (Ben-Gurion University of the Negev, Beer-Sheva/Israel)
Double Truth in Jewish Averroist Philosophy

Panel: Averroistic Maimonideanism
Resianne Smidt van Gelder-Fontaine (Universiteit van Amsterdam/Netherlands)
Between Scepticism and Anti-Scepticism in Judah ha-Cohen's Presentation of Averroes in the *Midraš ha-Ḥokhmah*

Yehuda Halper (Bar-Ilan University, Ramat-Gan/Israel)
Socratic Scepticism in Hebrew: Al-Ḥarizi and Shem Ṭob Falaquera and Their Influence

Shira Weiss (Herzl Institute, Jerusalem/Israel)
Averroes's Influence upon Theological Responses to Scepticism in Late Medieval Jewish Philosophy

Panel: Averroistic Maimonideanism II
Rebecca Kneller-Rowe (Independent Scholar)
Scepticism and Anti-Scepticism in Samuel ibn Tibbon's Work

Yonatan Shemesh (University of Chicago/USA)
Averroes's *Incoherence of the Incoherence* and Narboni's Commentary on the *Guide of the Perplexed*

Racheli Haliva (Universität Hamburg/Germany)
Isaac Polqar's Anti-Sceptical Approach towards Miracles

Panel: Physics
Alexander Green (University at Buffalo/USA)
Gersonides and Ibn Kaspi on Scepticism about the Future

Maud Kozodoy (Independent Scholar)
Sources of Knowledge about the Heavens: Profayt Duran and Averroes

Warren Zev Harvey (Hebrew University of Jerusalem/Israel)
Crescas's Attitude towards Averroes

Panel: Metaphysics and Theology
Michael Engel (Universität Hamburg/Germany)
Averroes's *Long Commentary on the Posterior Analytics* and Its Impact on Jewish and Latin Averroism during the Italian Renaissance

Esti Eisenmann (The Open University of Israel)
Gersonides's Critique of Averroes: Between Physics, Metaphysics, and Theology

Yoav Meyrav (Universität Hamburg/Germany)
Against Averroes: Moses Ha-Levi's Defence of Avicenna's Necessary Existent

Panel: Prophecy and Human Perfection
Giada Coppola (Universität Hamburg/Germany)
Certainty and Uncertainty Regarding Metaphysics in Sforno's *Or ʿAmmim*

Bakinaz Abdalla (McGill University, Montreal/Canada)
Reconsidering Isaac Albalag's Theory of Prophecy: A Sceptical Approach

Adrian Sackson (Tel Aviv University/Israel)
What is the Status of Prophets from Other Religions for the Jewish Averroists?

Keynote Lecture
Hanna Kasher (Bar-Ilan University, Ramat-Gan/Israel)
The Jewish Averroists—Linking Thoughts between Maimonides and Spinoza

23–25 September 2019
Simone Luzzatto's Scepticism in the Context of Early Modern Thought
Convenors: Giuseppe Veltri (Universität Hamburg/Germany), Michela Torbidoni (Universität Hamburg/Germany), Anna Lissa (Université Paris 8/France)

This conference aimed to call attention to the sceptical thought of Rabbi Simone Luzzatto (ca. 1583–1663) and its value within seventeenth-century philosophy and political thought. The three-day conference brought together established scholars and young researchers working in the field of the intellectual history of Judaism in the sixteenth and seventeenth centuries. The lectures offered the opportunity to explore the philosophical sources and strategies of Luzzatto's scepticism, his intense dialogue with Christian philosophy, and his relationship to Judaism.

The discussion of Luzzatto provided the chance to investigate the role of scepticism within an era marked by paradoxes and contrasts between religious devotion and scientific rationalism, between rabbinical-biblical Jewish tradition and the open tendency towards engagement with non-Jewish philosophical, literary, scientific, and theological culture.

The conference's programme included a round table on Rabbi Simone Luzzatto's newly published works, *Discourse on the State of the Jews* (1638) and *Socrates, Or On Human Knowledge* (1651), discussed with the authors and translators.

Programme and Abstracts

Panel 1: Liberty, Religion, and Scientific Thought
Cristiana Facchini (Università di Bologna/Italy)
Practices of Critical Thought: The Rise of New Science, the Bible, and Jewish Thought

Evelien Chayes (Radboud Universiteit/Netherlands)
A Rabbi's Marks in Early Modern Subversive Literature. Simone Luzzatto and his *Nachleben*

Panel 2: Scepticism and Religion
Anna Lissa (Université Paris 8/France)
The Scepticism of Simone Luzzatto and his Appraisal of the *Book of Job*

Mina Lee (University of Tokyo/Japan)
The Concepts of Religion in Simone Luzzatto's *Discorso*

Panel 3: Sceptical Philosophy
Michela Torbidoni (Universität Hamburg/Germany)
Simone Luzzatto's Concept of Nature as a Source of Morality and Religiosity

Josef Stern (University of Chicago/USA)
What Kind of a (Sceptical) work is Simone Luzzatto's *Socrate?*

Evening Session
Ensemble Accademia dei Carpìti (Anna Tarca, Silvia Valenti, Paolo Davolio, Lorenzo Ziller)
Musical Interlude: Sacred and Profane Polyphony by Salomone Rossi and Luca Marenzio

Giuseppe Veltri (Universität Hamburg/Germany)
Introduction

Warren Zev Harvey (Hebrew University of Jerusalem/Israel)
Keynote Lecture: The Image of King Solomon in Simone Luzzatto's Writings

Panel 4: Ancient and Medieval Sources
Joanna Weinberg (University of Oxford/UK)
The Jewish Debate over Philo in Early Modern Italy

Fabrizio Lelli (Università del Salento/Italy)
Simone Luzzatto's Scepticism in Light of Medieval Jewish Apologetics

Panel 5: Politics and Economics
Vasileios Syros (University of Jyväskylä/Finland)
A Venetian Rabbi and a Baroness from the Low Countries on the Condition of the Jews

Guido Bartolucci (Università della Calabria/Italy)
Simone Luzzatto's Political Thought: Between Scepticism and Reason of State

Panel 6: Politics and Economics
Luca Andreoni (Istituto Comprensivo "Caio Giulio Cesare"/Italy)
Simone Luzzatto and the Debate on the Economical Role of the Jews in Eighteenth Century Italy

Myriam Silvera (Università degli Studi di Roma "Tor Vergata"/Italy)
"Seek the Peace and Prosperity of the City...": The Political Use of Jer 29:7 by Simone Luzzatto and Other Jewish Sources

Panel 7: Anthropology, Society, and Tolerance
Giuseppe Veltri (Universität Hamburg/Germany)
Self-Love, Blood-Ties, and Amity in Simone Luzzatto's Philosophy

Book Launch
Scepticism and Tolerance: New Perspectives for the Jewish Intellectual History
Evelien Chayes (Radboud Universiteit/Netherlands)
Intellectual Life and Literature

Cristiana Facchini (Università di Bologna/Italy)
History

Guido Bartolucci (Università della Calabria/Italy)
History of Political Thought

Josef Stern (University of Chicago/USA)
Philosophy

Lecture Series "Feminism and Scepticism"
The Maimonides Centre for Advanced Studies established a lecture series which focuses on feminist approaches to and perspectives on scepticism. The Centre invites researchers to give a public lecture on this topic, and—following this lecture—to meet with the female researchers in order to share their experiences and career paths in their academic systems.

2 April 2019
Ruth Kara-Ivanov Kaniel (Shalom Hartman Institute in Jerusalem/Israel)
Scepticism and Gender. King David in the Image of the *Shekhinah*

King David is one of the most colourful heroes of Jewish myth; warrior and poet, sinner and penitent, conqueror and musician, adulterer and Messiah. The many facets of his character are rooted in biblical scripture and continued to develop in the literature of the Midrash and the Kabbalah. Each generation has added new layers to David's portrait, sketching him in a new light. Indeed, David's personality reflects the characters and hopes of his interpreters throughout generations. Embodying the hero "with a thousand faces" and representing the messianic idea, David is not merely a private character, but a collective entity, wearing many different forms. In the Zohar, David's collective image is identified with the *Shekhinah*, the Assembly of Israel [Knesset Israel] and the Divine Spouse.

Why was this figure of the warrior—the ultimate male, conqueror of cities and kingdoms, the redeemer, who was presented as a masculine hero in both Christian and Jewish literature—"converted" by the Zohar to signify a feminine image? In her lecture, Ruth Kara-Ivanov Kaniel tried to solve this riddle by using questions of scepticism, identity, and gender.

This year's lecture on scepticism and feminism was accompanied by a workshop.

1 April 2019
Biti Roi (Shalom Hartman Institute in Jerusalem/Israel)
Scepticism in Zoharic and Hasidic Literature

The Zoharic literature of twelfth- to fourteenth-century Spain, which has been defined as the most influential Kabbalistic movement in the Middle Ages, and its followers—the eighteenth-century Eastern European Hasidic movement—demonstrate an interesting window into the complicated relationship between scepticism and theological absolutism. In this workshop, the way in which scepticism defines affinity with God in the Jewish mystical tradition and the way non-philosophical texts deal with philosophical perceptions were explored. The role of gender issues in the Kabbalistic movement were also articulated. Texts which were examined included works of rabbinic, Zoharic, medieval, and Hasidic literature as well as the well-known liturgical foreword "Elijah opened and said" and its similarities to *Al-Fātiḥah*, the famous Islamic prayer.

Maimonides Lectures on Scepticism

Researchers focusing on various aspects of scepticism are invited to present and discuss their research in an evening lecture. The lectures are usually recorded and published on the webpage of the Centre in order to make them available to a larger audience.

22 November 2018
Richard Bett (Johns Hopkins University, Baltimore/USA)
Can we be Ancient Sceptics?

The lecture considered how far Pyrrhonism might be a viable outlook today. The answer depends in part on what range of issues suspension of judgment might plausibly be generated in today's intellectual climate. On many ethical, religious, and philosophical issues, the prospects seem just as good as in Sextus's day. Concerning natural science, the matter is more complicated. Here there are many issues where we know too much for suspension of judgment to be realistic, and where science has infiltrated ordinary life. On the other hand, suspension of judgment is possible on some issues, such as climate change, where there is a vocal popular opinion about a scientific question. In addition, it is possible on philosophical questions concerning the status of science itself, questions that did not occur to the ancients. A further issue is the value of suspension of judgment. Here Sextus seems overly ambitious, because of exaggerated claims about *ataraxia* as its outcome. Whether suspension of judgment would yield *ataraxia* depends on people's character and circumstances, regardless of the perceived importance of the topic. What remains is a worthwhile recommendation to be open-minded.

9 July 2019
Nadja El Kassar (ETH Zürich/Switzerland)
How to Deal with Ignorance? Some Historical Suggestions

Ever since its early days, philosophers have been fascinated by ignorance. In her talk, Nadja El Kassar discussed historical answers to the question of how it should be dealt with. The question itself is seldom raised explicitly, but a number of philosophers have addressed the issue of dealing with ignorance. The lecture focused in particular on Socrates's, John Locke's, and Immanuel Kant's suggestions on the subject. Socratic ignorance can be read as a suggestion that knowledge and wisdom are the best ways of dealing with ignorance. Locke suggests that four principles (the principle of indifference, the principle of evidence, the principle of appraisal, and the principle of proportionality) are the best ways of dealing with ignorance and also with error. Finally, Kant argues that orienting oneself is the key to dealing with ignorance. Nadja El Kassar discussed and compared these three historical proposals and concluded by examining their relevance to today's question of how to deal with ignorance.

Lecture Series "Skepsis und Toleranz. Moses Mendelssohn, Salomon Maimon und die jüdische Aufklärungsphilosophie"

Organised by the Institute for Jewish Philosophy and Religion, in cooperation with the Maimonides Centre for Advanced Studies

Convenors: Ze'ev Strauss (HfJS Heidelberg/Germany), José María Sánchez de León Serrano (Universität Hamburg/Germany), Libera Pisano (Universität Hamburg/Germany)

The lectures within the framework of the lecture series provided an introduction into the innovative thought of the two central philosophers of the German-Jewish Enlightenment: Moses Mendelssohn and Solomon Maimon. Two concepts of their philosophical systems—tolerance and scepticism—gained centre stage. The main question that these lectures addressed is the reciprocal relation between tolerance and scepticism, and how each of these thinkers takes a stance in this regard. To answer this question, special attention was devoted to Jewish religion, inasmuch as it constitutes an elemental component of their Weltanschauungen.

Programme

1 April 2019
Giuseppe Veltri (Universität Hamburg/Germany) and Anne Fiebig (Universität Hamburg/Germany)
Toleranz als gesellschaftliche Herausforderung in jüdischen Quellen und Vorstellung des Toleranzprojekts

6 May 2019
Shmuel Feiner (Bar-Ilan University, Ramat-Gan/Israel)
Moses Mendelssohn: The Myth, the History, and the Jewish Battle for Religious Tolerance

13 May 2019
Jason Yonover (Johns Hopkins University, Baltimore/USA)
Mendelssohn and Maimon on Superstition

20 May 2019
Timothy Franz (New School for Social Research in New York City/USA)
"They Go From Strength to Strength": Solomon Maimon's Way from Critique to Transcendental Philosophy

27 May 2019
José María Sánchez de León Serrano (Universität Hamburg/Germany)
Abgötterei bei Mendelssohn und Spinoza

3 June 2019
Warren Zev Harvey (Hebrew University of Jerusalem/Israel)
Mendelssohn and Maimon on the Garden of Eden

17 June 2019
Ze'ev Strauss (HfJS Heidelberg/Germany)
Das Judentum als Grundlage im Religionsgebäude

24 June 2019
Grit Schorch (Hebrew University of Jerusalem/Israel)
Keine Philosophie ohne Sprache – Eine neue Perspektive auf Moses Mendelssohns Denken

26 June 2019
Idit Chikurel (Universität Potsdam/Germany)
Scepticism and Certainty in Salomon Maimon's Theory of Invention

1 July 2019
George Y. Kohler (Bar-Ilan University, Ramat-Gan/Israel)
Skeptische Aufnahme – Kritik an Mendelssohns Religionsphilosophie im deutsch-jüdischen Denken des 19. Jahrhunderts

8 July 2019
Libera Pisano (Universität Hamburg/Germany)
"Ein Wohltäter aller Deutsch redenden Juden:" Zunz' Deutung von Moses Mendelssohn

10 July 2019
Michah Gottlieb (New York University/USA)
Mendelssohn and Minority Toleration

MCAS Participation in External Conferences
4–7 March 2019
Annual Conference of the European Academy of Religion
Panel: Jewish Scepticism as Philosophical Question

Research on Jewish philosophical scepticism is still in its very early stages. This also holds true for cultural expressions of scepticism, i.e., modes of sceptical strategies present in Jewish literature, cultural practices, history, the organisation of social groups, and especially education. This lack of interest in the intricacies of Jewish scepticism in Jewish studies may be rooted in the modality of Jewish philosophy as such, its being notoriously in-between, subsuming traditional wisdom, philosophy, theology, Jewish and general Weltanschauung, as well as cultural history, and representing a hazardous bridge between orthopraxy and orthodoxy. In this panel, the controversial binomial of Jewish scepticism was investigated from a philosophical perspective.

Chair: Libera Pisano (Universität Hamburg/Germany)

Giuseppe Veltri (Universität Hamburg/Germany)
Short introduction

Yoav Meyrav (Universität Hamburg/Germany)
The Danger of Perplexity: a Farabian Problem—and a Maimonidean Solution?

Michela Torbidoni (Universität Hamburg/Germany)
The Sceptical Socrates of Simone Luzzatto between Reason and Revelation

José María Sánchez de León Serrano (Universität Hamburg/Germany)
God's Transparency and Hyperbolic Doubt in Spinoza's Metaphysics

Silke Schaeper

Report on the Library of Jewish Scepticism

The Library of Jewish Scepticism is a special collection administered by a single librarian—a so-called one-person-library (OPL). During the Maimonides Centre's first period, from October 2015 to February 2019, the position of librarian was jointly funded by the German Research Foundation and Universität Hamburg (90 %). Between March and August 2019, the position of librarian was vacant due to lack of funding. Since 2019, the position of librarian is funded by Universität Hamburg part-time (50 %). Tasks and budget remain the same.

Student Assistants

The librarian was supported by Ms. Vanessa Zerwas, a sign-language student, until February 2019.

Catalogue Merger

German academic libraries contribute cataloguing records to regional library networks, for historic reasons. The meta search engine "Karlsruhe Virtual Catalog" serves as a virtual national union catalogue, drawing together the content of these regional catalogues. In March 2019, two very large German regional library networks (GBV, SWB) merged their catalogues to form **K10plus**, a data pool containing over 200 million holdings. While catalogue users will not notice a lot of changes, library professionals need to handle technical changes to cataloguing and retrieval patterns. The librarian attended **K10plus** training in February 2019. Development, deduplication, and debugging continues.

Specialised Information Service in Jewish Studies/Israel Studies

Since 2016, the "Fachinformationsdienst Jüdische Studien" (FID) provides managed access to electronic resources in Jewish Studies to senior academics and bona fide researchers in Germany. In February 2018, the Judaica team at Universitätsbibliothek Frankfurt a. M. led by Dr. Rachel Heuberger, which runs the FID, conducted a survey among the research and library community. Based on survey feedback and available financial resources, the FID has negotiated access to three additional e-resources in 2019:

https://doi.org/10.1515/9783110618839-017

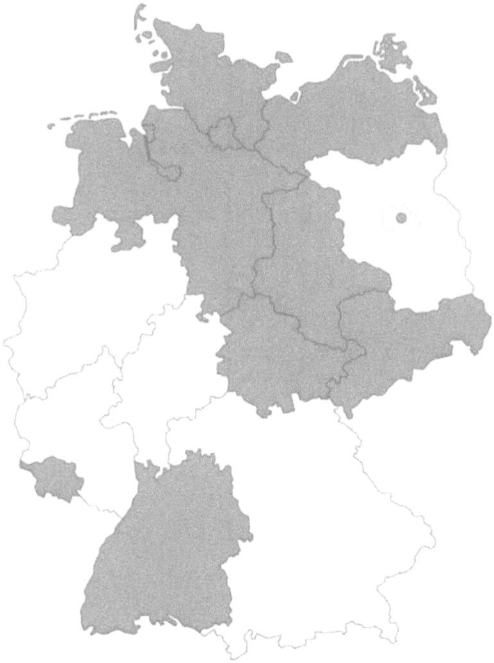

Coverage of K10plus. For further information, see https://www.bszgbv.de/.

COTAR online = Collection of Torah articles *New in 2019*
Index to Hebrew Periodicals
JSTOR Complete Jewish Studies Collection
Magnes Press e-Book collection Hebrew titles *New in 2019*
Otzar HaHochma
Rav Milim Dictionary Online
Journal of Ancient Judaism (Vandenhoeck & Ruprecht) *New in 2019*

In 2019, the Frankfurt FID team won the bid for another period of funding from the German Research Foundation (2019–2021). The librarian continues to manage FID access permissions for staff and fellows of the Maimonides Centre.

FID subscriptions July 2018 to September 2019

Staff	20
Fellows, post-docs	3

Faculty Library Committee

Since September 2019, the librarian is an advising member of the Library Committee of the Faculty of Humanities at Universität Hamburg.

Book Donations

We wish to express our sincere thanks for donations of books received from Daniel Boyarin (University of California), Moshe Idel (Hebrew University of Jerusalem), Ariel Malachi (Bar-Ilan University, Ramat-Gan), Yitzhak Melamed (Johns Hopkins University, Baltimore), Yoav Meyrav (Universität Hamburg), Dalyah Netser-Tsafrir (Moshav Kefar Yehoshu'a), Lilian Türk (Universität Hamburg), Giuseppe Veltri (Universität Hamburg), Irina Wandrey (Centre for the Study of Manuscript Cultures, Hamburg), and unnamed former fellows.

Statistics

July 2018 to September 2019

Institute for Jewish Philosophy and Religion

Acquisitions (physical volumes)	85
Items catalogued (incl. earlier acquisitions)	172

Maimonides Centre for Advanced Studies

Acquisitions (physical volumes)	225
Items catalogued (incl. earlier acquisitions)	157

Bei Fragen zur Produktsicherheit wenden Sie sich bitte an:
If you have any questions regarding product safety,
please contact:

Walter de Gruyter GmbH
Genthiner Straße 13
10785 Berlin
productsafety@degruyterbrill.com